PRAISE FOR *THE MANGA GUIDE TO DATABASES*

"Without a doubt, *The Manga Guide to Databases* was the most enjoyable tech book I've ever read . . . I loved this book and highly recommend it."
—RIKKI KITE, LINUX PRO MAGAZINE

"Whether you're new to the whole concept … geek, you need to have *The Manga Guide to Databa…*
—JOSH BERKUS, POSTGRESQL CORE TEAM

"For an American audience, this is certainly an off-the-beaten-path approach to technical training. However, its ability to effectively plunge into a topic that can be a morass of arcane theory is undeniable. Over the years, we've learned to expect the unexpected from No Starch Press."
—MACDIRECTORY

"This is a fun book, there is no denying that. If you're in the market to learn more about databases and you hate normal tech books this is a great pickup."
—THE CAFFINATION PODCAST

"If I was going to teach a class or introduce a non-technical person to the world of databases, I would most likely start here."
—BLOGCRITICS.ORG

PRAISE FOR *THE MANGA GUIDE TO STATISTICS*

"This is really what a good math text should be like. Unlike the majority of books on subjects like statistics, it doesn't just present the material as a dry series of pointless-seeming formulas. It presents statistics as something fun, and something enlightening."
—GOOD MATH, BAD MATH

"*The Manga Guide to Statistics* is a solid introduction to the world of statistical analysis done in a fun and accessible way."
—ACTIVE ANIME

"Inoue's art is clean, cute and simplified, and it works with machine-like efficiency—the artist not only knows and speaks the language of manga, but does so fluently."
—NEWSARAMA

"*The Manga Guide to Statistics* offers a visualization of statistics that can't be found in any mere textbook."
—ANIME 3000

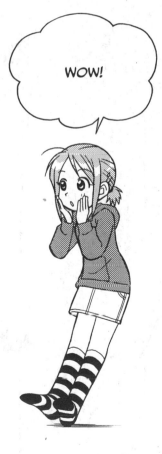

WOW!

THE MANGA GUIDE™ TO PHYSICS

THE MANGA GUIDE™ TO
PHYSICS

WITHDRAWN

HIDEO NITTA
KEITA TAKATSU
TREND-PRO CO., LTD.

Ohmsha

no starch
press

The Manga Guide to Physics is a translation of the Japanese original, *Manga de Wakaru Butsuri* [*Rikigaku hen*], published by Ohmsha, Ltd. of Tokyo, Japan, © 2006 by Hideo Nitta and TREND-PRO Co., Ltd.

The English edition is co-published by No Starch Press and Ohmsha, Ltd.

13 12 11 10 09   2 3 4 5 6 7 8 9

ISBN-10: 1-59327-196-4
ISBN-13: 978-1-59327-196-1

SUSTAINABLE FORESTRY INITIATIVE Certified Fiber Sourcing
Label applies to the text stock www.sfiprogram.org

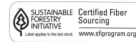

Publisher: William Pollock
Author: Hideo Nitta
Illustrator: Keita Takatsu
Producer: TREND-PRO Co., Ltd.
Production Editor: Megan Dunchak
Developmental Editor: Tyler Ortman
Technical Reviewers: Keith Brown and Whitney Ortman-Link
Compositor: Riley Hoffman
Proofreader: Cristina Chan
Indexer: Sarah Schott

For information on book distributors or translations, please contact No Starch Press, Inc. directly:
No Starch Press, Inc.
555 De Haro Street, Suite 250, San Francisco, CA 94107
phone: 415.863.9900; fax: 415.863.9950; info@nostarch.com; http://www.nostarch.com/

Library of Congress Cataloging-in-Publication Data

Nitta, Hideo, 1957-
[Manga de wakaru butsuri. Rikigaku hen. English]
The manga guide to physics / Hideo Nitta, Keita Takatsu ; Trend-pro Co., Ltd.
p. cm.
Includes index.
ISBN-13: 978-1-59327-196-1
ISBN-10: 1-59327-196-4
1. Physics--Comic books, strips, etc. 2. Physics--Popular works. I. Takatsu, Keita. II. Trend-pro Co. III. Title.
QC24.5.N5813 2009
530--dc22

2009012720

CONTENTS

PREFACE

It is essential to the understanding of physics to correctly "see" what you wish to examine. In classical mechanics, in particular, you need to understand how physical laws apply to transient, moving objects. But unfortunately, conventional textbooks rarely provide adequate images of such motion.

This book attempts to conquer the limits of those conventional textbooks by using cartoons. Cartoons are not just simple illustrations—they are an expressive and dynamic medium that can represent the flow of time. By using cartoons, it is possible to vividly express changes in motion. Cartoons can also transform seemingly dry laws and unreal scenarios into things that are familiar, friendly, and easy to understand. And it goes without saying—cartoons are fun. We have emphasized that in this book, as well.

As an author eager to know whether or not my intent has succeeded, I can only wait for readers to make their judgments. This work has been finished to my deep satisfaction, except for the omission of one chapter—due to page count constraints—featuring a trip to an amusement park to explain circular movement and the noninertial system.

The main character of this book is a high school student named Megumi Ninomiya who finds physics rather difficult. It is my sincere desire that this book reaches out to as many readers as possible who think "physics is tough" and who "don't like physics," helping them find pleasure in physics like Megumi does—even if it's only a little.

Last but not least, I would like to express my deep appreciation to the staff at the OHM Development Office, scenario writer re_akino, and illustrator Keita Takatsu—their combined efforts have resulted in this wonderful cartoon work that would have been impossible for one individual to complete.

HIDEO NITTA
NOVEMBER 2006

PROLOGUE

DOES PHYSICS BOTHER YOU?

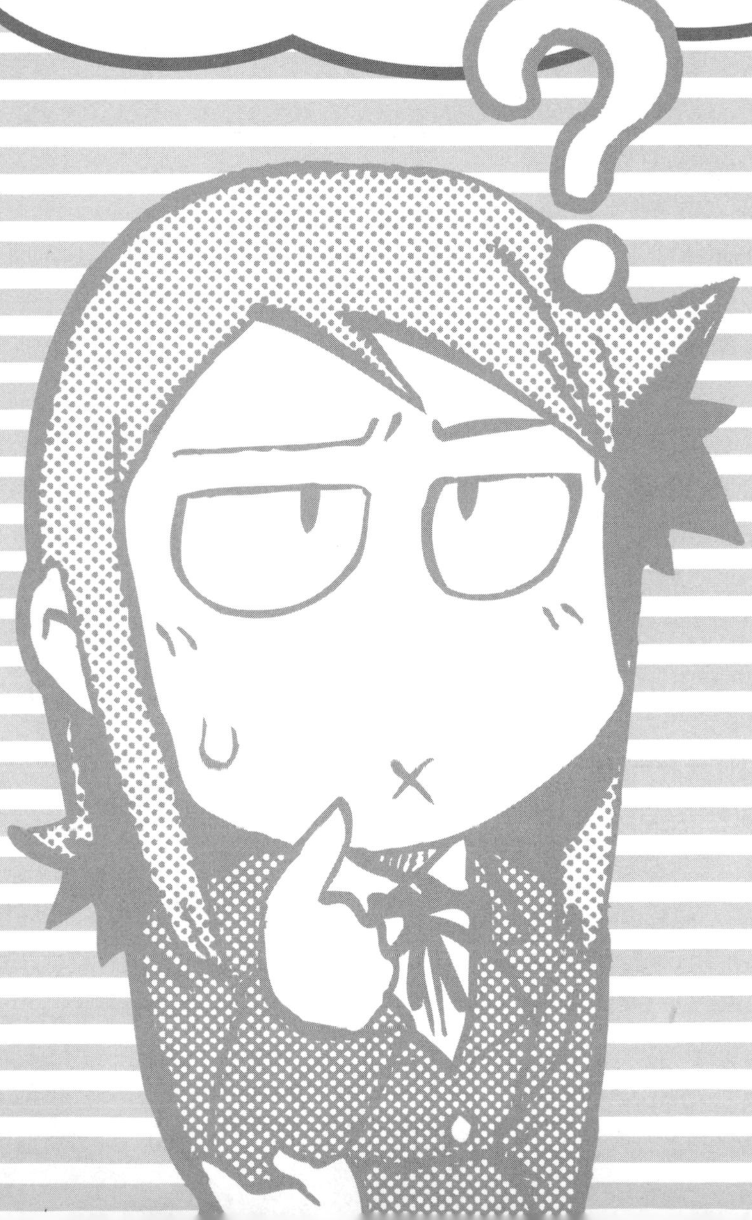

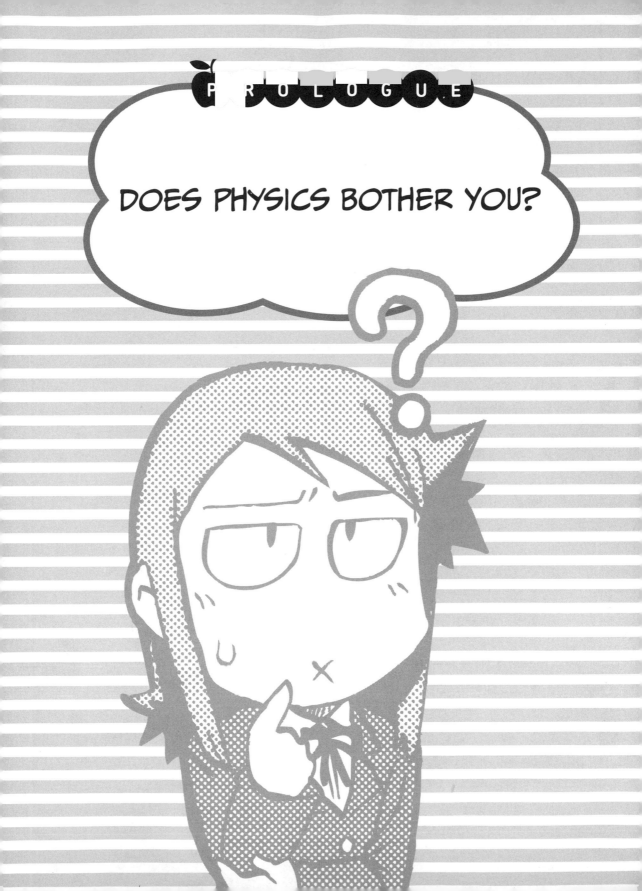

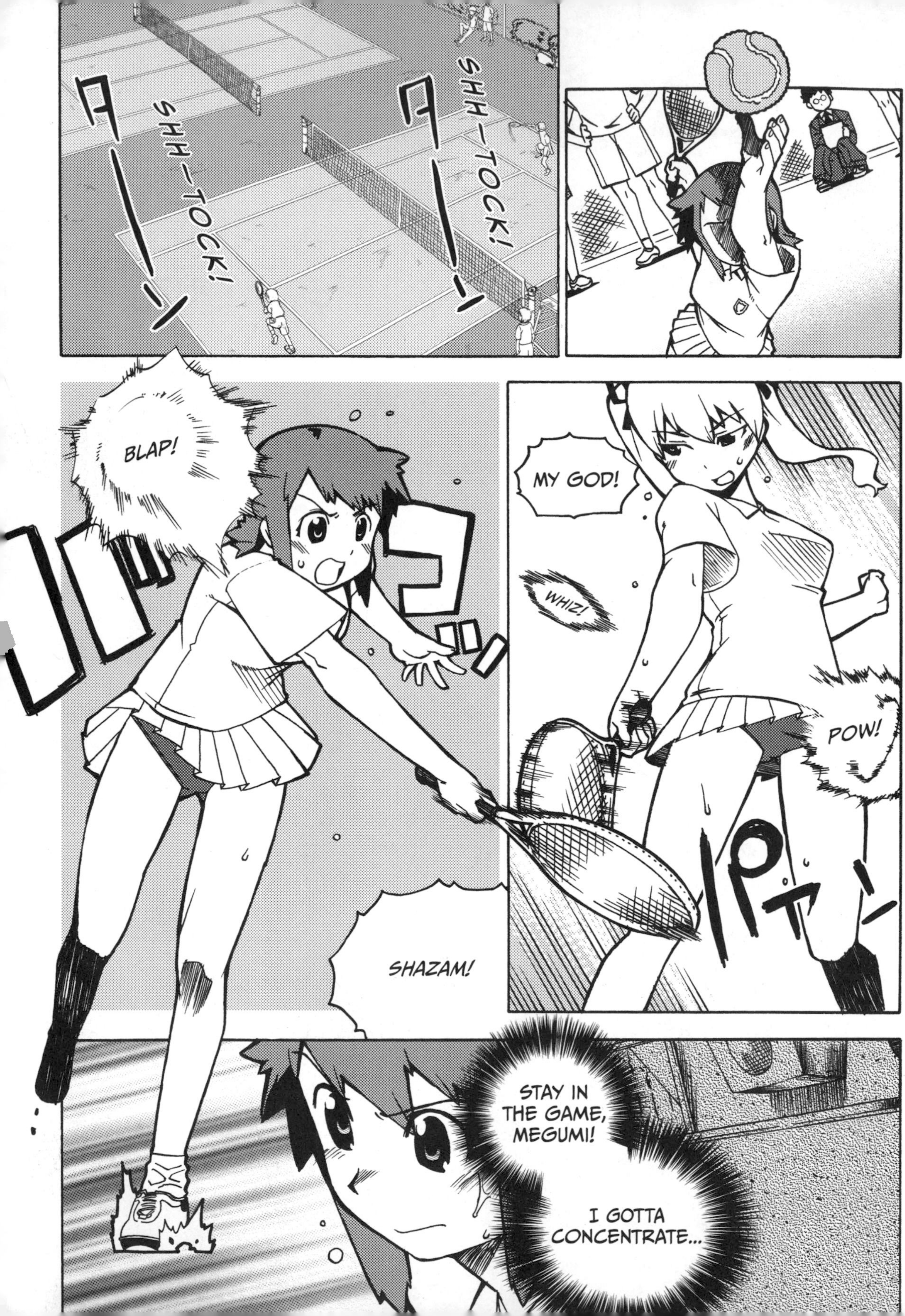
SHH-TOCK!
SHH-TOCK!
BLAP!
SHAZAM!
MY GOD!
WHIZ!
POW!
STAY IN THE GAME, MEGUMI!
I GOTTA CONCENTRATE...

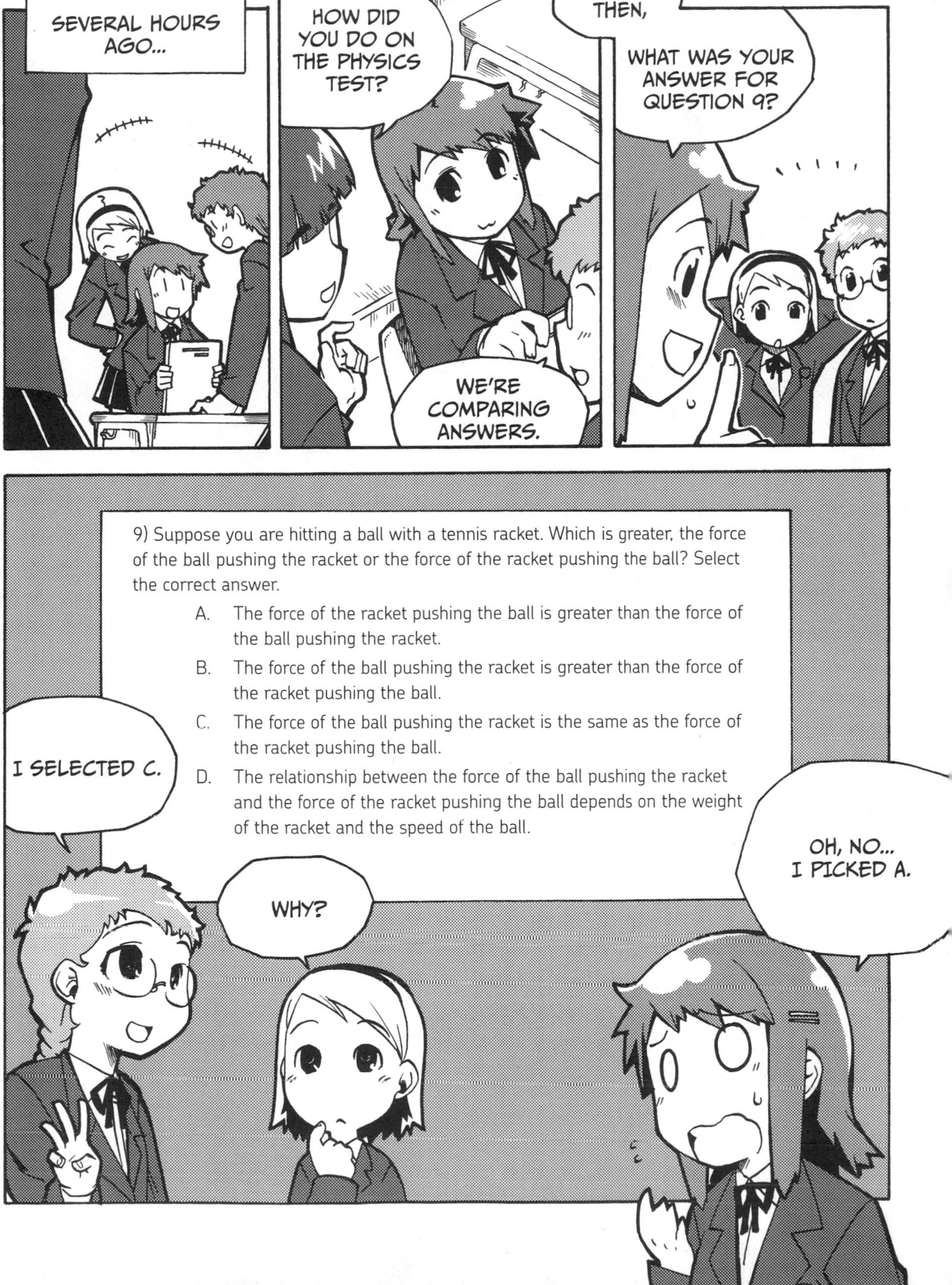
SEVERAL HOURS AGO...
HOW DID YOU DO ON THE PHYSICS TEST?
WE'RE COMPARING ANSWERS.
WELL, THEN,
WHAT WAS YOUR ANSWER FOR QUESTION 9?
9) Suppose you are hitting a ball with a tennis racket. Which is greater, the force of the ball pushing the racket or the force of the racket pushing the ball? Select the correct answer.
A. The force of the racket pushing the ball is greater than the force of the ball pushing the racket.
B. The force of the ball pushing the racket is greater than the force of the racket pushing the ball.
C. The force of the ball pushing the racket is the same as the force of the racket pushing the ball.
D. The relationship between the force of the ball pushing the racket and the force of the racket pushing the ball depends on the weight of the racket and the speed of the ball.
I SELECTED C.
WHY?
OH, NO... I PICKED A.

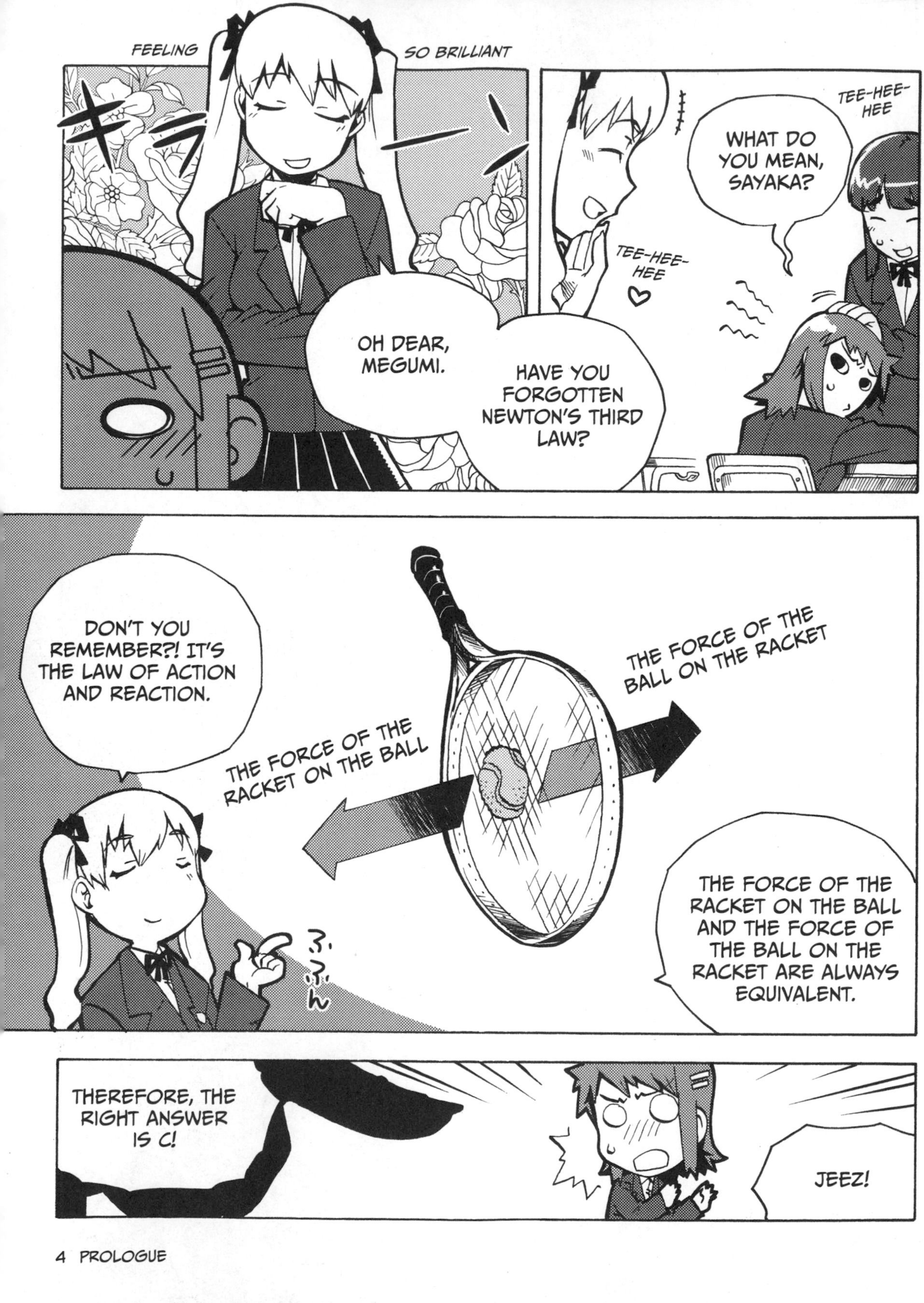
FEELING
SO BRILLIANT
OH DEAR, MEGUMI.
HAVE YOU FORGOTTEN NEWTON'S THIRD LAW?
TEE-HEE-HEE
WHAT DO YOU MEAN, SAYAKA?
TEE-HEE-HEE
DON'T YOU REMEMBER?! IT'S THE LAW OF ACTION AND REACTION.
THE FORCE OF THE RACKET ON THE BALL
THE FORCE OF THE BALL ON THE RACKET
ふふん
THE FORCE OF THE RACKET ON THE BALL AND THE FORCE OF THE BALL ON THE RACKET ARE ALWAYS EQUIVALENT.
THEREFORE, THE RIGHT ANSWER IS C!
JEEZ!

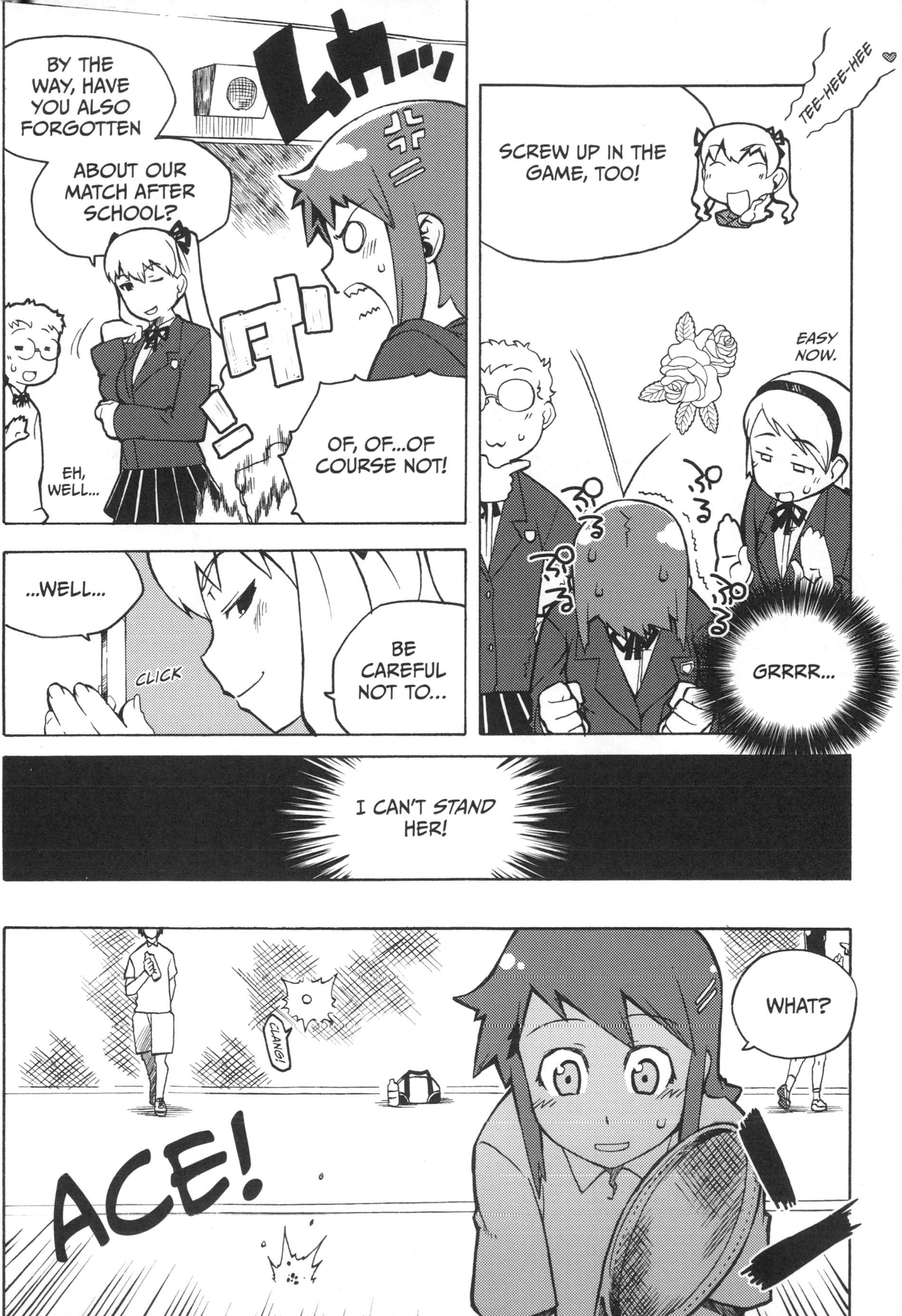

BY THE WAY, HAVE YOU ALSO FORGOTTEN
ABOUT OUR MATCH AFTER SCHOOL?
EH, WELL...
OF, OF...OF COURSE NOT!
...WELL...
CLICK
BE CAREFUL NOT TO...
SCREW UP IN THE GAME, TOO!
TEE-HEE-HEE
EASY NOW.
GRRRR...
I CAN'T STAND HER!
CLANG!
WHAT?
ACE!

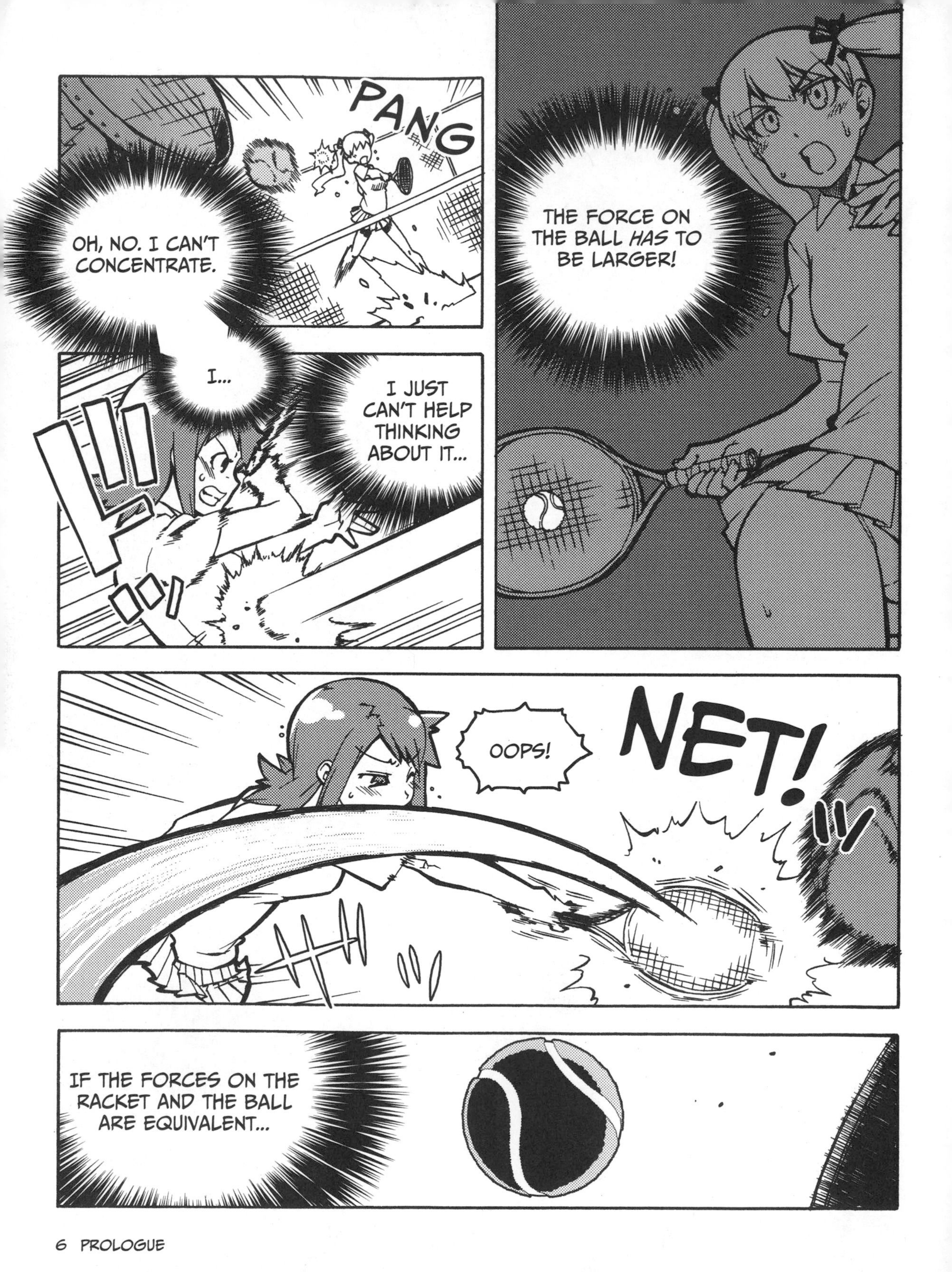
PANG
OH, NO. I CAN'T CONCENTRATE.
I...
I JUST CAN'T HELP THINKING ABOUT IT...
THE FORCE ON THE BALL HAS TO BE LARGER!
OOPS!
NET!
IF THE FORCES ON THE RACKET AND THE BALL ARE EQUIVALENT...

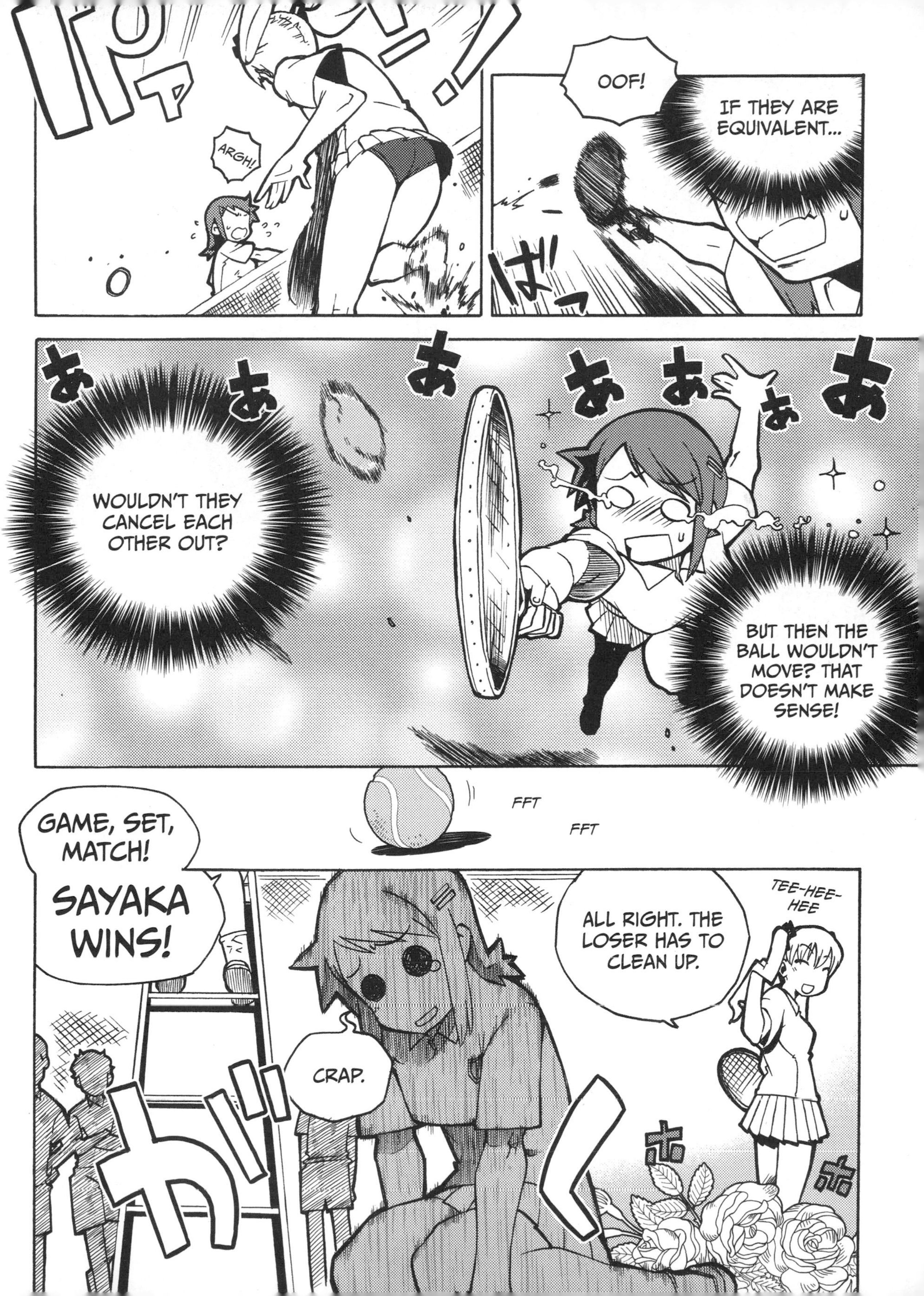
ARGH!
OOF!
IF THEY ARE EQUIVALENT...
WOULDN'T THEY CANCEL EACH OTHER OUT?
BUT THEN THE BALL WOULDN'T MOVE? THAT DOESN'T MAKE SENSE!
FFT
FFT
GAME, SET, MATCH!
SAYAKA WINS!
CRAP.
ALL RIGHT. THE LOSER HAS TO CLEAN UP.
TEE-HEE-HEE

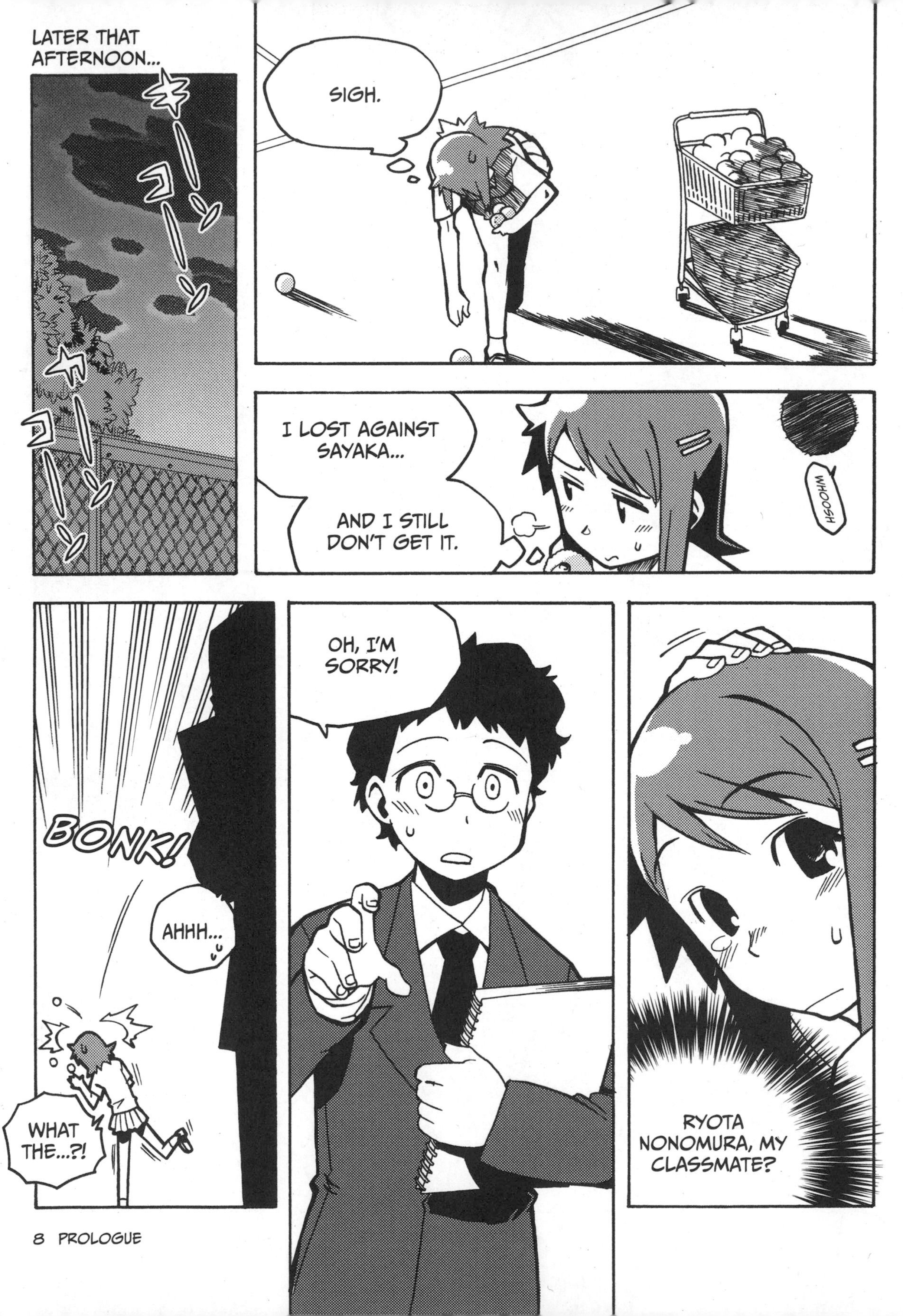
LATER THAT AFTERNOON...
SIGH.
I LOST AGAINST SAYAKA...
AND I STILL DON'T GET IT.
WHOOSH
BONK!
AHHH...
WHAT THE...?!
OH, I'M SORRY!
RYOTA NONOMURA, MY CLASSMATE?

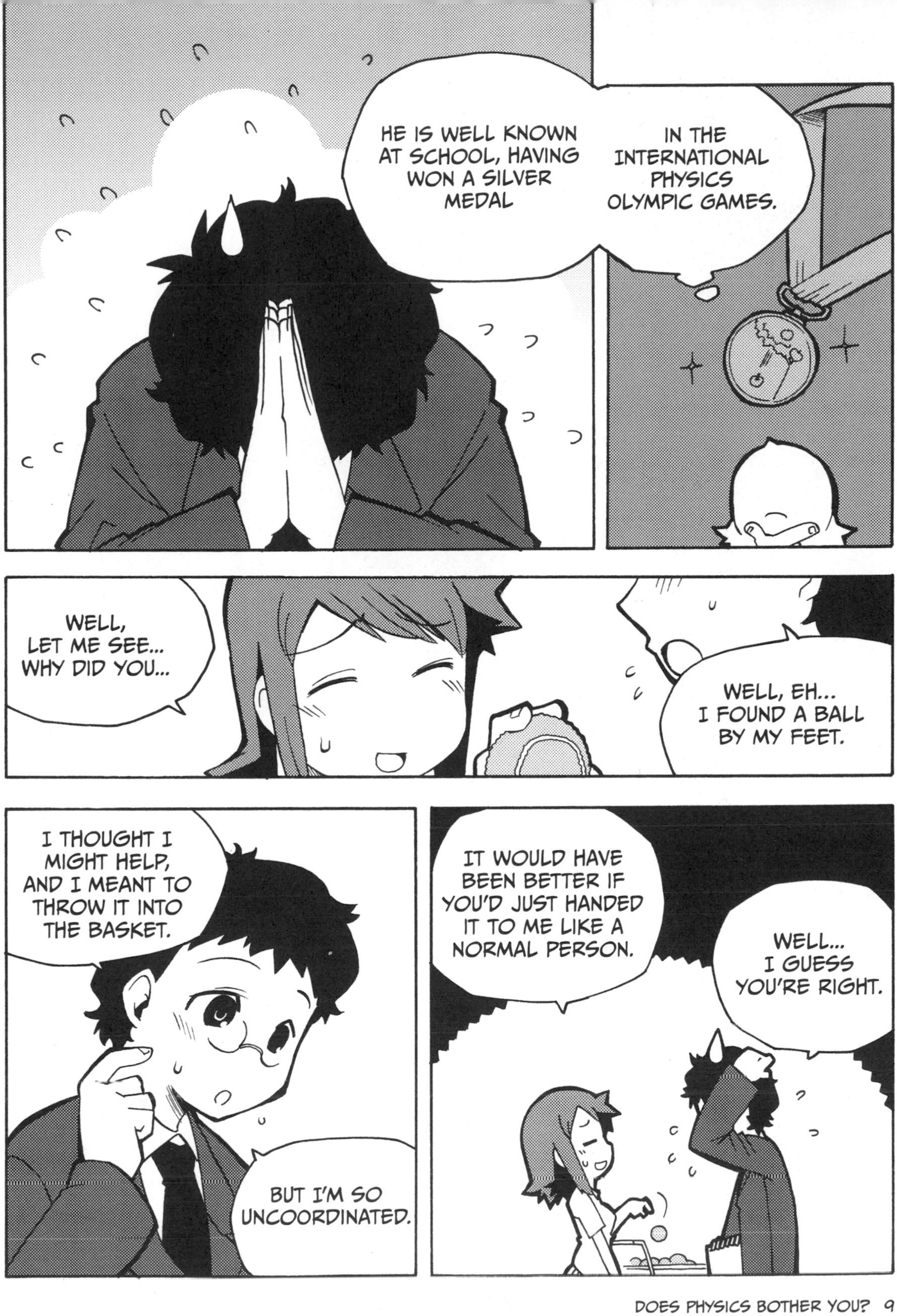
HE IS WELL KNOWN AT SCHOOL, HAVING WON A SILVER MEDAL
IN THE INTERNATIONAL PHYSICS OLYMPIC GAMES.
WELL, LET ME SEE... WHY DID YOU...
WELL, EH... I FOUND A BALL BY MY FEET.
I THOUGHT I MIGHT HELP, AND I MEANT TO THROW IT INTO THE BASKET.
BUT I'M SO UNCOORDINATED.
IT WOULD HAVE BEEN BETTER IF YOU'D JUST HANDED IT TO ME LIKE A NORMAL PERSON.
WELL... I GUESS YOU'RE RIGHT.

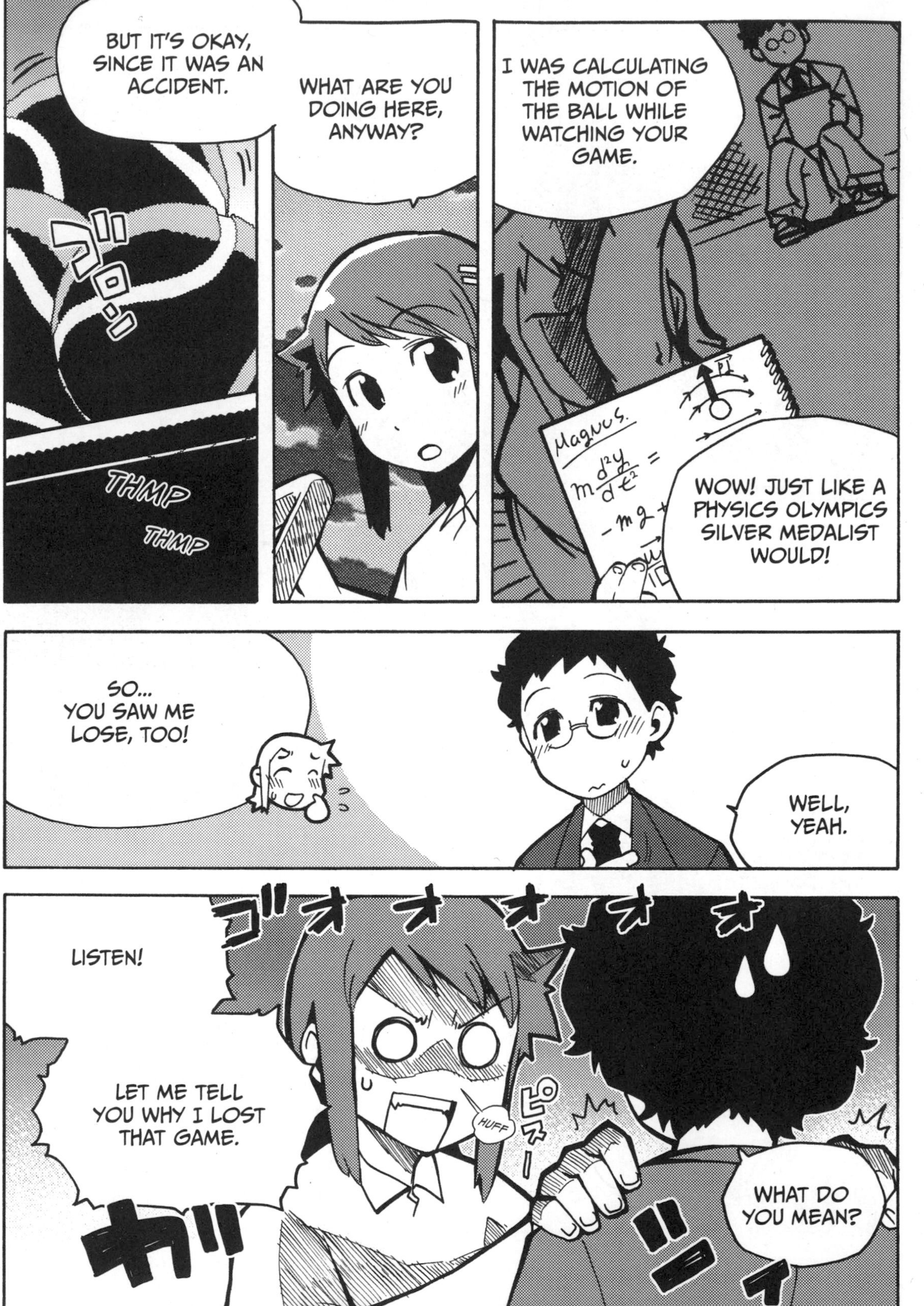
BUT IT'S OKAY, SINCE IT WAS AN ACCIDENT.
ゴロン
THMP
THMP
WHAT ARE YOU DOING HERE, ANYWAY?
I WAS CALCULATING THE MOTION OF THE BALL WHILE WATCHING YOUR GAME.
Magnus.
WOW! JUST LIKE A PHYSICS OLYMPICS SILVER MEDALIST WOULD!
SO... YOU SAW ME LOSE, TOO!
WELL, YEAH.
ゴオオオオオ
LISTEN!
LET ME TELL YOU WHY I LOST THAT GAME.
HUFF
ピスー
ガッ
WHAT DO YOU MEAN?

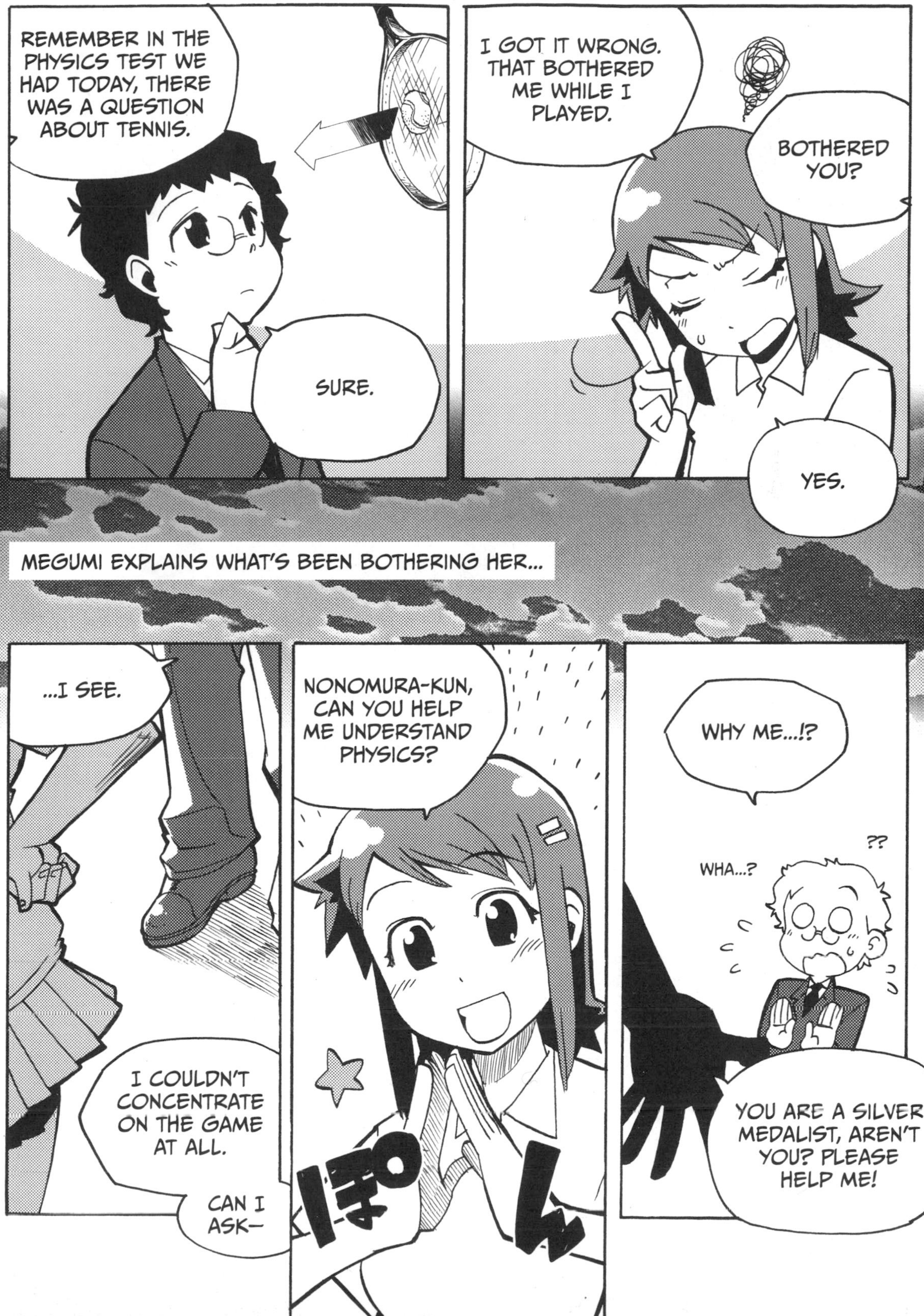
REMEMBER IN THE PHYSICS TEST WE HAD TODAY, THERE WAS A QUESTION ABOUT TENNIS.
SURE.
I GOT IT WRONG. THAT BOTHERED ME WHILE I PLAYED.
BOTHERED YOU?
YES.
MEGUMI EXPLAINS WHAT'S BEEN BOTHERING HER...
...I SEE.
I COULDN'T CONCENTRATE ON THE GAME AT ALL.
CAN I ASK–
NONOMURA-KUN, CAN YOU HELP ME UNDERSTAND PHYSICS?
ぽーん
WHY ME...!?
WHA...?
??
YOU ARE A SILVER MEDALIST, AREN'T YOU? PLEASE HELP ME!

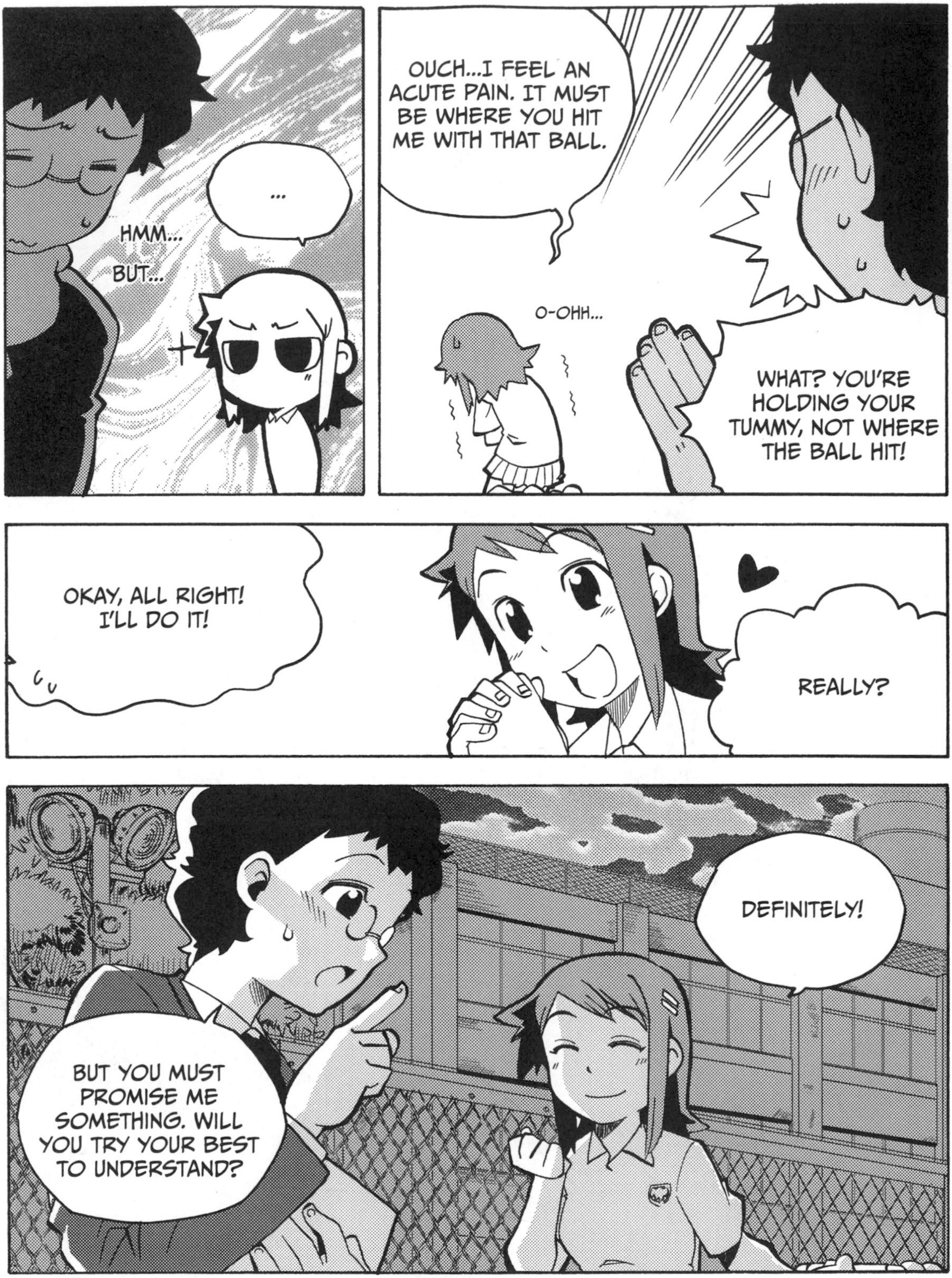

HMM...
BUT...
...
OUCH...I FEEL AN ACUTE PAIN. IT MUST BE WHERE YOU HIT ME WITH THAT BALL.
O-OHH...
WHAT? YOU'RE HOLDING YOUR TUMMY, NOT WHERE THE BALL HIT!
OKAY, ALL RIGHT! I'LL DO IT!
REALLY?
DEFINITELY!
BUT YOU MUST PROMISE ME SOMETHING. WILL YOU TRY YOUR BEST TO UNDERSTAND?

1

LAW OF ACTION AND REACTION

LAW OF ACTION AND REACTION

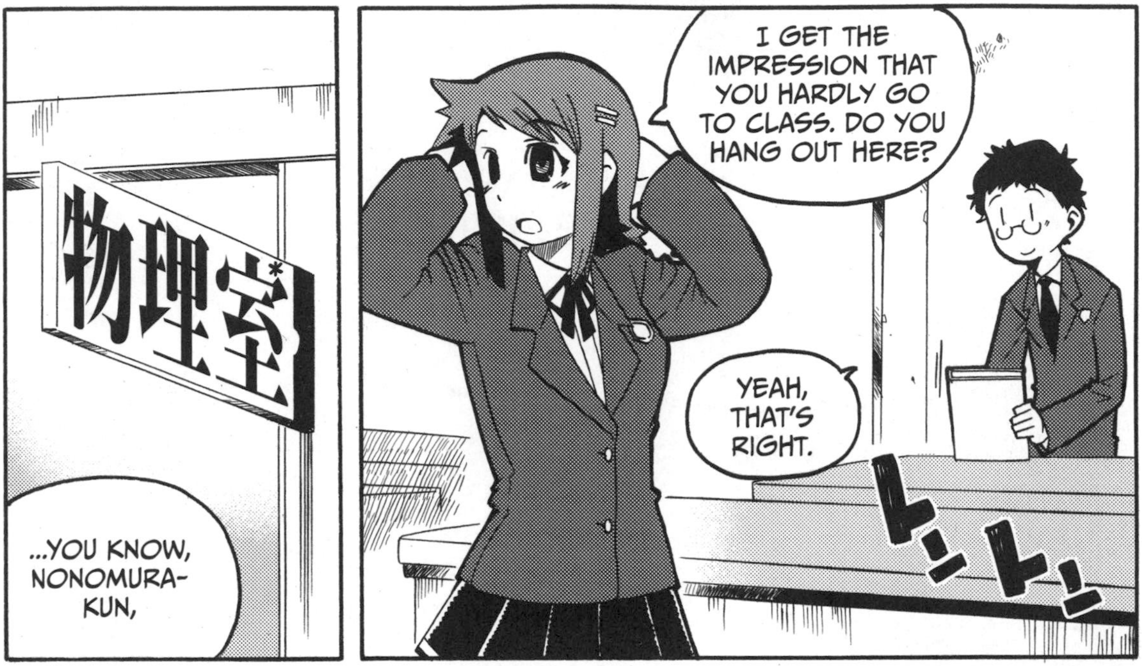
...YOU KNOW, NONOMURA-KUN,
物理室*
I GET THE IMPRESSION THAT YOU HARDLY GO TO CLASS. DO YOU HANG OUT HERE?
YEAH, THAT'S RIGHT.
トン トン

* LABORATORY

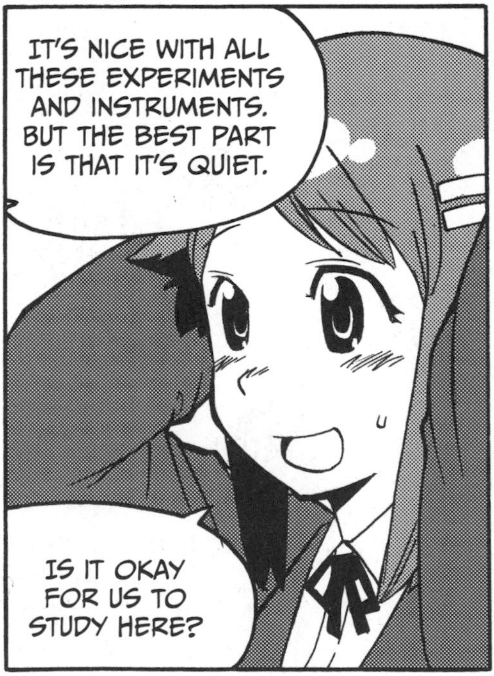
IT'S NICE WITH ALL THESE EXPERIMENTS AND INSTRUMENTS. BUT THE BEST PART IS THAT IT'S QUIET.
IS IT OKAY FOR US TO STUDY HERE?

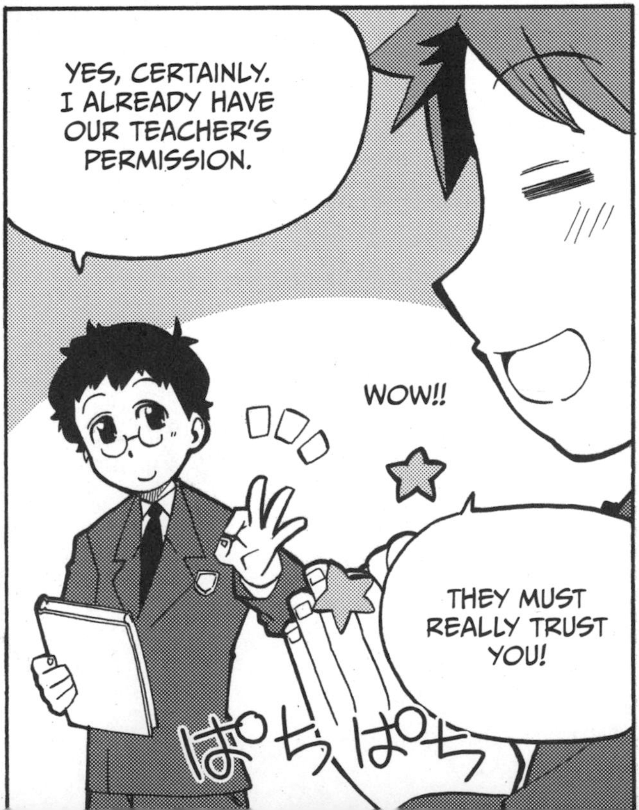
YES, CERTAINLY. I ALREADY HAVE OUR TEACHER'S PERMISSION.
WOW!!
THEY MUST REALLY TRUST YOU!
ぱち ぱち

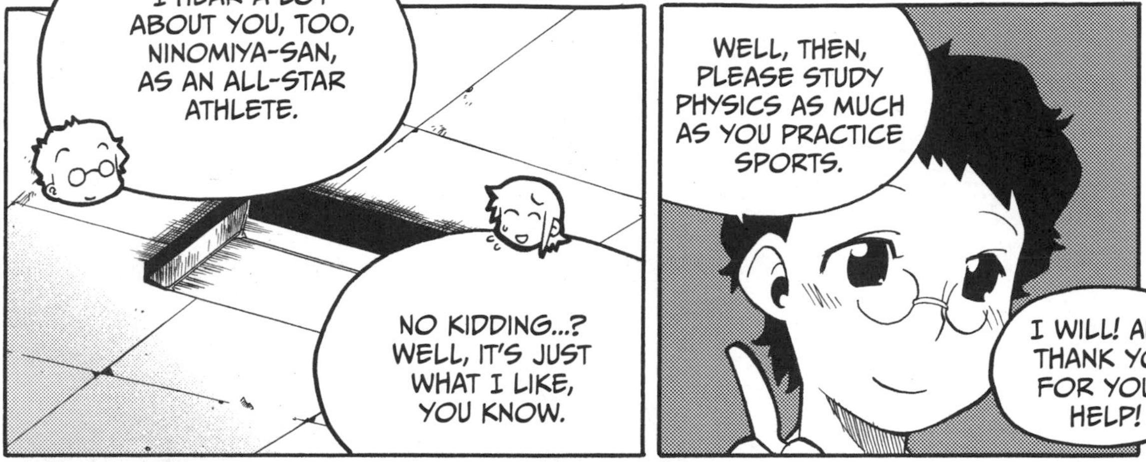
I HEAR A LOT ABOUT YOU, TOO, NINOMIYA-SAN, AS AN ALL-STAR ATHLETE.
NO KIDDING...? WELL, IT'S JUST WHAT I LIKE, YOU KNOW.
WELL, THEN, PLEASE STUDY PHYSICS AS MUCH AS YOU PRACTICE SPORTS.
I WILL! AND THANK YOU FOR YOUR HELP!

HOW THE LAW OF ACTION AND REACTION WORKS

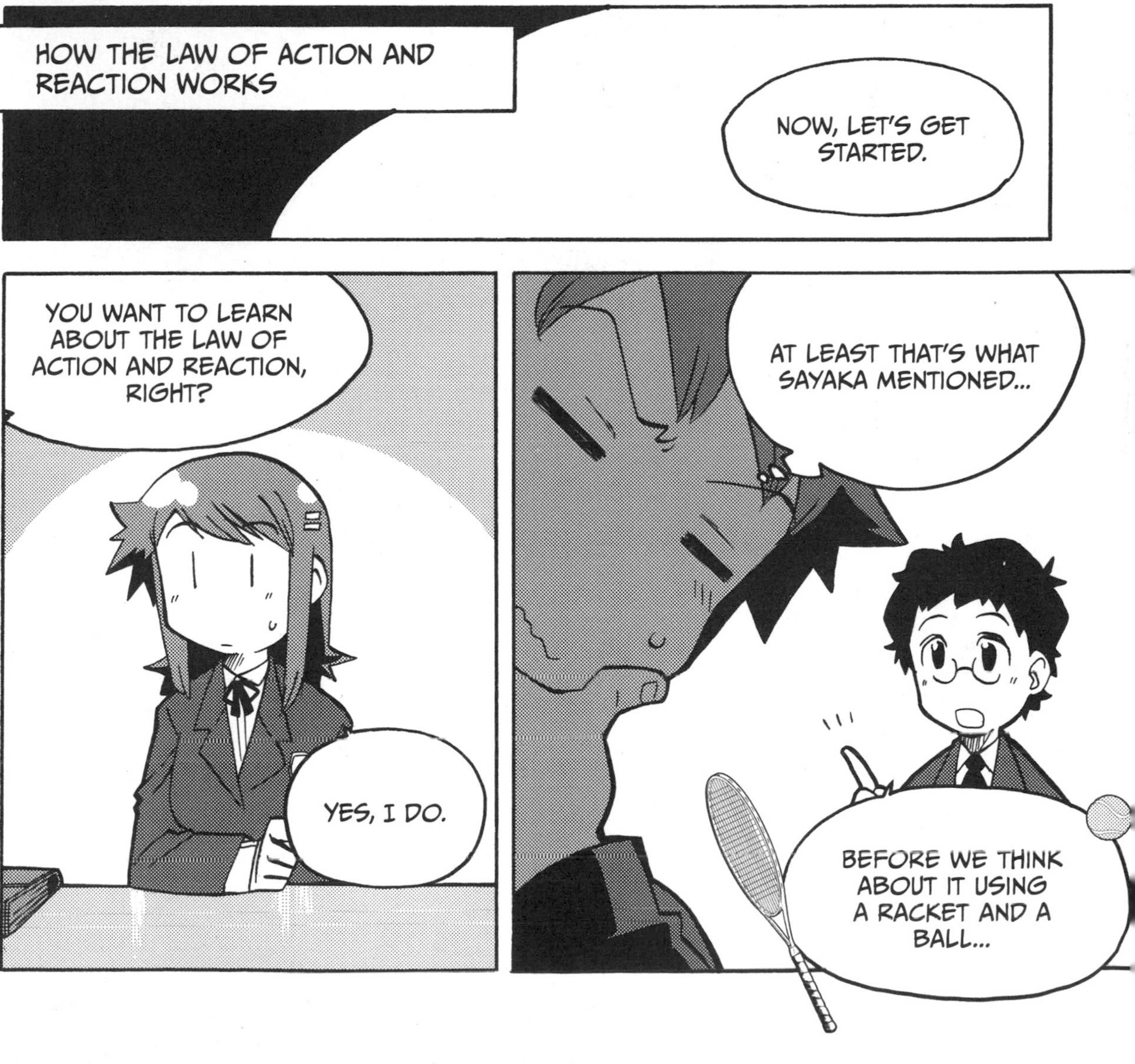
NOW, LET'S GET STARTED.
YOU WANT TO LEARN ABOUT THE LAW OF ACTION AND REACTION, RIGHT?
YES, I DO.
AT LEAST THAT'S WHAT SAYAKA MENTIONED...
BEFORE WE THINK ABOUT IT USING A RACKET AND A BALL...

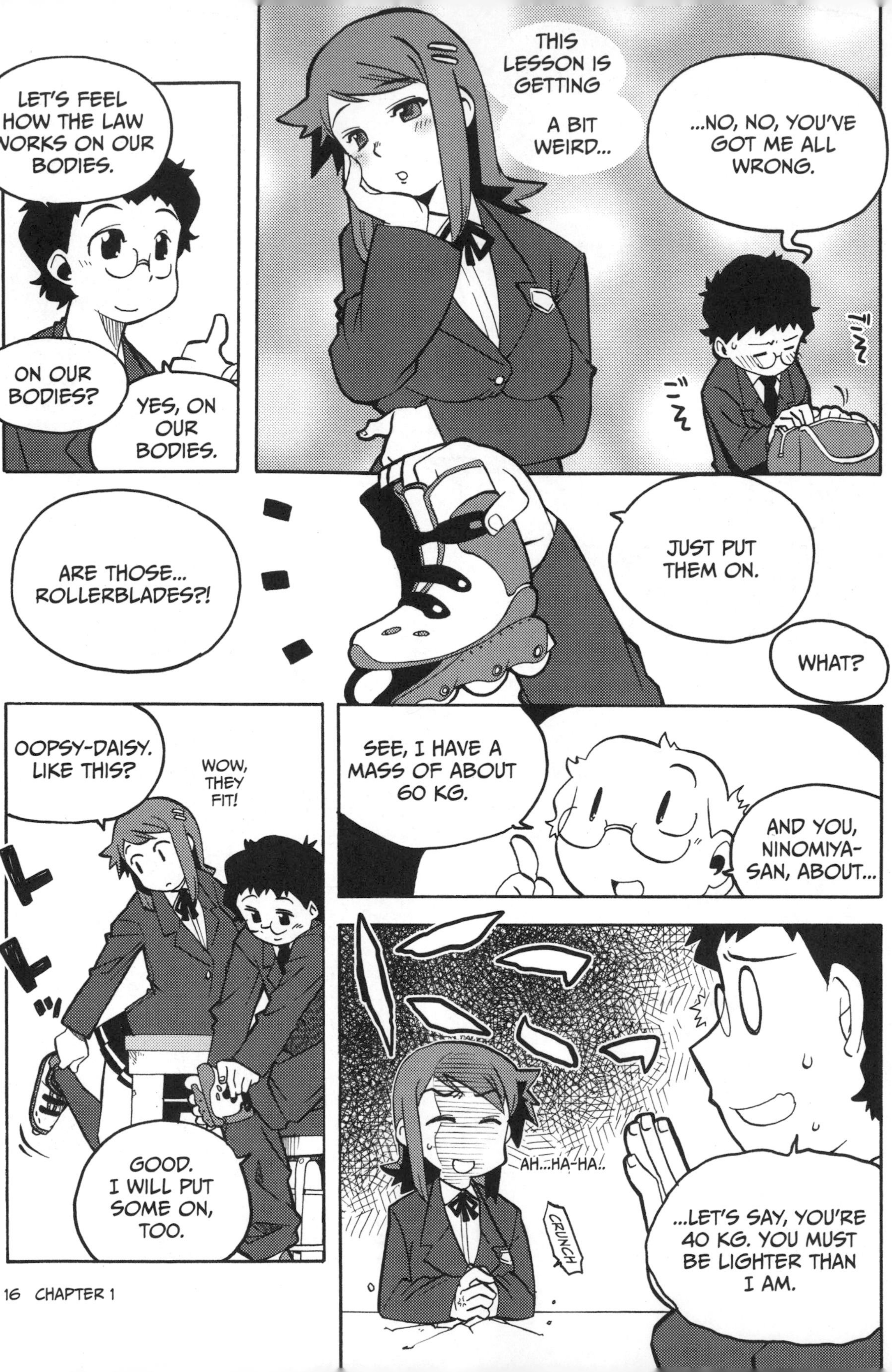
LET'S FEEL HOW THE LAW WORKS ON OUR BODIES.
ON OUR BODIES?
YES, ON OUR BODIES.
THIS LESSON IS GETTING
A BIT WEIRD...
...NO, NO, YOU'VE GOT ME ALL WRONG.
ARE THOSE... ROLLERBLADES?!
JUST PUT THEM ON.
WHAT?
OOPSY-DAISY. LIKE THIS?
WOW, THEY FIT!
GOOD. I WILL PUT SOME ON, TOO.
SEE, I HAVE A MASS OF ABOUT 60 KG.
AND YOU, NINOMIYA-SAN, ABOUT...
AH...HA-HA..
CRUNCH
...LET'S SAY, YOU'RE 40 KG. YOU MUST BE LIGHTER THAN I AM.

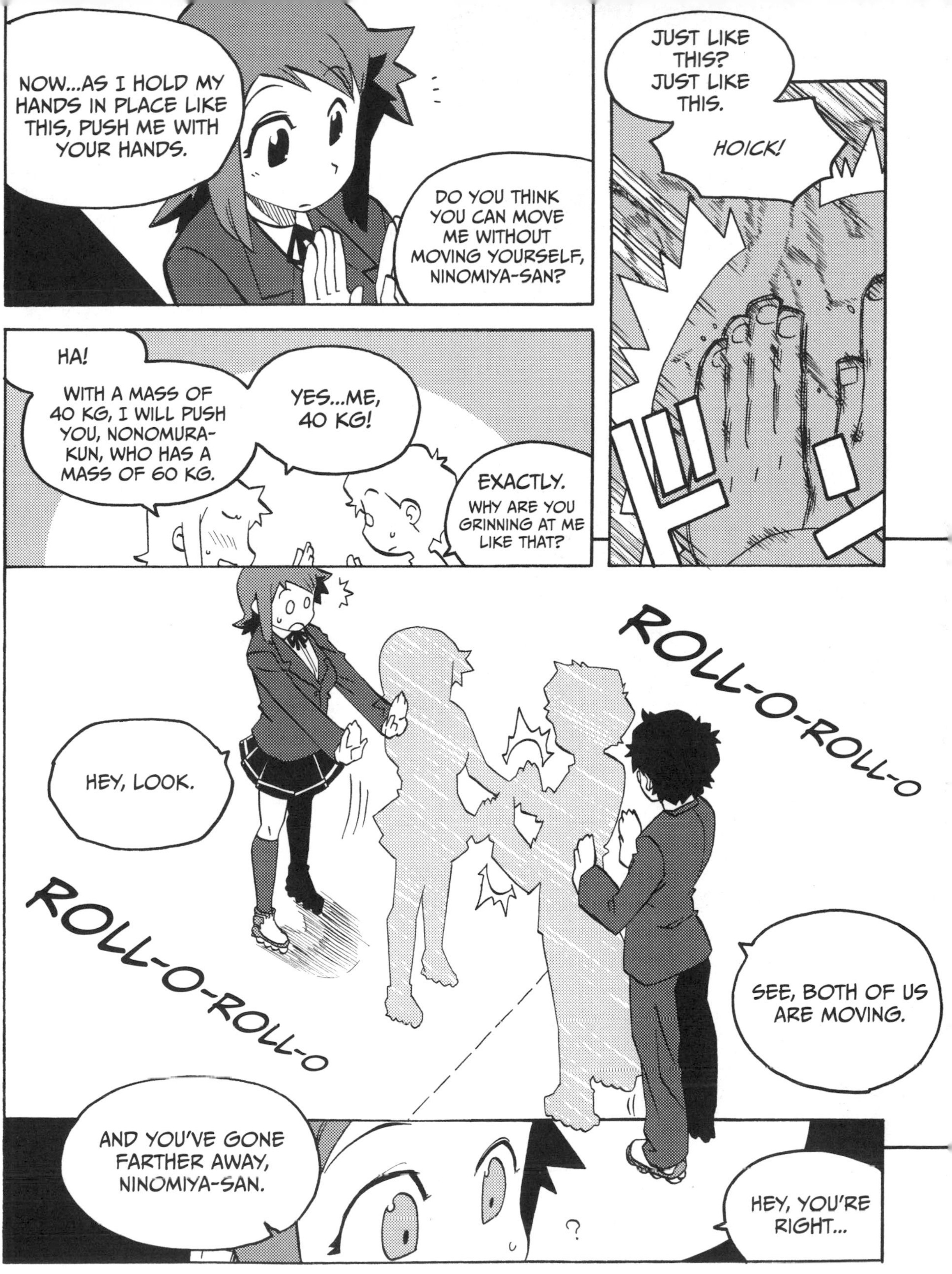
NOW...AS I HOLD MY HANDS IN PLACE LIKE THIS, PUSH ME WITH YOUR HANDS.
DO YOU THINK YOU CAN MOVE ME WITHOUT MOVING YOURSELF, NINOMIYA-SAN?
JUST LIKE THIS? JUST LIKE THIS.
HOICK!
HA!
WITH A MASS OF 40 KG, I WILL PUSH YOU, NONOMURA-KUN, WHO HAS A MASS OF 60 KG.
YES...ME, 40 KG!
EXACTLY.
WHY ARE YOU GRINNING AT ME LIKE THAT?
ROLL-O-ROLL-O
HEY, LOOK.
ROLL-O-ROLL-O
SEE, BOTH OF US ARE MOVING.
AND YOU'VE GONE FARTHER AWAY, NINOMIYA-SAN.
HEY, YOU'RE RIGHT...

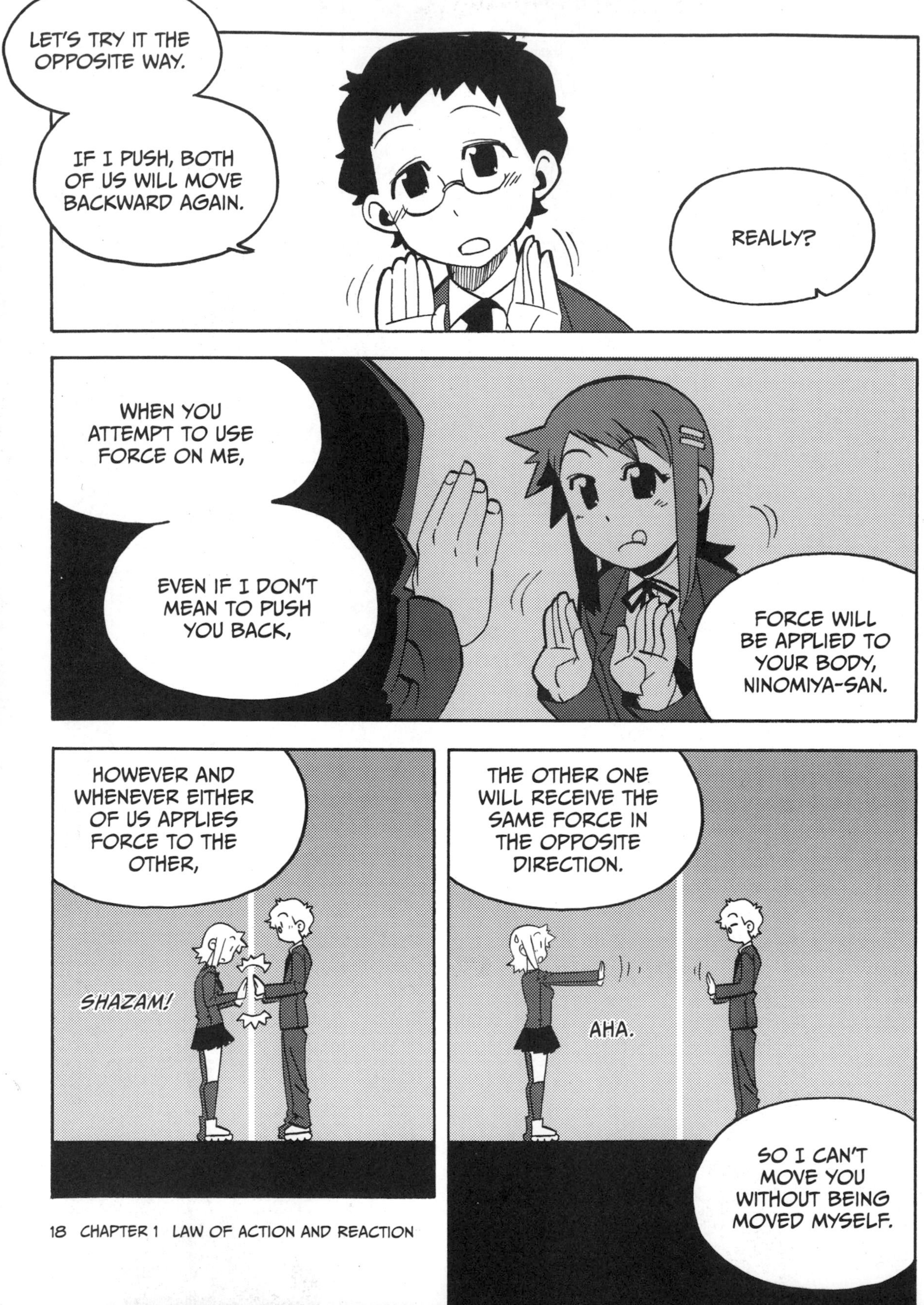
LET'S TRY IT THE OPPOSITE WAY.
IF I PUSH, BOTH OF US WILL MOVE BACKWARD AGAIN.
REALLY?
WHEN YOU ATTEMPT TO USE FORCE ON ME,
EVEN IF I DON'T MEAN TO PUSH YOU BACK,
FORCE WILL BE APPLIED TO YOUR BODY, NINOMIYA-SAN.
HOWEVER AND WHENEVER EITHER OF US APPLIES FORCE TO THE OTHER,
SHAZAM!
THE OTHER ONE WILL RECEIVE THE SAME FORCE IN THE OPPOSITE DIRECTION.
AHA.
SO I CAN'T MOVE YOU WITHOUT BEING MOVED MYSELF.

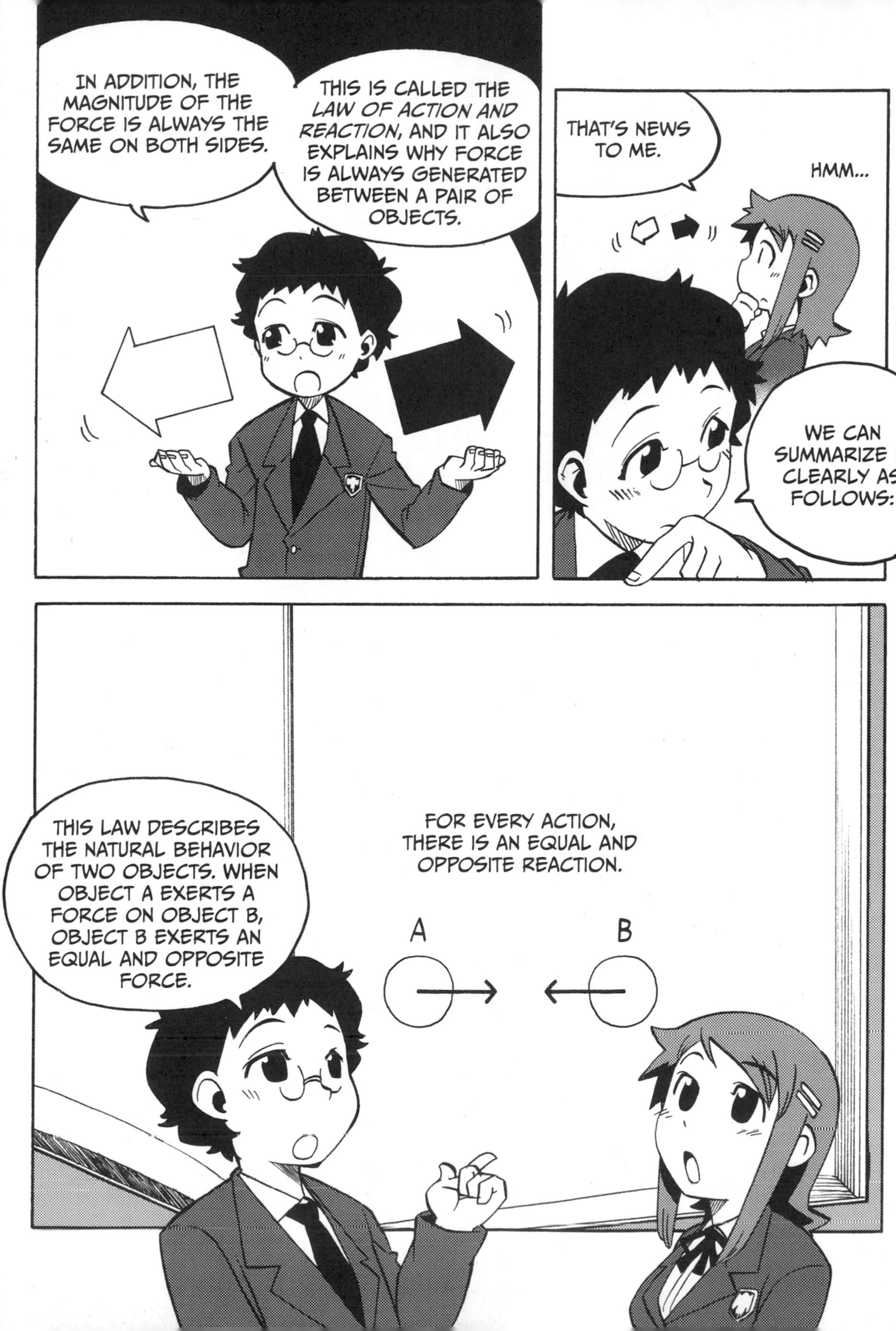
IN ADDITION, THE MAGNITUDE OF THE FORCE IS ALWAYS THE SAME ON BOTH SIDES.
THIS IS CALLED THE *LAW OF ACTION AND REACTION*, AND IT ALSO EXPLAINS WHY FORCE IS ALWAYS GENERATED BETWEEN A PAIR OF OBJECTS.
THAT'S NEWS TO ME.
HMM...
WE CAN SUMMARIZE IT CLEARLY AS FOLLOWS:
THIS LAW DESCRIBES THE NATURAL BEHAVIOR OF TWO OBJECTS. WHEN OBJECT A EXERTS A FORCE ON OBJECT B, OBJECT B EXERTS AN EQUAL AND OPPOSITE FORCE.
FOR EVERY ACTION, THERE IS AN EQUAL AND OPPOSITE REACTION.
A
B

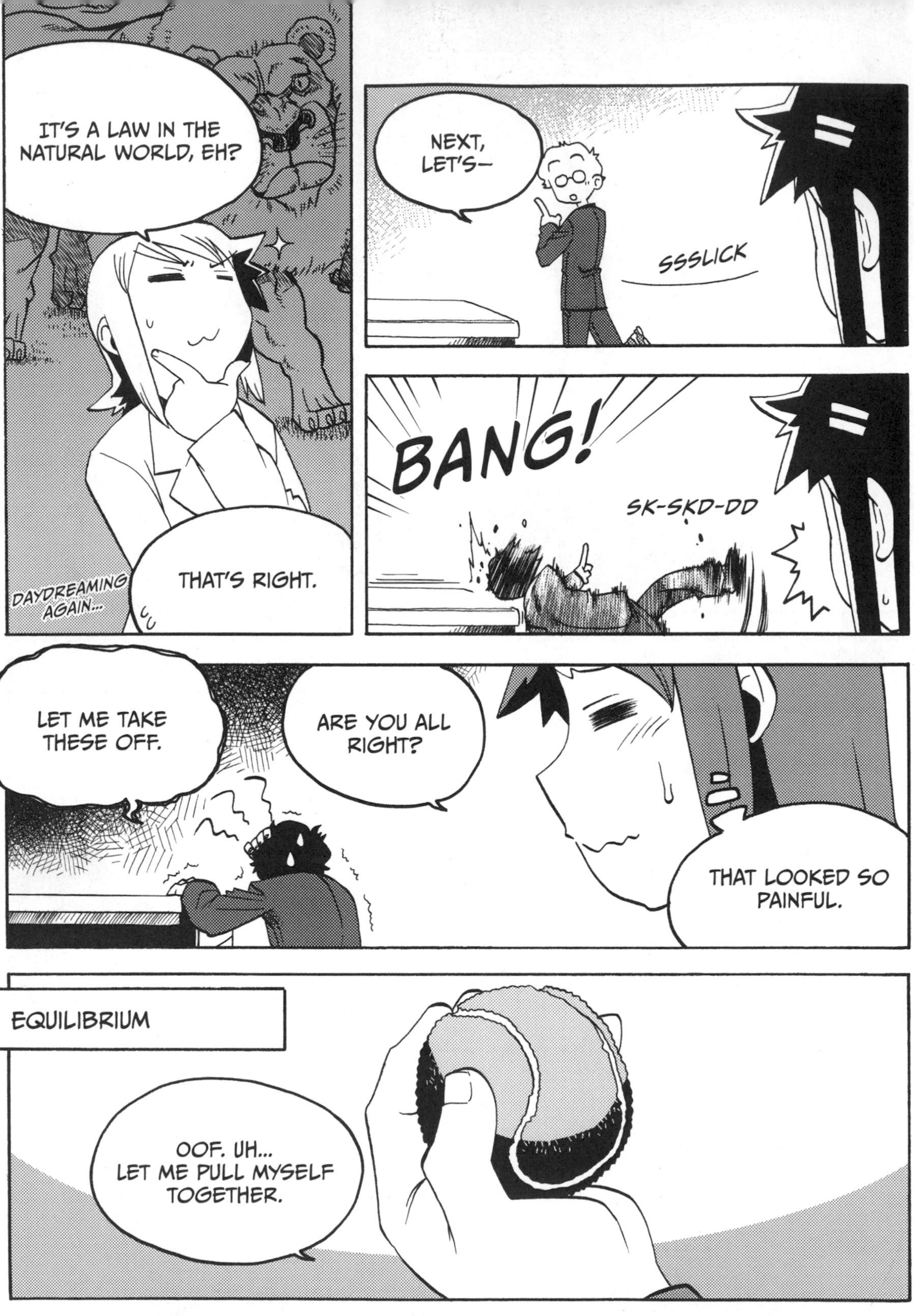
IT'S A LAW IN THE NATURAL WORLD, EH?
DAYDREAMING AGAIN...
THAT'S RIGHT.
NEXT, LET'S—
SSSLICK
BANG!
SK-SKD-DD
LET ME TAKE THESE OFF.
ARE YOU ALL RIGHT?
THAT LOOKED SO PAINFUL.
EQUILIBRIUM
OOF. UH... LET ME PULL MYSELF TOGETHER.

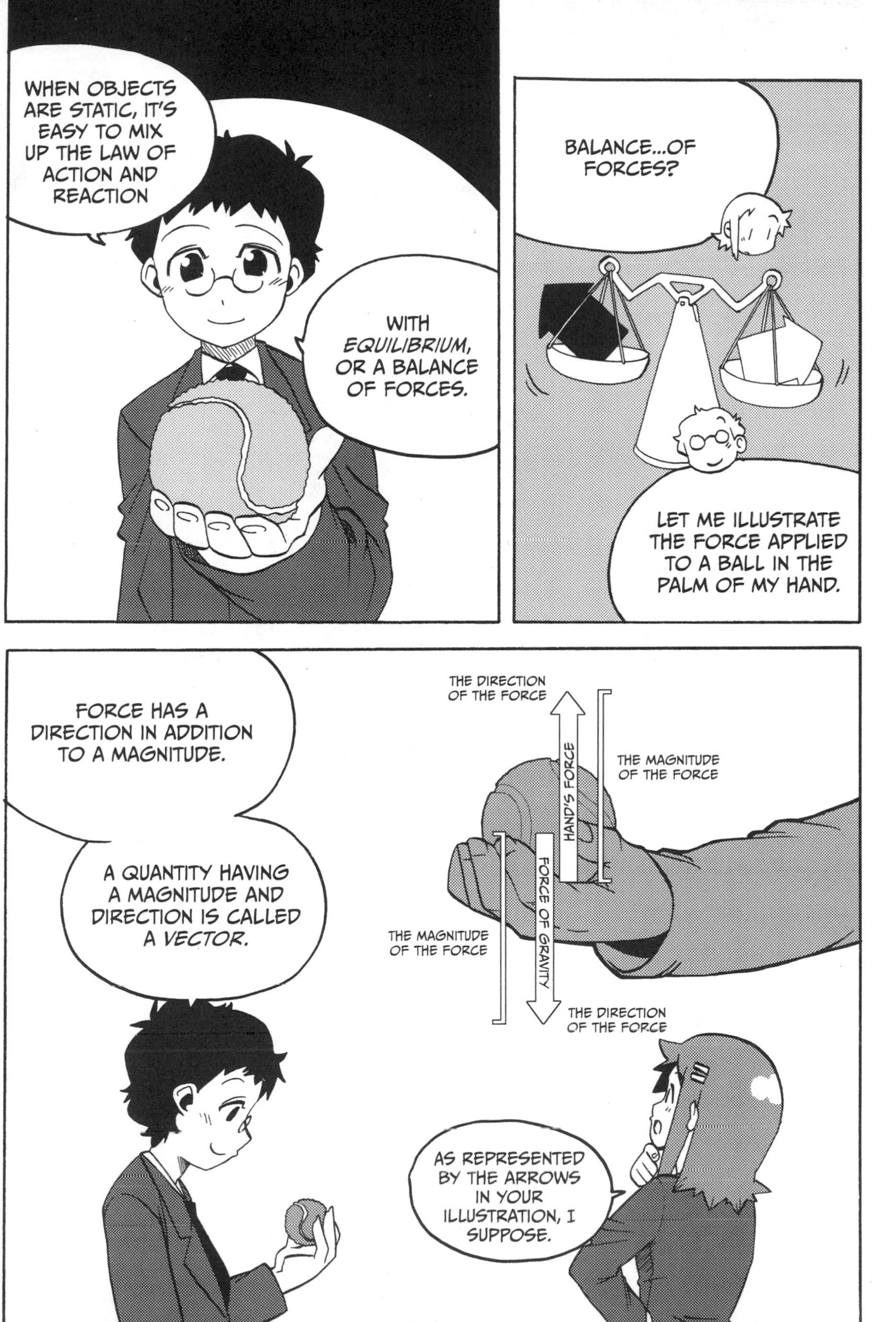
WHEN OBJECTS ARE STATIC, IT'S EASY TO MIX UP THE LAW OF ACTION AND REACTION
WITH *EQUILIBRIUM*, OR A BALANCE OF FORCES.
BALANCE...OF FORCES?
LET ME ILLUSTRATE THE FORCE APPLIED TO A BALL IN THE PALM OF MY HAND.
FORCE HAS A DIRECTION IN ADDITION TO A MAGNITUDE.
A QUANTITY HAVING A MAGNITUDE AND DIRECTION IS CALLED A *VECTOR*.
THE DIRECTION OF THE FORCE
HAND'S FORCE
THE MAGNITUDE OF THE FORCE
FORCE OF GRAVITY
THE MAGNITUDE OF THE FORCE
THE DIRECTION OF THE FORCE
AS REPRESENTED BY THE ARROWS IN YOUR ILLUSTRATION, I SUPPOSE.

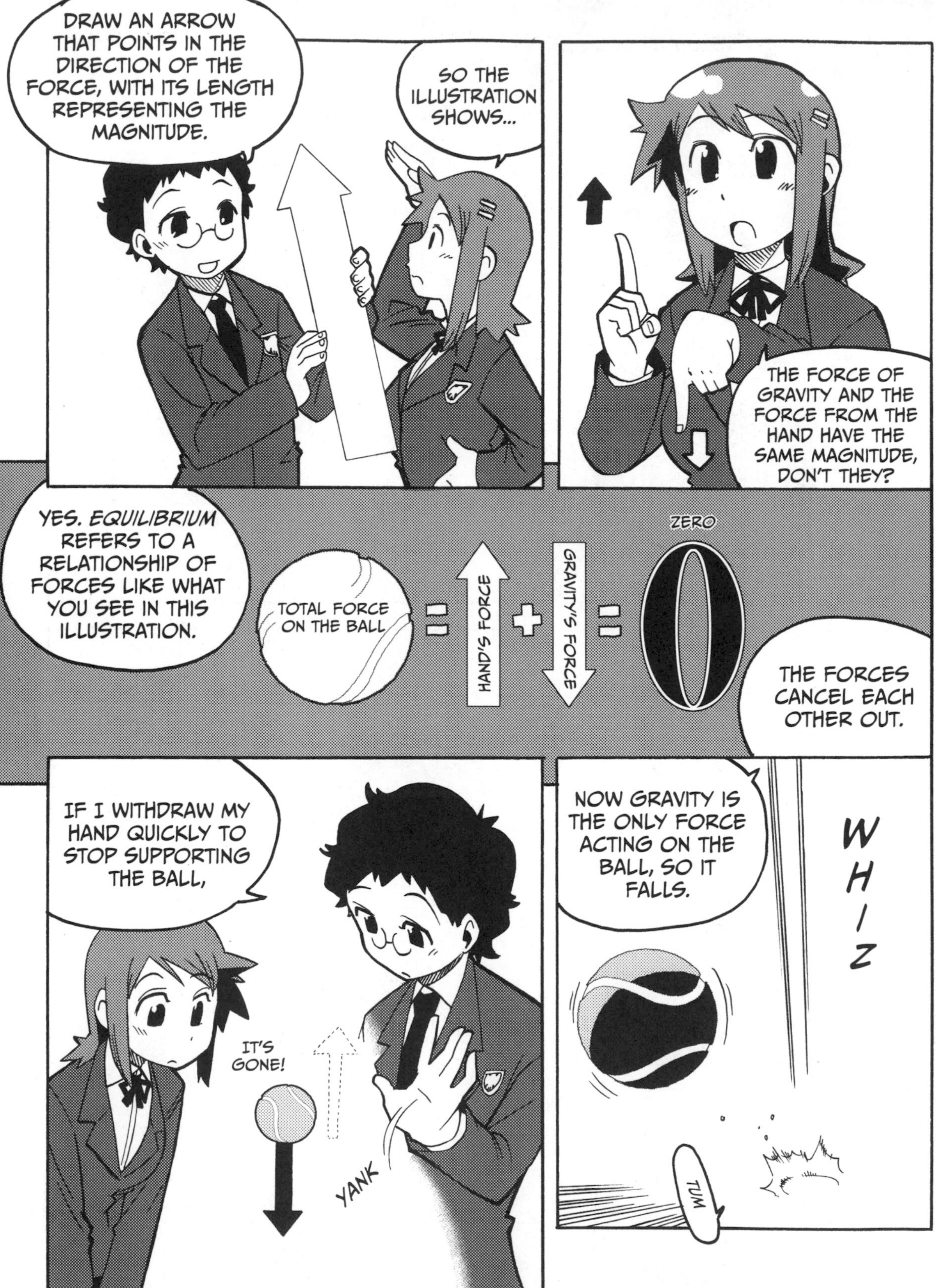
DRAW AN ARROW THAT POINTS IN THE DIRECTION OF THE FORCE, WITH ITS LENGTH REPRESENTING THE MAGNITUDE.
SO THE ILLUSTRATION SHOWS...
THE FORCE OF GRAVITY AND THE FORCE FROM THE HAND HAVE THE SAME MAGNITUDE, DON'T THEY?
YES. *EQUILIBRIUM* REFERS TO A RELATIONSHIP OF FORCES LIKE WHAT YOU SEE IN THIS ILLUSTRATION.
TOTAL FORCE ON THE BALL
=
HAND'S FORCE
+
GRAVITY'S FORCE
=
ZERO
0
THE FORCES CANCEL EACH OTHER OUT.
IF I WITHDRAW MY HAND QUICKLY TO STOP SUPPORTING THE BALL,
IT'S GONE!
YANK
NOW GRAVITY IS THE ONLY FORCE ACTING ON THE BALL, SO IT FALLS.
WHIZ
TUM

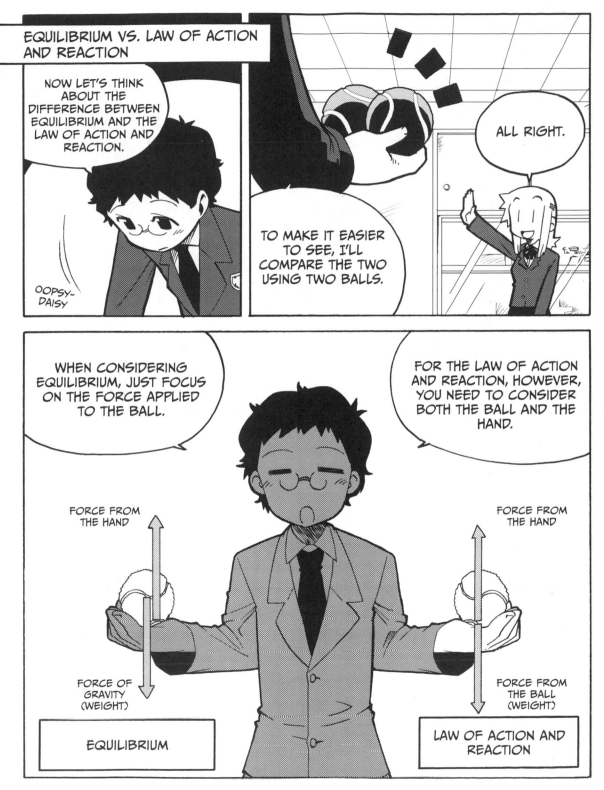

EQUILIBRIUM VS. LAW OF ACTION AND REACTION

NOW LET'S THINK ABOUT THE DIFFERENCE BETWEEN EQUILIBRIUM AND THE LAW OF ACTION AND REACTION.

OOPSY-DAISY

ALL RIGHT.

TO MAKE IT EASIER TO SEE, I'LL COMPARE THE TWO USING TWO BALLS.

WHEN CONSIDERING EQUILIBRIUM, JUST FOCUS ON THE FORCE APPLIED TO THE BALL.

FOR THE LAW OF ACTION AND REACTION, HOWEVER, YOU NEED TO CONSIDER BOTH THE BALL AND THE HAND.

FORCE FROM THE HAND

FORCE FROM THE HAND

FORCE OF GRAVITY (WEIGHT)

FORCE FROM THE BALL (WEIGHT)

EQUILIBRIUM

LAW OF ACTION AND REACTION

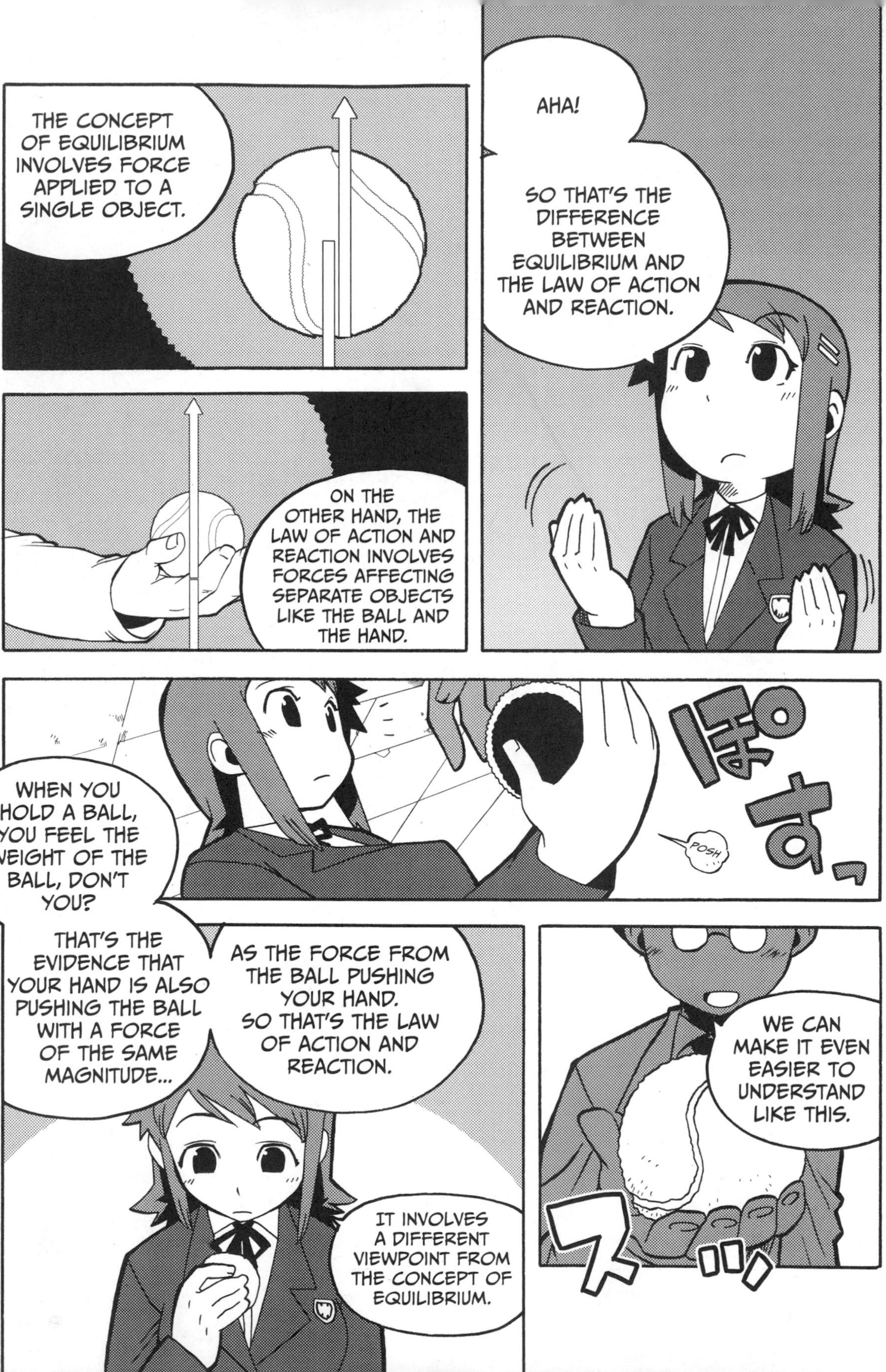
THE CONCEPT OF EQUILIBRIUM INVOLVES FORCE APPLIED TO A SINGLE OBJECT.
ON THE OTHER HAND, THE LAW OF ACTION AND REACTION INVOLVES FORCES AFFECTING SEPARATE OBJECTS LIKE THE BALL AND THE HAND.
AHA!
SO THAT'S THE DIFFERENCE BETWEEN EQUILIBRIUM AND THE LAW OF ACTION AND REACTION.
WHEN YOU HOLD A BALL, YOU FEEL THE WEIGHT OF THE BALL, DON'T YOU?
ぽす
POSH
THAT'S THE EVIDENCE THAT YOUR HAND IS ALSO PUSHING THE BALL WITH A FORCE OF THE SAME MAGNITUDE...
AS THE FORCE FROM THE BALL PUSHING YOUR HAND. SO THAT'S THE LAW OF ACTION AND REACTION.
IT INVOLVES A DIFFERENT VIEWPOINT FROM THE CONCEPT OF EQUILIBRIUM.
WE CAN MAKE IT EVEN EASIER TO UNDERSTAND LIKE THIS.
ス

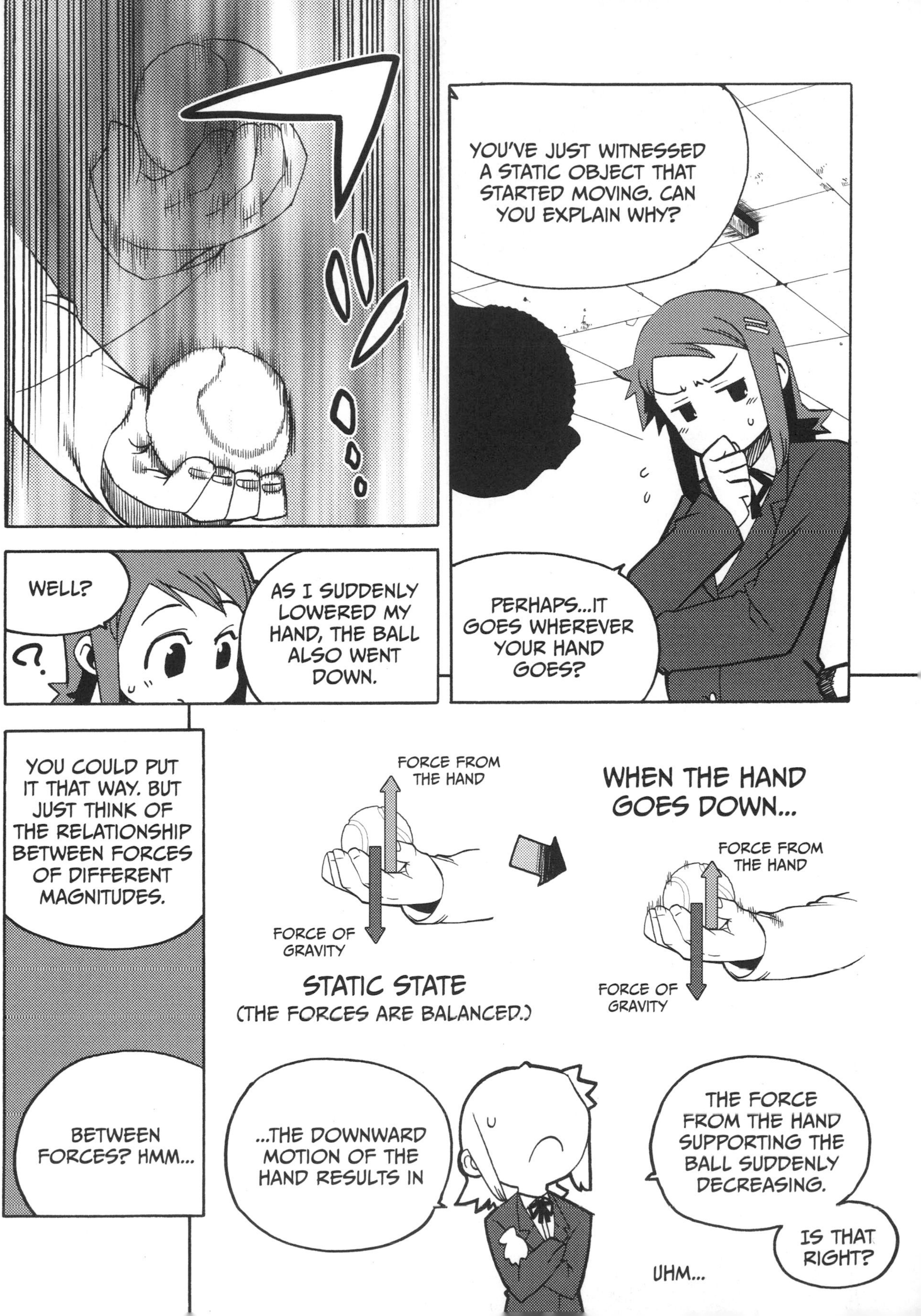
YOU'VE JUST WITNESSED A STATIC OBJECT THAT STARTED MOVING. CAN YOU EXPLAIN WHY?
WELL?
AS I SUDDENLY LOWERED MY HAND, THE BALL ALSO WENT DOWN.
PERHAPS...IT GOES WHEREVER YOUR HAND GOES?
YOU COULD PUT IT THAT WAY. BUT JUST THINK OF THE RELATIONSHIP BETWEEN FORCES OF DIFFERENT MAGNITUDES.
FORCE FROM THE HAND
FORCE OF GRAVITY
STATIC STATE
(THE FORCES ARE BALANCED.)
WHEN THE HAND GOES DOWN...
FORCE FROM THE HAND
FORCE OF GRAVITY
BETWEEN FORCES? HMM...
...THE DOWNWARD MOTION OF THE HAND RESULTS IN
THE FORCE FROM THE HAND SUPPORTING THE BALL SUDDENLY DECREASING.
IS THAT RIGHT?
UHM...

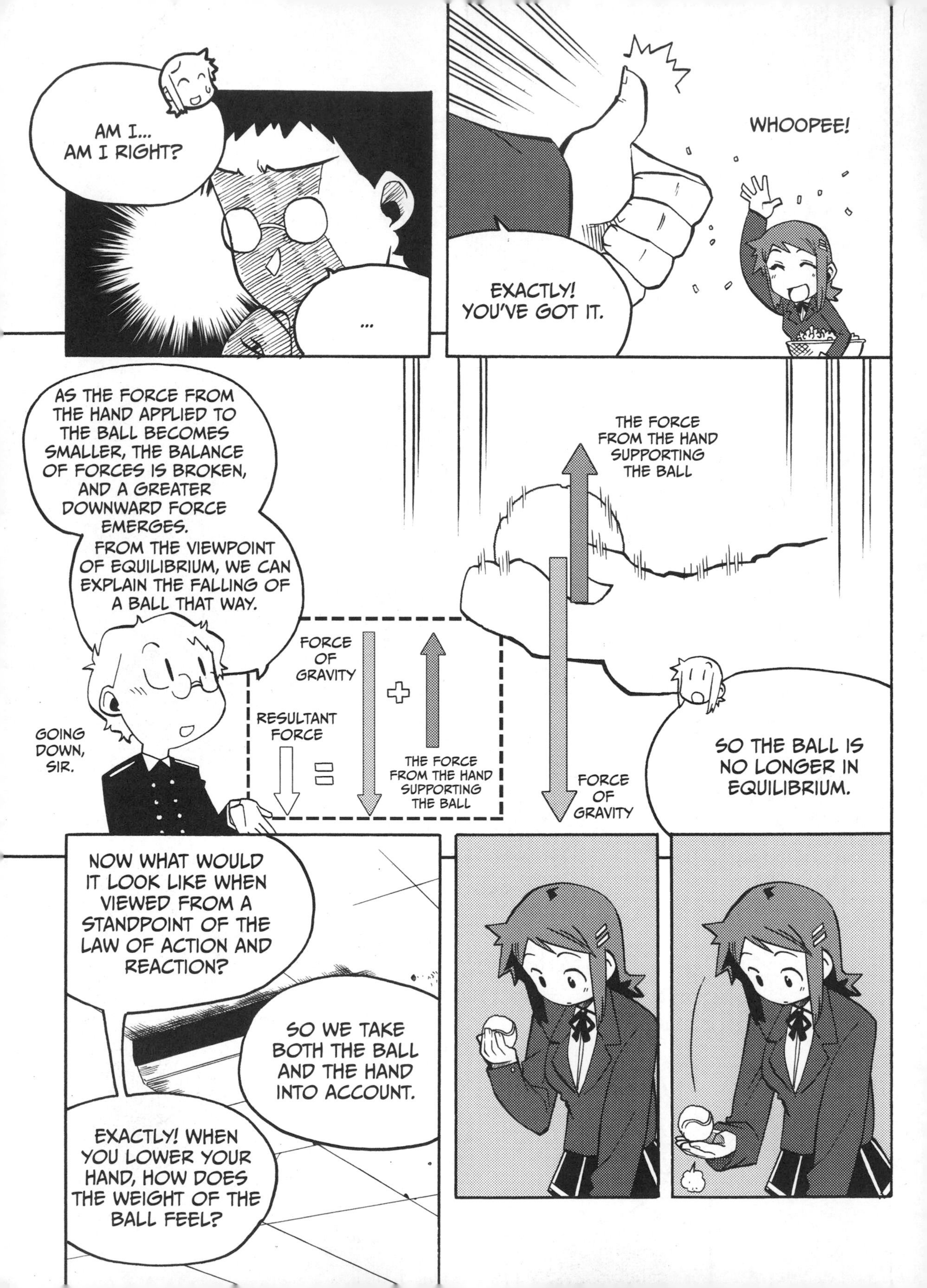
AM I... AM I RIGHT?
...
EXACTLY! YOU'VE GOT IT.
WHOOPEE!
AS THE FORCE FROM THE HAND APPLIED TO THE BALL BECOMES SMALLER, THE BALANCE OF FORCES IS BROKEN, AND A GREATER DOWNWARD FORCE EMERGES.
FROM THE VIEWPOINT OF EQUILIBRIUM, WE CAN EXPLAIN THE FALLING OF A BALL THAT WAY.
GOING DOWN, SIR.
FORCE OF GRAVITY
RESULTANT FORCE
THE FORCE FROM THE HAND SUPPORTING THE BALL
THE FORCE FROM THE HAND SUPPORTING THE BALL
FORCE OF GRAVITY
SO THE BALL IS NO LONGER IN EQUILIBRIUM.
NOW WHAT WOULD IT LOOK LIKE WHEN VIEWED FROM A STANDPOINT OF THE LAW OF ACTION AND REACTION?
SO WE TAKE BOTH THE BALL AND THE HAND INTO ACCOUNT.
EXACTLY! WHEN YOU LOWER YOUR HAND, HOW DOES THE WEIGHT OF THE BALL FEEL?

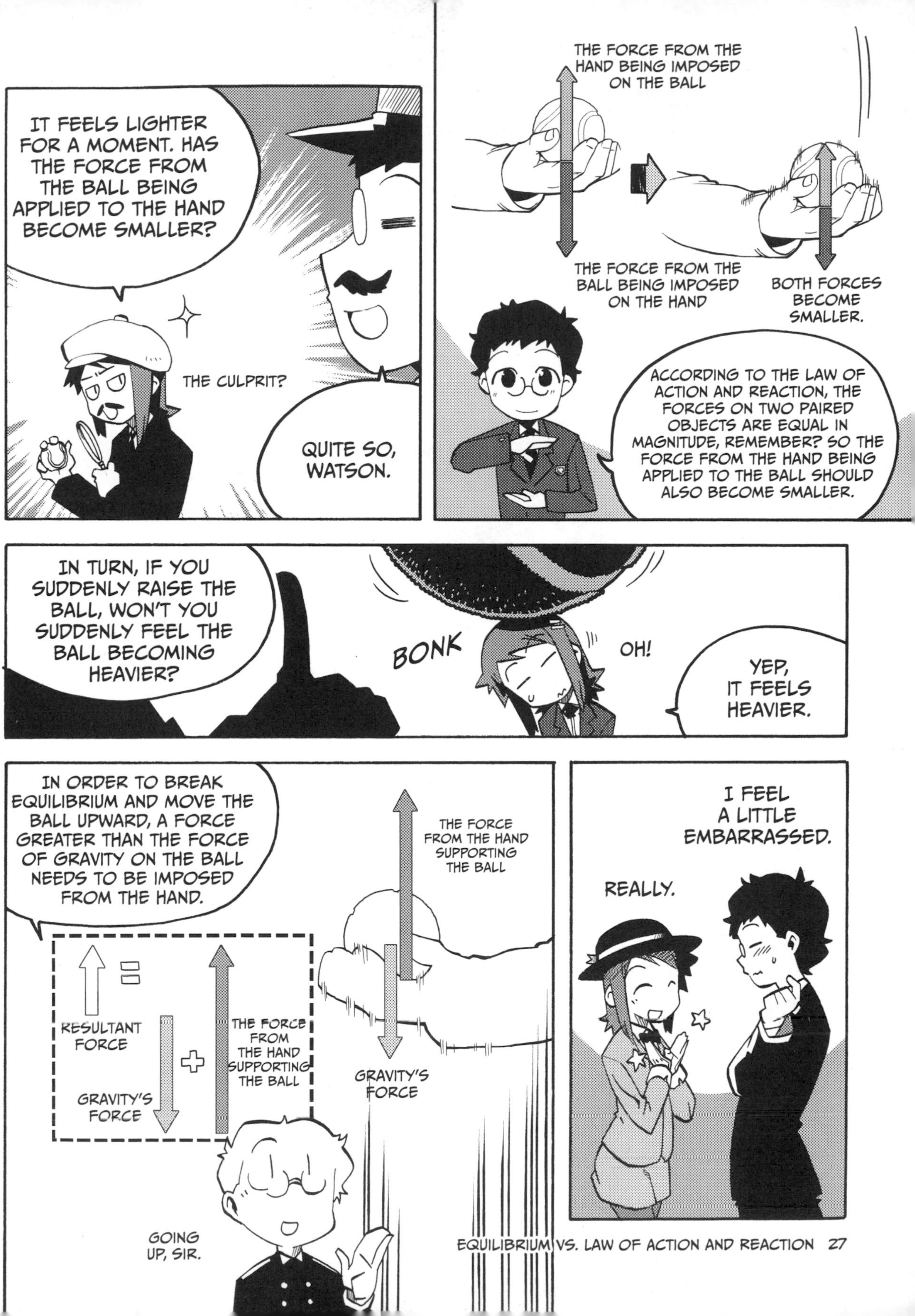
IT FEELS LIGHTER FOR A MOMENT. HAS THE FORCE FROM THE BALL BEING APPLIED TO THE HAND BECOME SMALLER?
THE CULPRIT?
QUITE SO, WATSON.
THE FORCE FROM THE HAND BEING IMPOSED ON THE BALL
THE FORCE FROM THE BALL BEING IMPOSED ON THE HAND
BOTH FORCES BECOME SMALLER.
ACCORDING TO THE LAW OF ACTION AND REACTION, THE FORCES ON TWO PAIRED OBJECTS ARE EQUAL IN MAGNITUDE, REMEMBER? SO THE FORCE FROM THE HAND BEING APPLIED TO THE BALL SHOULD ALSO BECOME SMALLER.
IN TURN, IF YOU SUDDENLY RAISE THE BALL, WON'T YOU SUDDENLY FEEL THE BALL BECOMING HEAVIER?
BONK
OH!
YEP, IT FEELS HEAVIER.
IN ORDER TO BREAK EQUILIBRIUM AND MOVE THE BALL UPWARD, A FORCE GREATER THAN THE FORCE OF GRAVITY ON THE BALL NEEDS TO BE IMPOSED FROM THE HAND.
RESULTANT FORCE
=
GRAVITY'S FORCE
+
THE FORCE FROM THE HAND SUPPORTING THE BALL
THE FORCE FROM THE HAND SUPPORTING THE BALL
GRAVITY'S FORCE
GOING UP, SIR.
I FEEL A LITTLE EMBARRASSED.
REALLY.

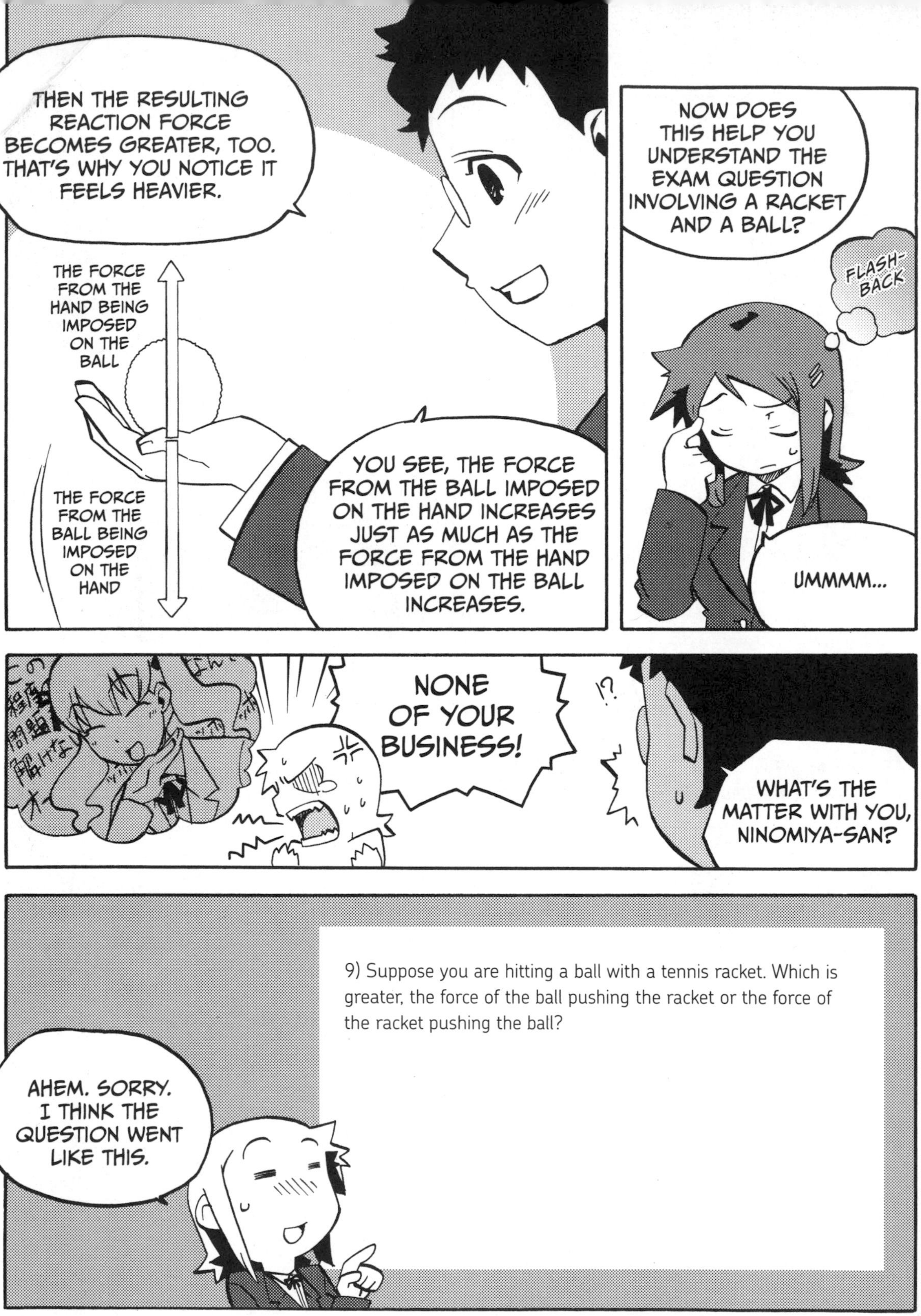
THEN THE RESULTING REACTION FORCE BECOMES GREATER, TOO. THAT'S WHY YOU NOTICE IT FEELS HEAVIER.
THE FORCE FROM THE HAND BEING IMPOSED ON THE BALL
THE FORCE FROM THE BALL BEING IMPOSED ON THE HAND
YOU SEE, THE FORCE FROM THE BALL IMPOSED ON THE HAND INCREASES JUST AS MUCH AS THE FORCE FROM THE HAND IMPOSED ON THE BALL INCREASES.
NOW DOES THIS HELP YOU UNDERSTAND THE EXAM QUESTION INVOLVING A RACKET AND A BALL?
FLASH-BACK
UMMMM...
NONE OF YOUR BUSINESS!
!?
WHAT'S THE MATTER WITH YOU, NINOMIYA-SAN?
AHEM. SORRY. I THINK THE QUESTION WENT LIKE THIS.
9) Suppose you are hitting a ball with a tennis racket. Which is greater, the force of the ball pushing the racket or the force of the racket pushing the ball?

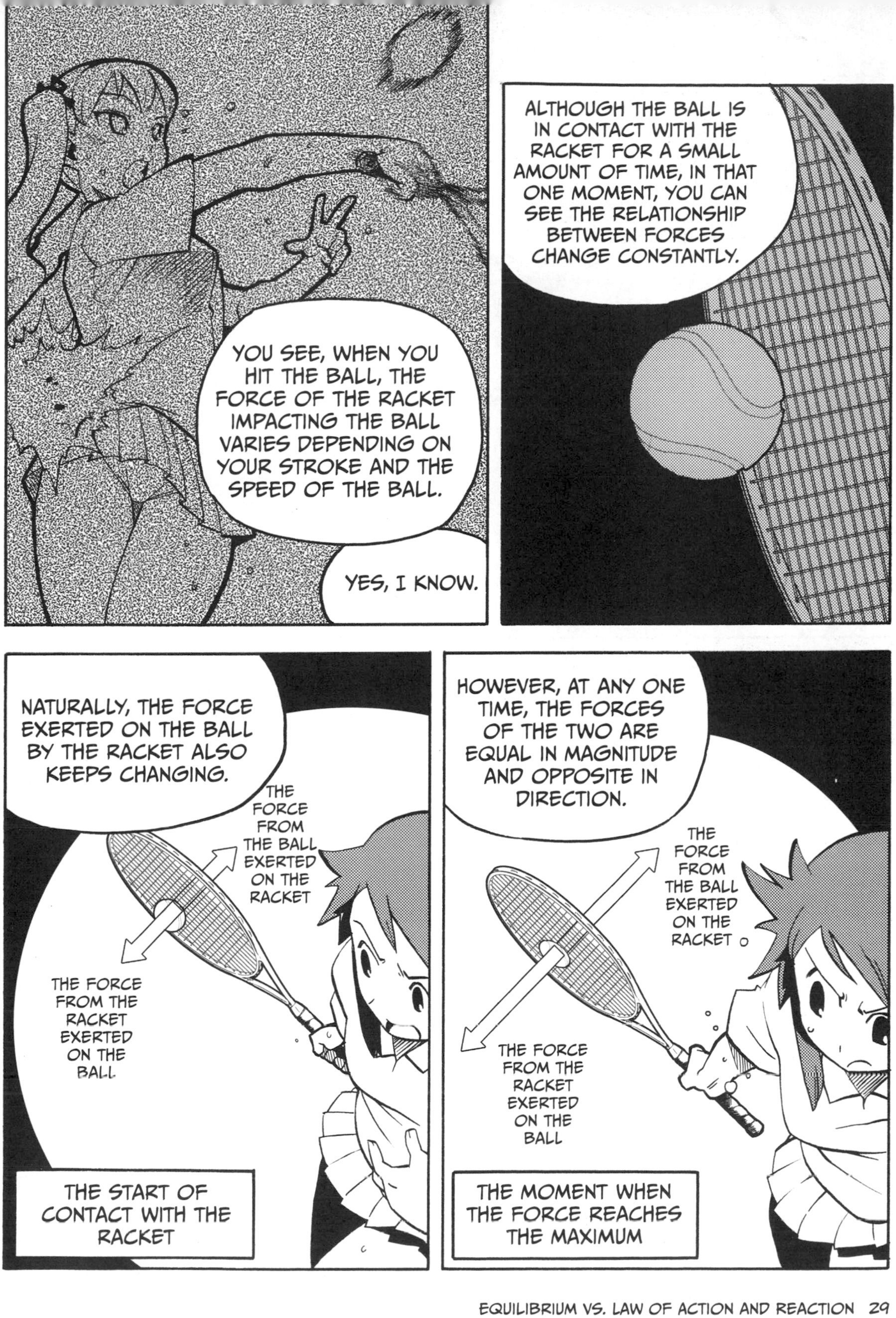
YOU SEE, WHEN YOU HIT THE BALL, THE FORCE OF THE RACKET IMPACTING THE BALL VARIES DEPENDING ON YOUR STROKE AND THE SPEED OF THE BALL.
YES, I KNOW.
ALTHOUGH THE BALL IS IN CONTACT WITH THE RACKET FOR A SMALL AMOUNT OF TIME, IN THAT ONE MOMENT, YOU CAN SEE THE RELATIONSHIP BETWEEN FORCES CHANGE CONSTANTLY.
NATURALLY, THE FORCE EXERTED ON THE BALL BY THE RACKET ALSO KEEPS CHANGING.
THE FORCE FROM THE BALL EXERTED ON THE RACKET
THE FORCE FROM THE RACKET EXERTED ON THE BALL
THE START OF CONTACT WITH THE RACKET
HOWEVER, AT ANY ONE TIME, THE FORCES OF THE TWO ARE EQUAL IN MAGNITUDE AND OPPOSITE IN DIRECTION.
THE FORCE FROM THE BALL EXERTED ON THE RACKET
THE FORCE FROM THE RACKET EXERTED ON THE BALL
THE MOMENT WHEN THE FORCE REACHES THE MAXIMUM

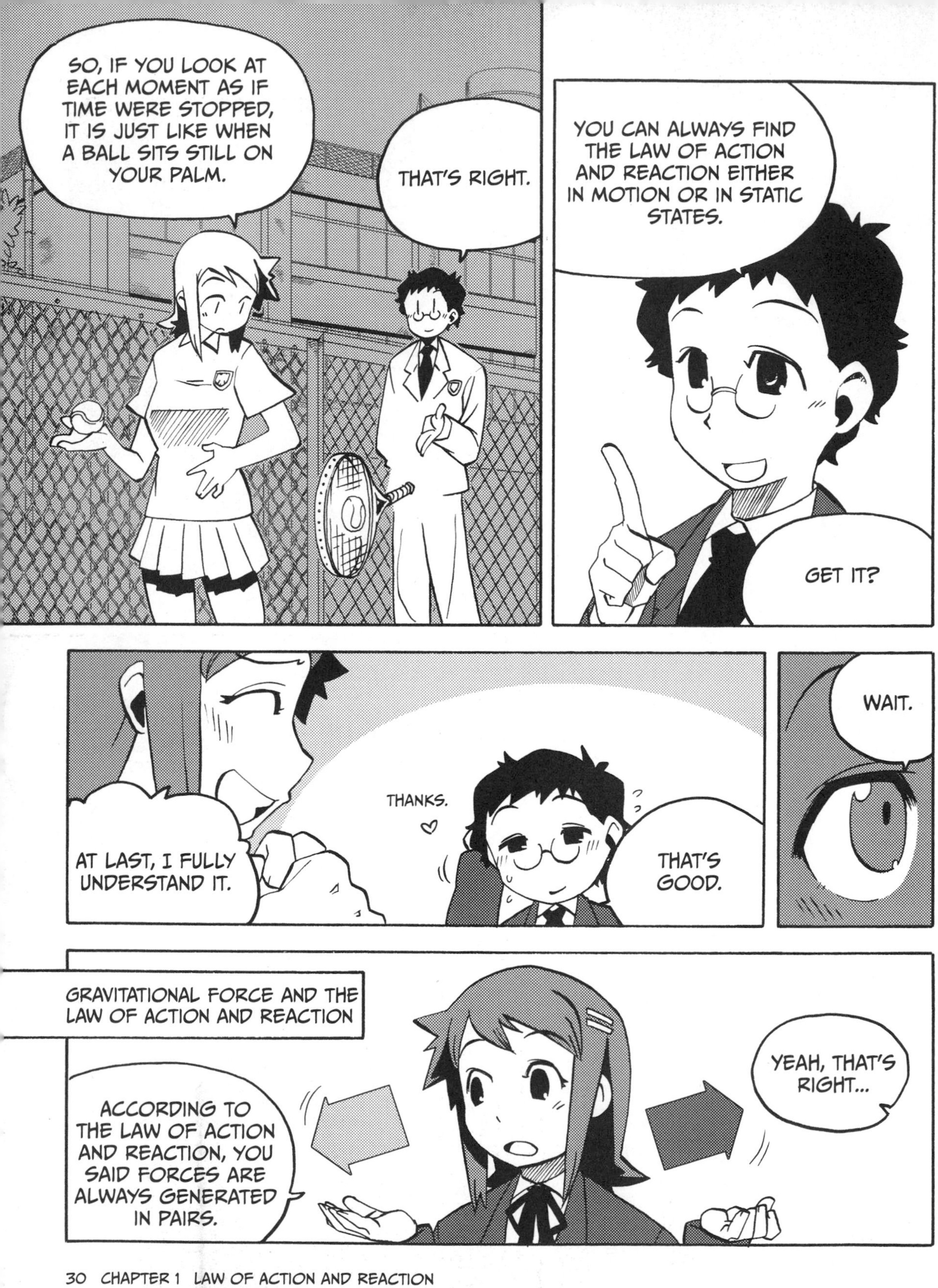
SO, IF YOU LOOK AT EACH MOMENT AS IF TIME WERE STOPPED, IT IS JUST LIKE WHEN A BALL SITS STILL ON YOUR PALM.
THAT'S RIGHT.
YOU CAN ALWAYS FIND THE LAW OF ACTION AND REACTION EITHER IN MOTION OR IN STATIC STATES.
GET IT?
AT LAST, I FULLY UNDERSTAND IT.
THANKS.
THAT'S GOOD.
WAIT.
GRAVITATIONAL FORCE AND THE LAW OF ACTION AND REACTION
ACCORDING TO THE LAW OF ACTION AND REACTION, YOU SAID FORCES ARE ALWAYS GENERATED IN PAIRS.
YEAH, THAT'S RIGHT...

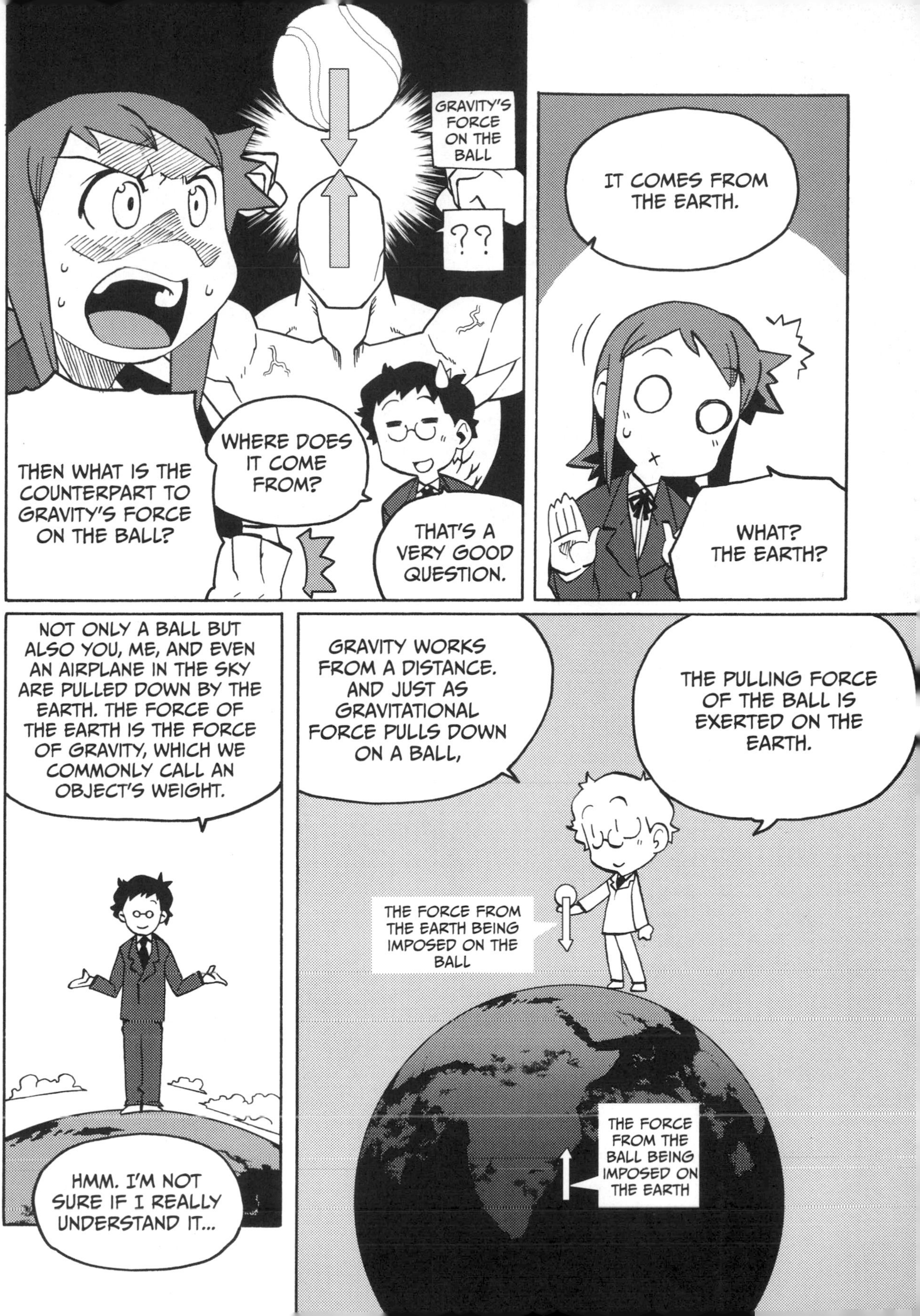
GRAVITY'S FORCE ON THE BALL
??
THEN WHAT IS THE COUNTERPART TO GRAVITY'S FORCE ON THE BALL?
WHERE DOES IT COME FROM?
THAT'S A VERY GOOD QUESTION.
IT COMES FROM THE EARTH.
WHAT? THE EARTH?
NOT ONLY A BALL BUT ALSO YOU, ME, AND EVEN AN AIRPLANE IN THE SKY ARE PULLED DOWN BY THE EARTH. THE FORCE OF THE EARTH IS THE FORCE OF GRAVITY, WHICH WE COMMONLY CALL AN OBJECT'S WEIGHT.
HMM. I'M NOT SURE IF I REALLY UNDERSTAND IT...
GRAVITY WORKS FROM A DISTANCE. AND JUST AS GRAVITATIONAL FORCE PULLS DOWN ON A BALL,
THE PULLING FORCE OF THE BALL IS EXERTED ON THE EARTH.
THE FORCE FROM THE EARTH BEING IMPOSED ON THE BALL
THE FORCE FROM THE BALL BEING IMPOSED ON THE EARTH

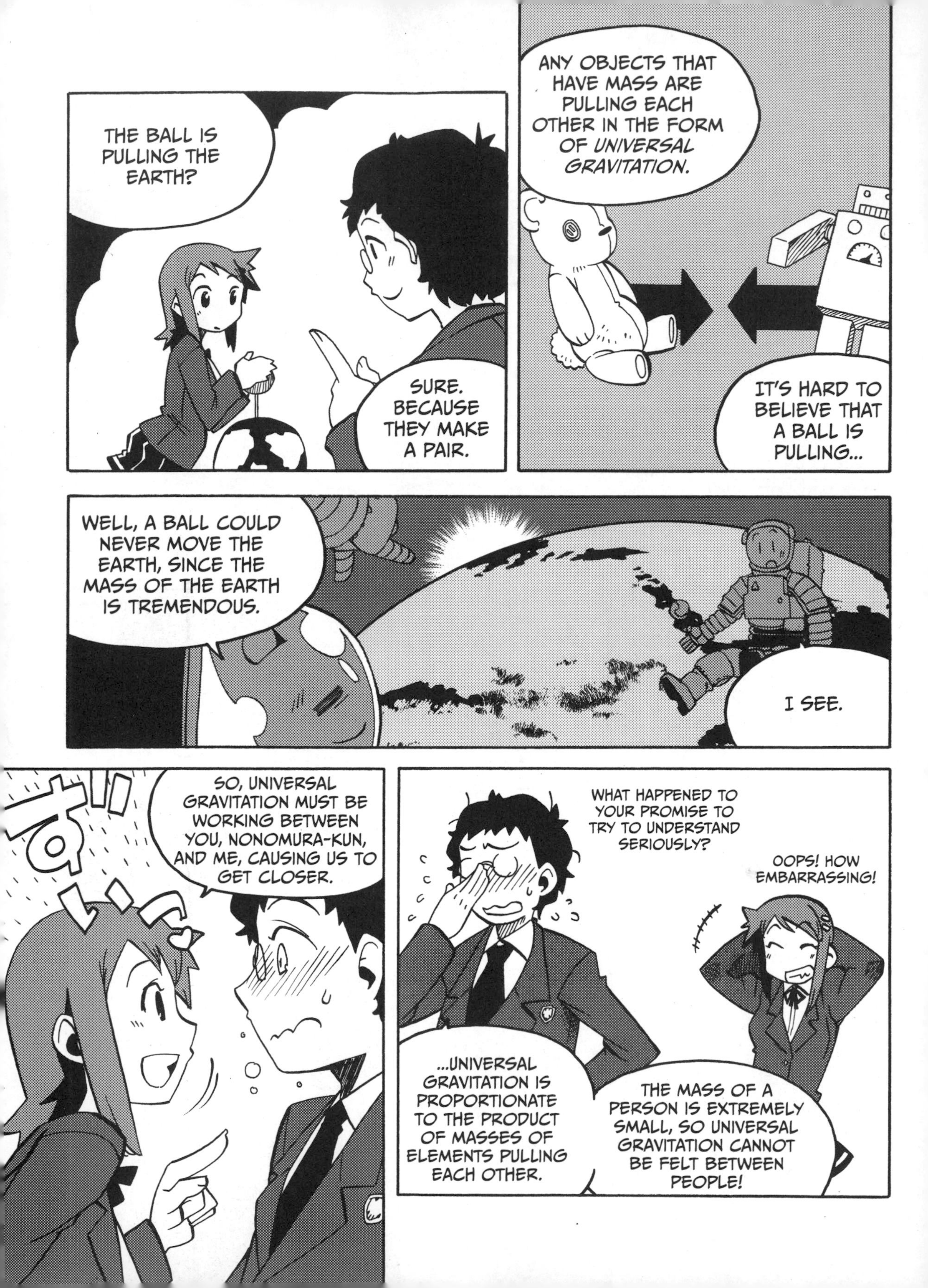
THE BALL IS PULLING THE EARTH?
SURE. BECAUSE THEY MAKE A PAIR.
ANY OBJECTS THAT HAVE MASS ARE PULLING EACH OTHER IN THE FORM OF *UNIVERSAL GRAVITATION*.
IT'S HARD TO BELIEVE THAT A BALL IS PULLING...
WELL, A BALL COULD NEVER MOVE THE EARTH, SINCE THE MASS OF THE EARTH IS TREMENDOUS.
I SEE.
SO, UNIVERSAL GRAVITATION MUST BE WORKING BETWEEN YOU, NONOMURA-KUN, AND ME, CAUSING US TO GET CLOSER.
WHAT HAPPENED TO YOUR PROMISE TO TRY TO UNDERSTAND SERIOUSLY?
OOPS! HOW EMBARRASSING!
...UNIVERSAL GRAVITATION IS PROPORTIONATE TO THE PRODUCT OF MASSES OF ELEMENTS PULLING EACH OTHER.
THE MASS OF A PERSON IS EXTREMELY SMALL, SO UNIVERSAL GRAVITATION CANNOT BE FELT BETWEEN PEOPLE!

NEWTON'S THREE LAWS OF MOTION

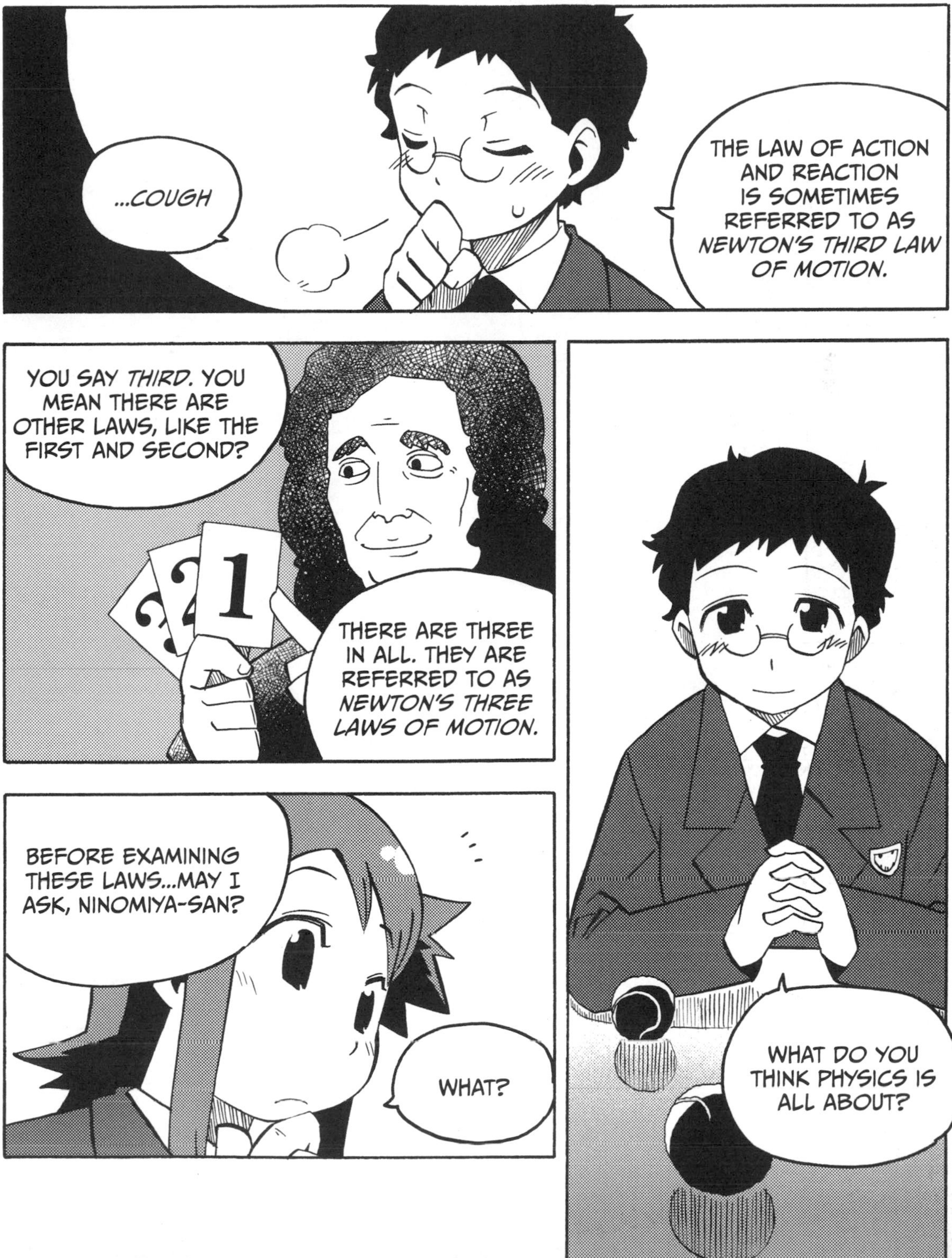
...COUGH
THE LAW OF ACTION AND REACTION IS SOMETIMES REFERRED TO AS *NEWTON'S THIRD LAW OF MOTION.*
YOU SAY *THIRD*. YOU MEAN THERE ARE OTHER LAWS, LIKE THE FIRST AND SECOND?
THERE ARE THREE IN ALL. THEY ARE REFERRED TO AS *NEWTON'S THREE LAWS OF MOTION.*
BEFORE EXAMINING THESE LAWS...MAY I ASK, NINOMIYA-SAN?
WHAT?
WHAT DO YOU THINK PHYSICS IS ALL ABOUT?

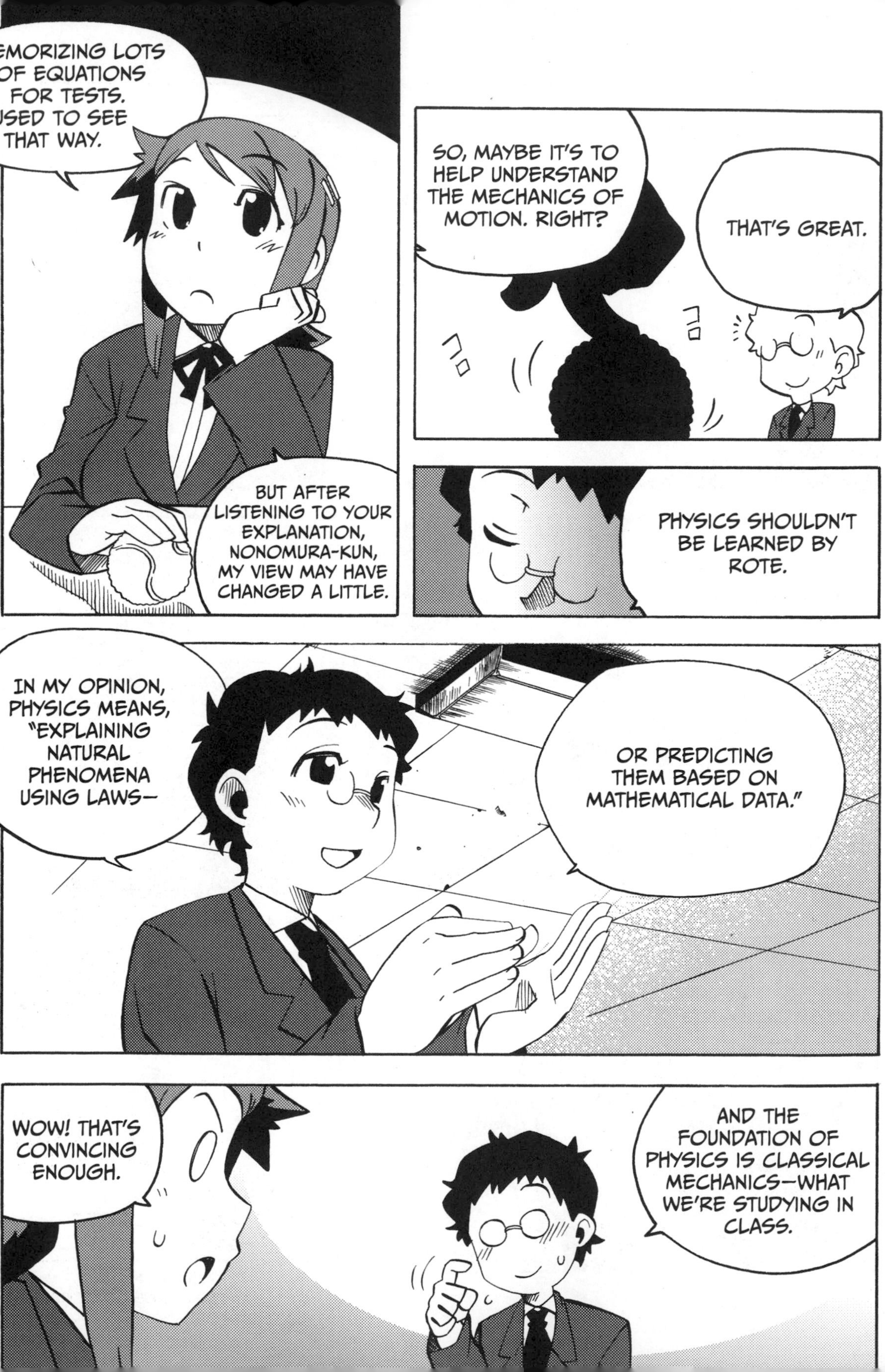
MEMORIZING LOTS OF EQUATIONS FOR TESTS. I USED TO SEE IT THAT WAY.
BUT AFTER LISTENING TO YOUR EXPLANATION, NONOMURA-KUN, MY VIEW MAY HAVE CHANGED A LITTLE.
SO, MAYBE IT'S TO HELP UNDERSTAND THE MECHANICS OF MOTION. RIGHT?
THAT'S GREAT.
PHYSICS SHOULDN'T BE LEARNED BY ROTE.
IN MY OPINION, PHYSICS MEANS, "EXPLAINING NATURAL PHENOMENA USING LAWS—
OR PREDICTING THEM BASED ON MATHEMATICAL DATA."
WOW! THAT'S CONVINCING ENOUGH.
AND THE FOUNDATION OF PHYSICS IS CLASSICAL MECHANICS—WHAT WE'RE STUDYING IN CLASS.

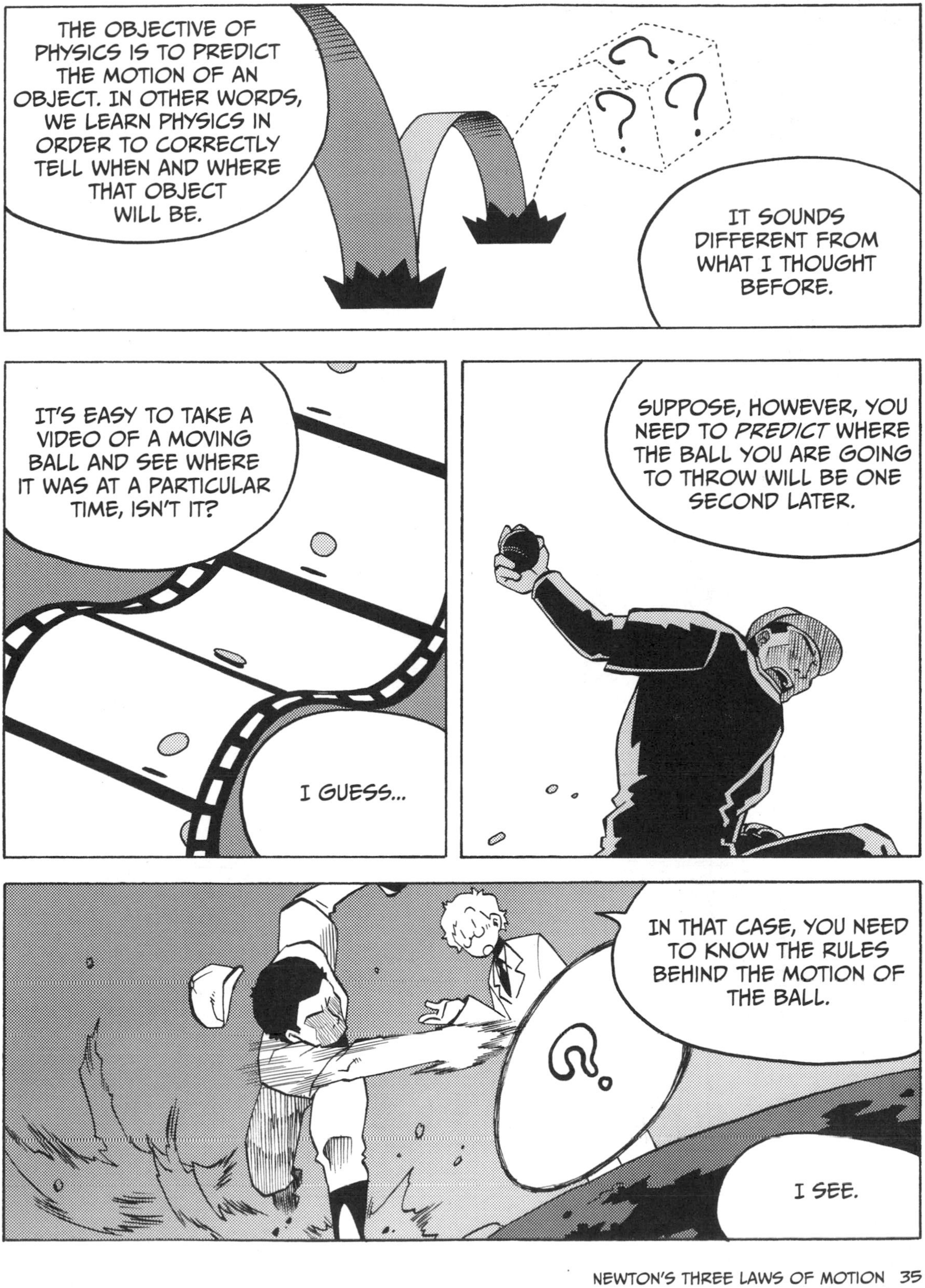
THE OBJECTIVE OF PHYSICS IS TO PREDICT THE MOTION OF AN OBJECT. IN OTHER WORDS, WE LEARN PHYSICS IN ORDER TO CORRECTLY TELL WHEN AND WHERE THAT OBJECT WILL BE.
IT SOUNDS DIFFERENT FROM WHAT I THOUGHT BEFORE.
IT'S EASY TO TAKE A VIDEO OF A MOVING BALL AND SEE WHERE IT WAS AT A PARTICULAR TIME, ISN'T IT?
I GUESS...
SUPPOSE, HOWEVER, YOU NEED TO *PREDICT* WHERE THE BALL YOU ARE GOING TO THROW WILL BE ONE SECOND LATER.
IN THAT CASE, YOU NEED TO KNOW THE RULES BEHIND THE MOTION OF THE BALL.
I SEE.

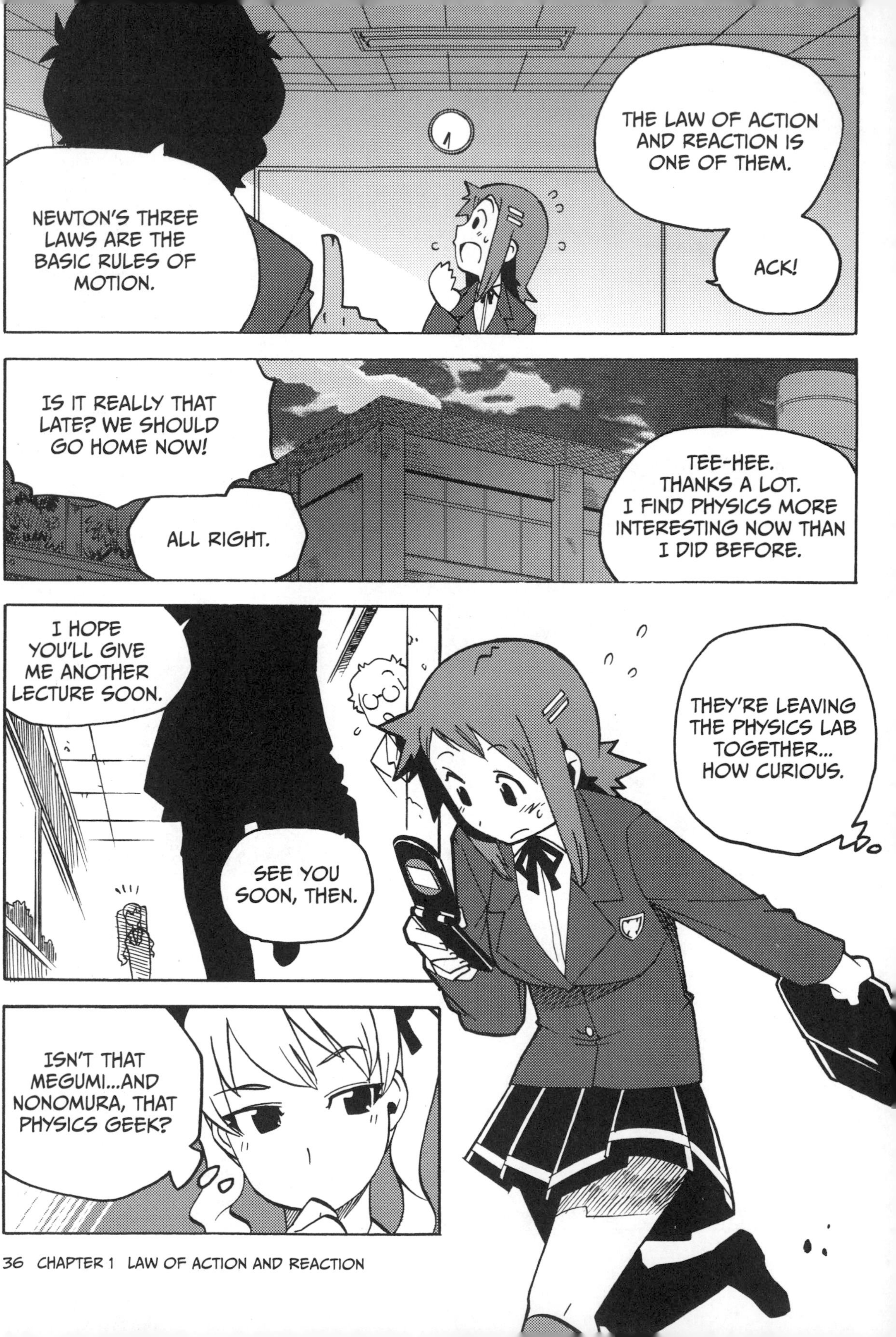
NEWTON'S THREE LAWS ARE THE BASIC RULES OF MOTION.
THE LAW OF ACTION AND REACTION IS ONE OF THEM.
ACK!
IS IT REALLY THAT LATE? WE SHOULD GO HOME NOW!
ALL RIGHT.
TEE-HEE. THANKS A LOT. I FIND PHYSICS MORE INTERESTING NOW THAN I DID BEFORE.
I HOPE YOU'LL GIVE ME ANOTHER LECTURE SOON.
SEE YOU SOON, THEN.
THEY'RE LEAVING THE PHYSICS LAB TOGETHER... HOW CURIOUS.
ISN'T THAT MEGUMI...AND NONOMURA, THAT PHYSICS GEEK?

SCALAR QUANTITIES VS. VECTOR QUANTITIES

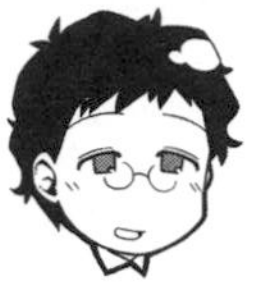

Physics involves measuring and predicting various quantities (or physical values) like force, mass, and velocity. These values can be classified into those having only magnitude and those having both magnitude and direction. A quantity that has magnitude without a direction is referred to as a *scalar* quantity. Mass is a scalar quantity. Energy and work, which we'll learn about in Chapter 4, are also scalar quantities.

On the other hand, force is a value with a direction. You can see that from the fact that the motion of an object changes if you apply force from a different direction. A quantity that has a direction is called a *vector*. Velocity and acceleration (which are introduced in Chapter 2) and momentum (discussed in Chapter 3) are also vector quantities, as they have direction. You may forget the terms *scalar* and *vector*, but you should keep in mind that there are two types of values in physics: those with just a magnitude and those with both a magnitude and a direction.

VECTOR BASICS

A vector is represented using an arrow. The length of the arrow represents the magnitude of the vector, and the point represents its orientation, or direction. Two vectors with the same magnitude and direction are equivalent to one another, even if they do not have the same origin.

Orientation
$\vec{a}$
Magnitude $|\vec{a}|$

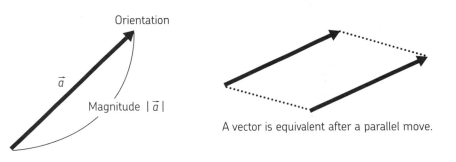

A vector is equivalent after a parallel move.

Also note that the magnitude of a vector (represented by the length of the arrow) can be noted with absolute value symbols, like $|\vec{a}|$, or simply as a.

$\vec{a} + \vec{b}$
$\vec{b}$
$\vec{a}$

$\vec{a}$
$\vec{b} + \vec{a}$
$\vec{b}$

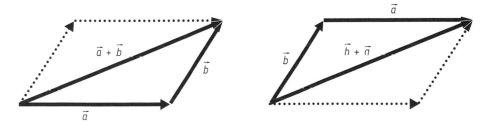

The sum of two vectors ($\vec{a} + \vec{b}$) is shown by joining the head of vector $\vec{a}$ to the tail of vector $\vec{b}$, and then extending a line from the tail of $\vec{a}$ to the head of $\vec{b}$, as shown in the figure on the left here. As this vector is a diagonal of the parallelogram in the figure, it is obvious that it is also equivalent to $\vec{b} + \vec{a}$. Therefore, we know that the following is true:

COMMUTATIVE LAW: $\vec{a} + \vec{b} = \vec{b} + \vec{a}$

The order in which you add vectors doesn't matter! You can find the sum of three or more vectors in the same way.

NEGATIVE VECTORS

Vector $-\vec{a}$, or $\vec{a}$ preceded by a minus sign, yields a sum of zero when added to vector $\vec{a}$. In an equation, the relationship looks like this:

$$\vec{a} + (-\vec{a}) = 0$$

In terms of geometry, $-\vec{a}$ is simply a vector of the same magnitude as $\vec{a}$, but in the exact opposite direction. The 0 on the right side of this equation represents zero as a vector, referred to as a *zero vector*. When vectors cancel each other out in this way, an object is said to be in *equilibrium*.

$\vec{a}$
$-\vec{a}$

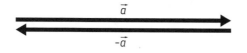

DIFFERENCE BETWEEN TWO VECTORS

The difference between two vectors ($\vec{a} - \vec{b}$) can be defined as follows:

$$\vec{a} - \vec{b} = \vec{a} + (-\vec{b})$$

Thus, we can find the result of the equation using the same process for summing vectors.

$(-\vec{b}) + \vec{a} = \vec{a} - \vec{b}$
$\vec{b}$
$-\vec{b}$
$\vec{a}$

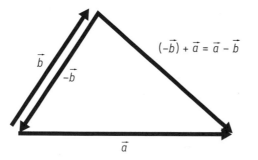

MULTIPLYING VECTORS BY SCALARS

Doubling vector $\vec{a}$ means doubling its magnitude without changing its direction. The result is represented as $2\vec{a}$.

$\vec{a}$
$\vec{a}$
$2\vec{a}$

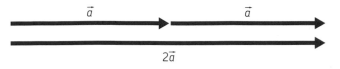

Generally, k multiplied by $\vec{a}$ ($k\vec{a}$) represents a vector with a magnitude k times greater than $\vec{a}$ but in the same direction.

EQUILIBRIUM AND VECTOR FORCES

In discussing the total forces on the tennis ball on page 22, we saw the following equation:

total force on the ball = force of gravity + force from the hand = 0

Do you think the plus sign is an error and there should be a minus sign there instead? It's not an error! Remember that forces are vectors—this equation is true as is. Considered as vectors, the total forces working on an object must equal the sum of all the forces.

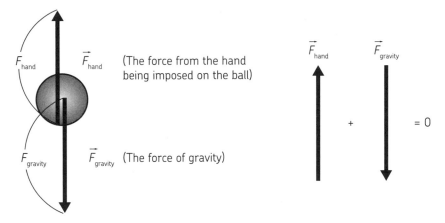

F_{hand}
$\vec{F}_{hand}$
(The force from the hand being imposed on the ball)
$F_{gravity}$
$\vec{F}_{gravity}$
(The force of gravity)
$\vec{F}_{hand}$
$\vec{F}_{gravity}$
+
= 0

Let's look into the balance between the forces on the ball and the hand holding it. Let's call the force from the hand on the ball $\vec{F}_{hand}$ and the force of gravity on the ball $\vec{F}_{gravity}$. The resultant force ($\vec{F}_{resultant}$) acting on the ball is expressed as follows:

$$\vec{F}_{resultant} = \vec{F}_{hand} + \vec{F}_{gravity}$$

The resultant force is also called the *net force*. If the forces on a ball are balanced, it means the resultant force has reached zero:

$$\vec{F}_{resultant} = 0 \text{ or, to put it another way, } \vec{F}_{hand} + \vec{F}_{gravity} = 0$$

Yes, that's exactly it. In short, $\vec{F}_{\text{hand}}$ and $\vec{F}_{\text{gravity}}$ are vectors of the same magnitude in opposite directions, resulting in zero when added:

the force exerted by the hand on the ball + force of gravity on the ball = zero

Now, let's look at forces in terms of magnitude only, not as vectors with direction. As explained on page 37, the magnitude of a force is expressed as $|\vec{F}_{\text{hand}}|$ or $|\vec{F}_{\text{gravity}}|$, using absolute value symbols. Developing these expressions further, you get equations like $|\vec{F}_{\text{hand}}| = F_{\text{hand}}$ and $|\vec{F}_{\text{gravity}}| = F_{\text{gravity}}$. Now you know the two forces have an equivalent magnitude, which can be expressed as follows in an equation for a subtraction:

$$F_{\text{hand}} = F_{\text{gravity}} \text{ or } F_{\text{hand}} - F_{\text{gravity}} = 0$$

Note that these forces are represented without arrows, which indicates that they are magnitudes. When giving equations for balanced forces, we need to make a clear distinction between cases where forces are considered to be vectors and cases where they are considered to be simple magnitudes without direction (scalars).

NEWTON'S THREE LAWS OF MOTION

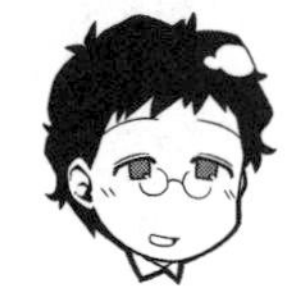

Isaac Newton was an English physicist born in 1643. Based on his observations of motion, he developed the following laws.

The first law (law of inertia): A body at rest tends to stay at rest unless acted upon by an outside net force. A body in motion tends to stay in motion at a constant velocity unless acted on by an outside net force.

The second law (law of acceleration): The net force on an object is equal to the mass of the object multiplied by its acceleration.

The third law (law of action and reaction): For every action there is an equal and opposite reaction.

Let me explain in terms of the ball held in my hand in this chapter. (We'll discuss this further in Chapter 2.)

Given the first law, we can tell that the total forces on a static object have reached zero in magnitude. Because the ball is in a state of equilibrium, it is static and remains so; this is the first law of motion in action. Because the ball is not moving, there must be no resultant force from the sum of the force of my hand and the force of gravity.

As we learned in this chapter, the law of action and reaction is the third law of motion. This law tells us that the force from the hand acting on the ball and the force from the ball acting on the hand are equivalent in magnitude and opposite in direction. The law of action and reaction is always present. It is also working when you keep a ball in motion by moving your hand.

The second law of motion tells us that an object receiving a net force begins moving with acceleration. If you suddenly lower your hand while holding a ball, the force from the hand on the ball (F_{hand}) suddenly decreases in magnitude, but the force of gravity on the ball ($F_{gravity}$) stays the same. Therefore, the balance of forces is broken, and the sum of $F_{gravity}$ and F_{hand} on the ball attains a nonzero value while the ball is moving. Thinking in terms of magnitude:

$$F_{net} = F_{gravity} - F_{hand} > 0$$

The above equation represents the magnitude of the force applied downward. At this time, given the second law of motion, which states that an object receiving a force attains acceleration proportional to that net force, the ball should begin accelerating, or start moving. This is how mechanics explains the motion of a ball caused when the hand retaining it is suddenly lowered. This same idea can be applied when a ball is suddenly lifted.

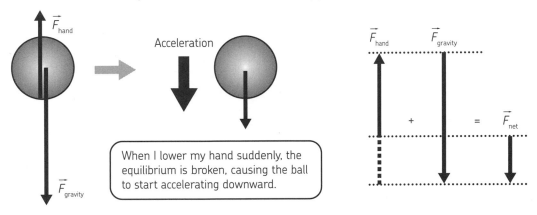

$\vec{F}_{hand}$
Acceleration
$\vec{F}_{gravity}$
When I lower my hand suddenly, the equilibrium is broken, causing the ball to start accelerating downward.

$\vec{F}_{hand}$
$\vec{F}_{gravity}$
+
=
$\vec{F}_{net}$

There is one thing you need to keep in mind. When a ball moves up and down at a constant speed, you should note that the net force (resultant force) on the ball remains zero, as the forces are balanced; the ball is not accelerating. The first law of motion tells us that. A nonzero net force is acting on the ball when the speed of the object's motion varies or any acceleration occurs. When the object moves at a constant speed, the acceleration is zero, and so is the net force. In other words, the forces applied are balanced, even though the ball is moving.

A force must be applied to an object for it to begin moving from a static state. Starting motion means the object transitions from a zero-velocity state to one with a velocity. When this occurs, the object has accelerated.

DRAWING A FREE-BODY DIAGRAM

In the figure showing vectors of forces acting on a ball in the previous section, $\vec{F}_{hand}$ and $\vec{F}_{gravity}$ have different starting points. Physicists call drawings like this *free-body diagrams*. When you draw a vector to represent the force of the hand on the ball, you start it at the point where the two come into contact with each other. That's not so confusing, but why do you think the starting point for gravity is located at the center of the ball?

In basic physics, an object is treated as a point of mass without a volume; it doesn't matter where the vector starts. We draw that mass point as an object with a certain volume simply because it is easier to see that way in a figure or illustration.

$\vec{F}_{hand}$
Ball
$\vec{F}_{gravity}$
$\vec{F}_{hand}$
Ball
$\vec{F}_{gravity}$

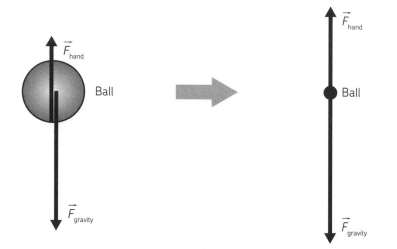

Let's consider an object with volume, and how we can represent forces imposed on it. In the case of a ball on my hand, the force of gravity is imposed on the center of the ball's mass (which also called the *center of gravity*). You can see in the diagram above that this is where the force vector is acting. However, the upward force of my hand is acting on the outside of the ball, as that is the point of contact. We'll draw the force vector starting there in our diagram.

But to simplify our calculations, we'll treat this object as a mass without volume—that is, a single point with mass. We'll simplify all objects with volume similarly, as the calculations for objects with volume can become very complicated. A diagram that represents this simplified free-body diagram is on the right. Bear in mind that we will simplify all the examples in this book in this way in our calculations, even if our diagrams appear more complex.

EXPRESSING NEWTON'S THIRD LAW WITH AN EQUATION

To express the law of action and reaction in correct wording, we need a lengthy phrase like "When an object impacts another object, both objects receive a force of the same magnitude but in opposite directions." So let's try to express the law of action and reaction as a simple equation instead. When object A imposes force $\vec{F}_{A \to B}$ on object B and object B imposes force $\vec{F}_{B \to A}$ on object A, the law of action and reaction is expressed as follows:

$$\vec{F}_{A \to B} = -\vec{F}_{B \to A}$$

So, you can express the law in a single equation, as shown above. In fact, comparing the elements in this equation in terms of absolute values, you get:

$$|\vec{F}_{A \to B}| = |-\vec{F}_{B \to A}|$$

Now you can see that the action and the reaction are equivalent in magnitude, and the minus sign tells you that their directions are opposite. Using equations can help you express Newton's laws in a simpler and more precise manner than verbal expression.

GRAVITY AND UNIVERSAL GRAVITATION

In the narrowest sense of the term, gravity is the force of the earth attracting objects toward itself. But the force of the earth comes from universal gravitation between all objects of mass, not just the earth itself. Between two objects, there is an attractive force proportional to the product of the objects' masses and inversely proportional to the distance between them, raised to the second power. This attractive force is *universal gravitation*, as discovered by Newton. It's called *universal* gravitation because it works on *all* objects with mass—it's not affected by the type of object. Its value only depends on the mass of objects affecting each other and the distance between them.

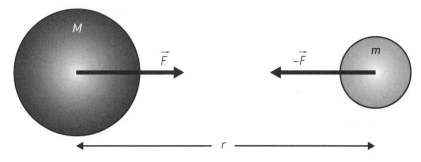

As shown in the figure, when two objects with mass M and mass m are separated from each other by distance r, a force of F attracts the two objects. The equation is as follows:

$$F = G \, \frac{mM}{r^2}$$

G is a universal constant referred to as the *universal gravitational constant*:

$$G = 6.67 \times 10^{-11} \; (\text{N} \times \text{m}^2/\text{kg}^2)$$

For an explanation of the unit newton (N), see page 92.

Universal gravitation satisfies the law of action and reaction, as it exerts a force on both masses M and m. The equation above can be used to calculate the force on either object. As their directions are obviously opposite, they satisfy the law of action and reaction. Thus, we should note that forces working between objects at a distance from each other (not just objects that come into contact with each other) also satisfy the law of action and reaction.

Universal gravitation is a very small force compared to electromagnetic force. While electromagnetic forces may be attractive or repulsive depending on a combination of positive and negative charges, universal gravitation always works as attractive force—that is, objects are always drawn closer to each other.

Because of universal gravitation, cosmic dust in outer space gathers into giant masses over time—such as the earth or the other planets.

2

FORCE AND MOTION

VELOCITY AND ACCELERATION

SIMPLE MOTION

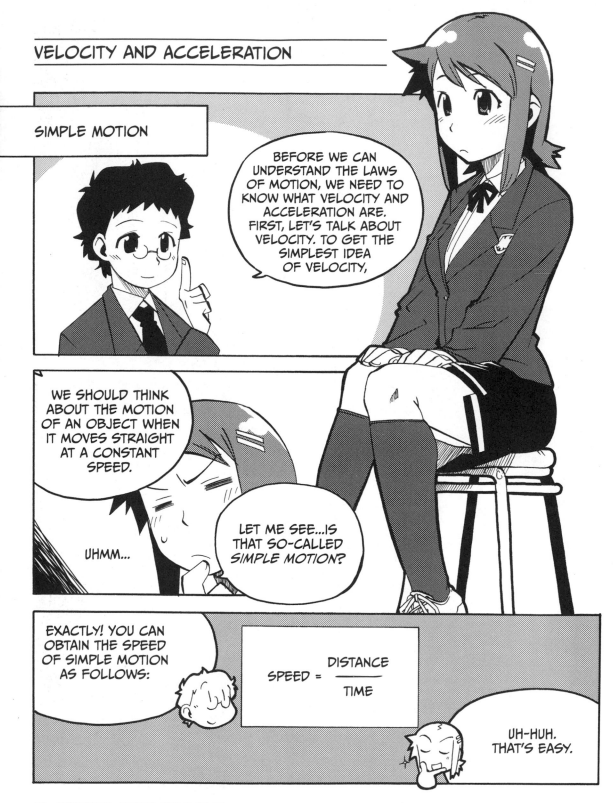

BEFORE WE CAN UNDERSTAND THE LAWS OF MOTION, WE NEED TO KNOW WHAT VELOCITY AND ACCELERATION ARE. FIRST, LET'S TALK ABOUT VELOCITY. TO GET THE SIMPLEST IDEA OF VELOCITY,

WE SHOULD THINK ABOUT THE MOTION OF AN OBJECT WHEN IT MOVES STRAIGHT AT A CONSTANT SPEED.

UHMM...

LET ME SEE...IS THAT SO-CALLED SIMPLE MOTION?

EXACTLY! YOU CAN OBTAIN THE SPEED OF SIMPLE MOTION AS FOLLOWS:

$$\text{SPEED} = \frac{\text{DISTANCE}}{\text{TIME}}$$

UH-HUH. THAT'S EASY.

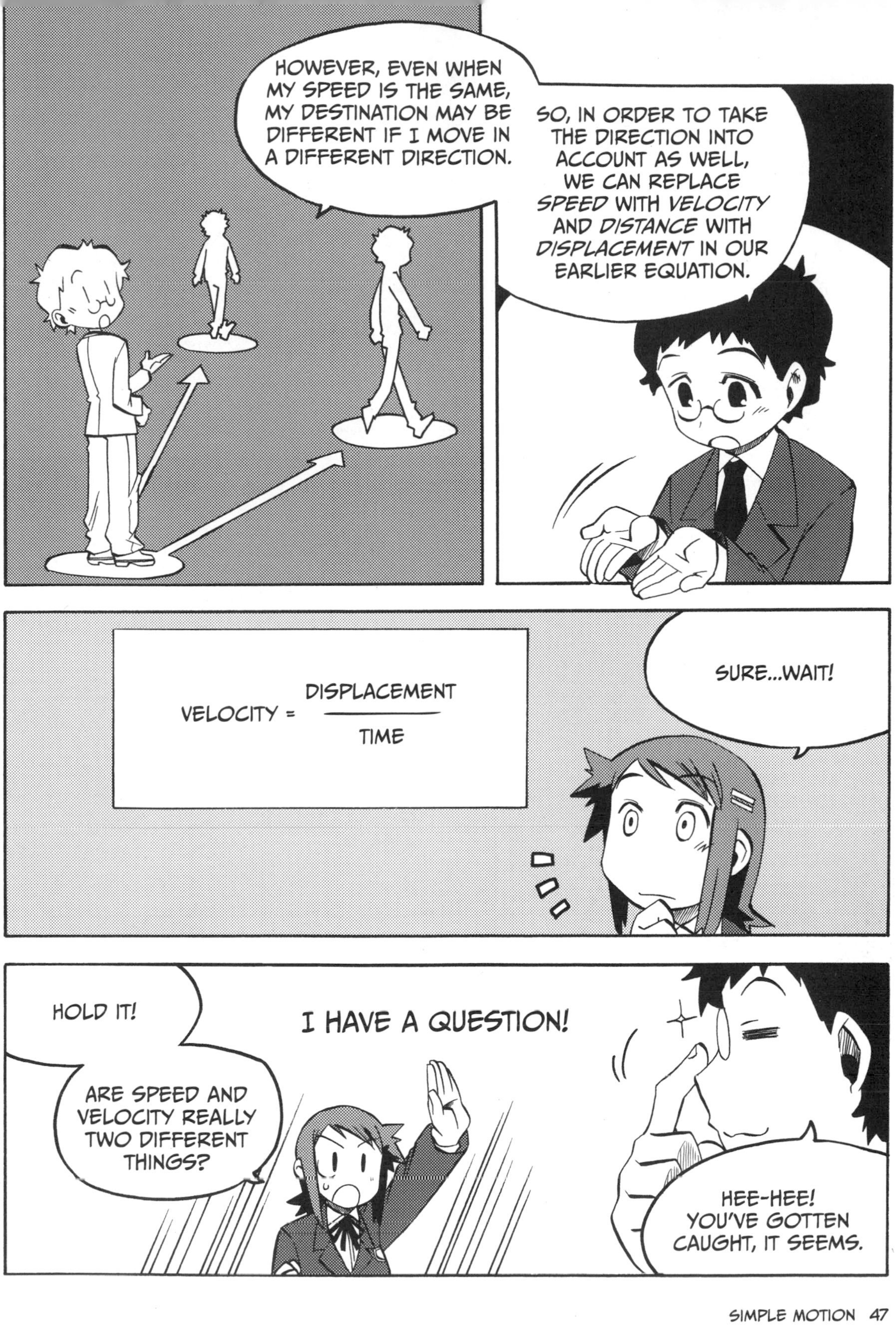
HOWEVER, EVEN WHEN MY SPEED IS THE SAME, MY DESTINATION MAY BE DIFFERENT IF I MOVE IN A DIFFERENT DIRECTION.
SO, IN ORDER TO TAKE THE DIRECTION INTO ACCOUNT AS WELL, WE CAN REPLACE *SPEED* WITH *VELOCITY* AND *DISTANCE* WITH *DISPLACEMENT* IN OUR EARLIER EQUATION.
VELOCITY = DISPLACEMENT / TIME
SURE...WAIT!
HOLD IT!
I HAVE A QUESTION!
ARE SPEED AND VELOCITY REALLY TWO DIFFERENT THINGS?
HEE-HEE! YOU'VE GOTTEN CAUGHT, IT SEEMS.

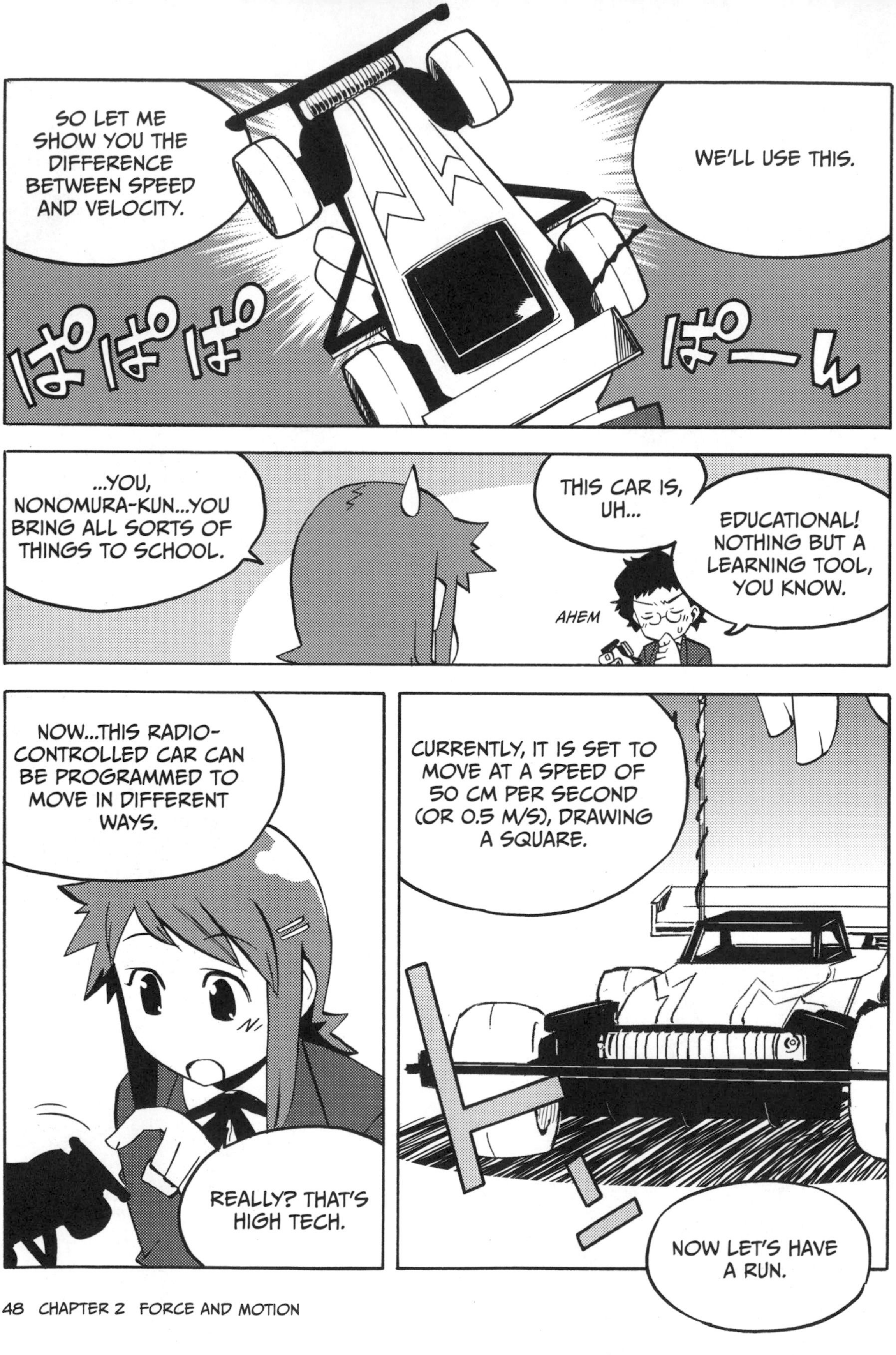
SO LET ME SHOW YOU THE DIFFERENCE BETWEEN SPEED AND VELOCITY.
WE'LL USE THIS.
ぱぱぱ
ぱーん
...YOU, NONOMURA-KUN...YOU BRING ALL SORTS OF THINGS TO SCHOOL.
THIS CAR IS, UH...
AHEM
EDUCATIONAL! NOTHING BUT A LEARNING TOOL, YOU KNOW.
NOW...THIS RADIO-CONTROLLED CAR CAN BE PROGRAMMED TO MOVE IN DIFFERENT WAYS.
REALLY? THAT'S HIGH TECH.
CURRENTLY, IT IS SET TO MOVE AT A SPEED OF 50 CM PER SECOND (OR 0.5 M/S), DRAWING A SQUARE.
NOW LET'S HAVE A RUN.

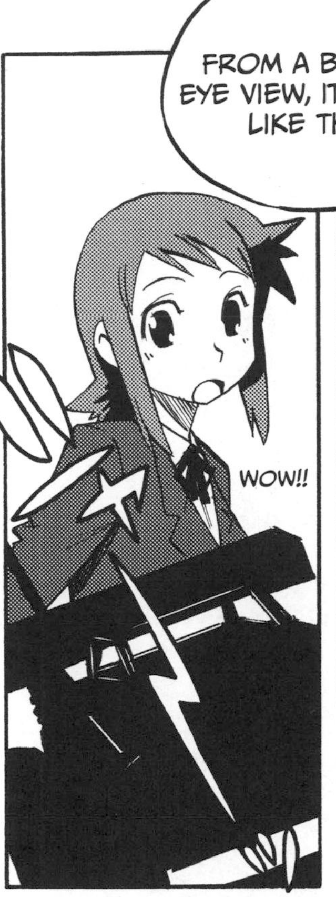
FROM A BIRD'S-EYE VIEW, IT LOOKS LIKE THIS.
WOW!!

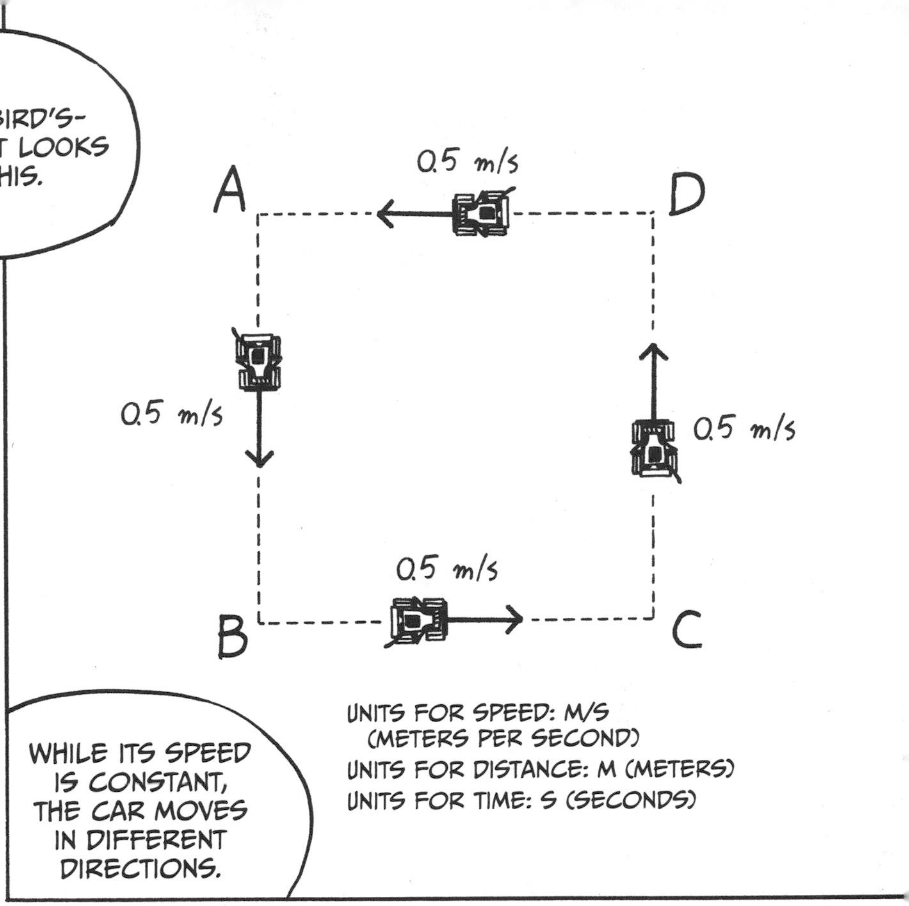
A
D
B
C
0.5 m/s
0.5 m/s
0.5 m/s
0.5 m/s
UNITS FOR SPEED: M/S (METERS PER SECOND)
UNITS FOR DISTANCE: M (METERS)
UNITS FOR TIME: S (SECONDS)
WHILE ITS SPEED IS CONSTANT, THE CAR MOVES IN DIFFERENT DIRECTIONS.

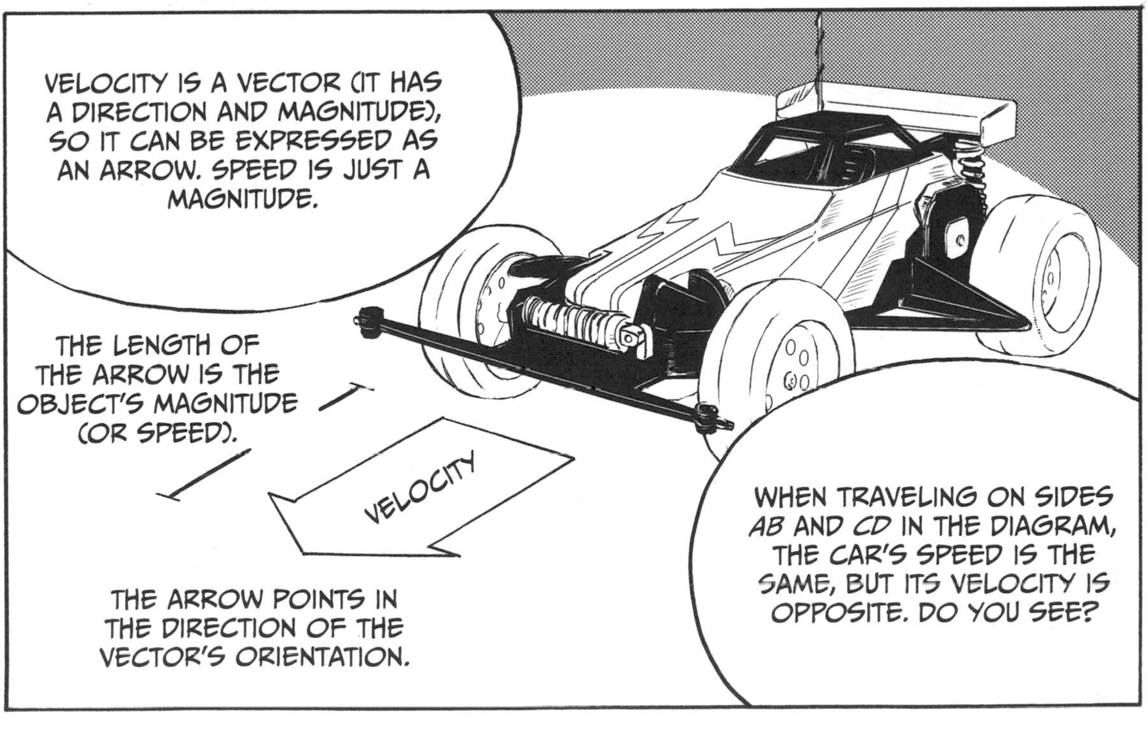
VELOCITY IS A VECTOR (IT HAS A DIRECTION AND MAGNITUDE), SO IT CAN BE EXPRESSED AS AN ARROW. SPEED IS JUST A MAGNITUDE.
THE LENGTH OF THE ARROW IS THE OBJECT'S MAGNITUDE (OR SPEED).
VELOCITY
THE ARROW POINTS IN THE DIRECTION OF THE VECTOR'S ORIENTATION.
WHEN TRAVELING ON SIDES AB AND CD IN THE DIAGRAM, THE CAR'S SPEED IS THE SAME, BUT ITS VELOCITY IS OPPOSITE. DO YOU SEE?

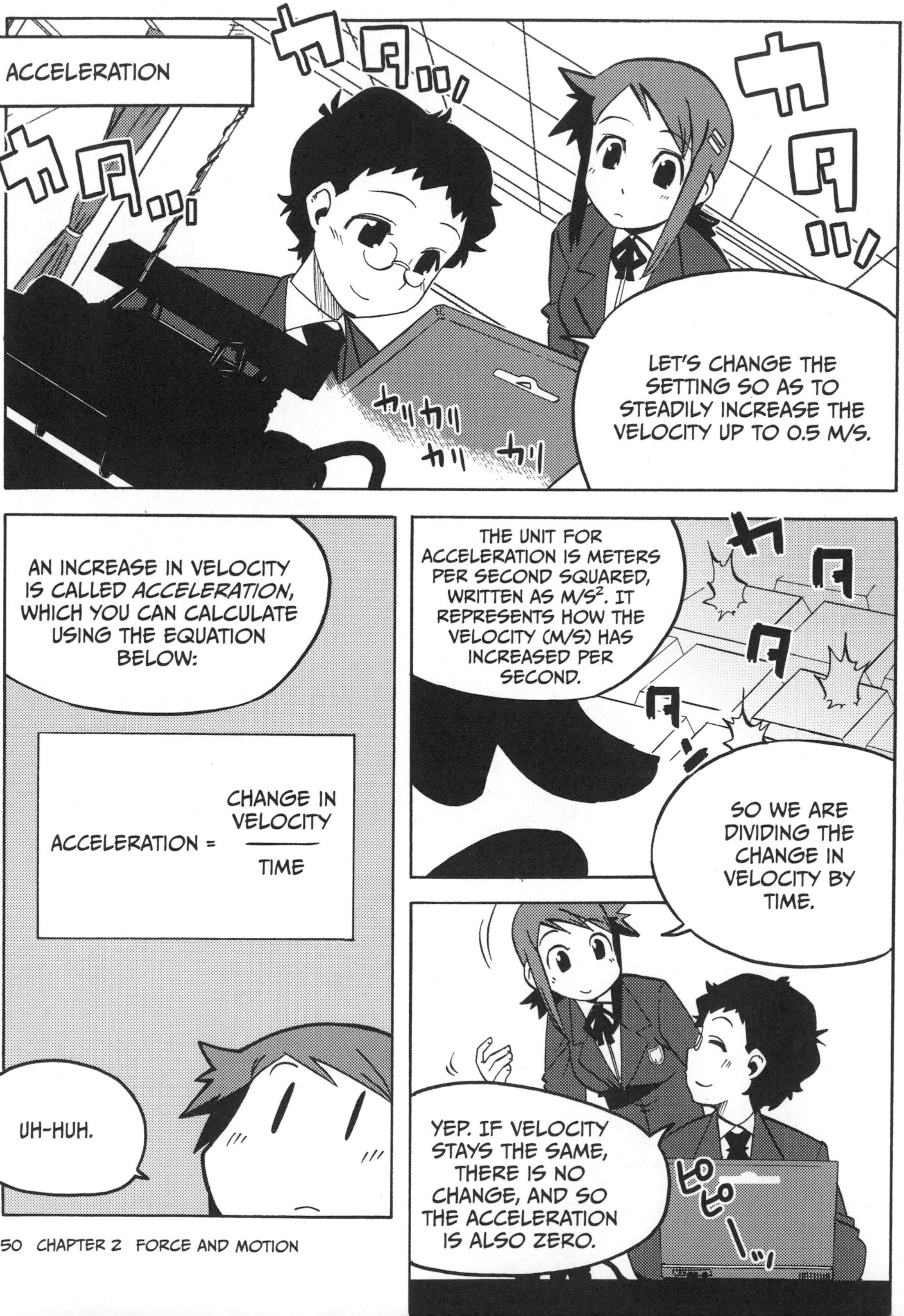
ACCELERATION
カタ…
カタ…
カタ…
LET'S CHANGE THE SETTING SO AS TO STEADILY INCREASE THE VELOCITY UP TO 0.5 M/S.
カリカリ
カリ カリ
AN INCREASE IN VELOCITY IS CALLED *ACCELERATION*, WHICH YOU CAN CALCULATE USING THE EQUATION BELOW:
ACCELERATION = CHANGE IN VELOCITY / TIME
UH-HUH.
THE UNIT FOR ACCELERATION IS METERS PER SECOND SQUARED, WRITTEN AS M/S². IT REPRESENTS HOW THE VELOCITY (M/S) HAS INCREASED PER SECOND.
SO WE ARE DIVIDING THE CHANGE IN VELOCITY BY TIME.
YEP. IF VELOCITY STAYS THE SAME, THERE IS NO CHANGE, AND SO THE ACCELERATION IS ALSO ZERO.

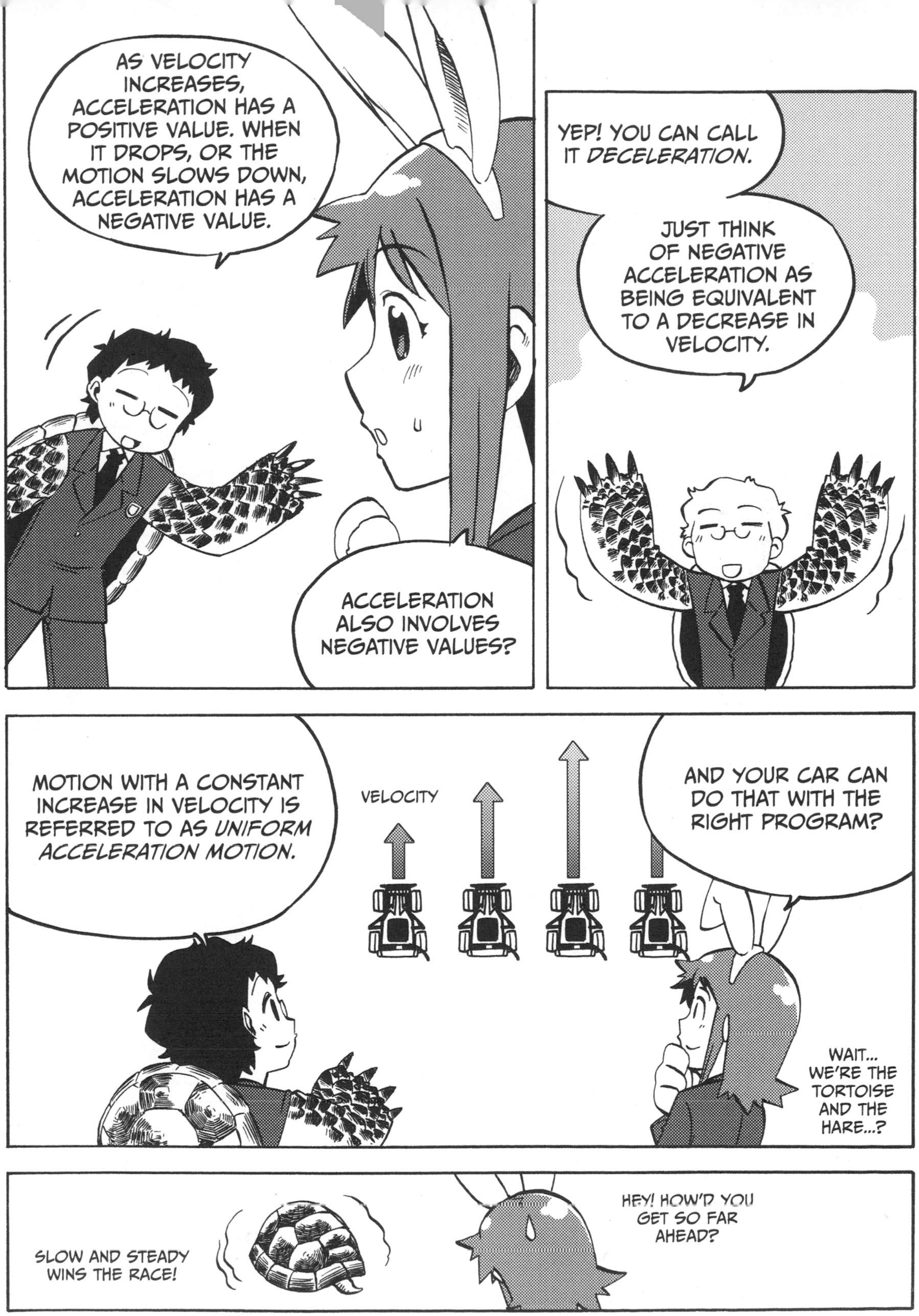
AS VELOCITY INCREASES, ACCELERATION HAS A POSITIVE VALUE. WHEN IT DROPS, OR THE MOTION SLOWS DOWN, ACCELERATION HAS A NEGATIVE VALUE.
ACCELERATION ALSO INVOLVES NEGATIVE VALUES?
YEP! YOU CAN CALL IT *DECELERATION.*
JUST THINK OF NEGATIVE ACCELERATION AS BEING EQUIVALENT TO A DECREASE IN VELOCITY.
MOTION WITH A CONSTANT INCREASE IN VELOCITY IS REFERRED TO AS *UNIFORM ACCELERATION MOTION.*
VELOCITY
AND YOUR CAR CAN DO THAT WITH THE RIGHT PROGRAM?
WAIT... WE'RE THE TORTOISE AND THE HARE...?
SLOW AND STEADY WINS THE RACE!
HEY! HOW'D YOU GET SO FAR AHEAD?

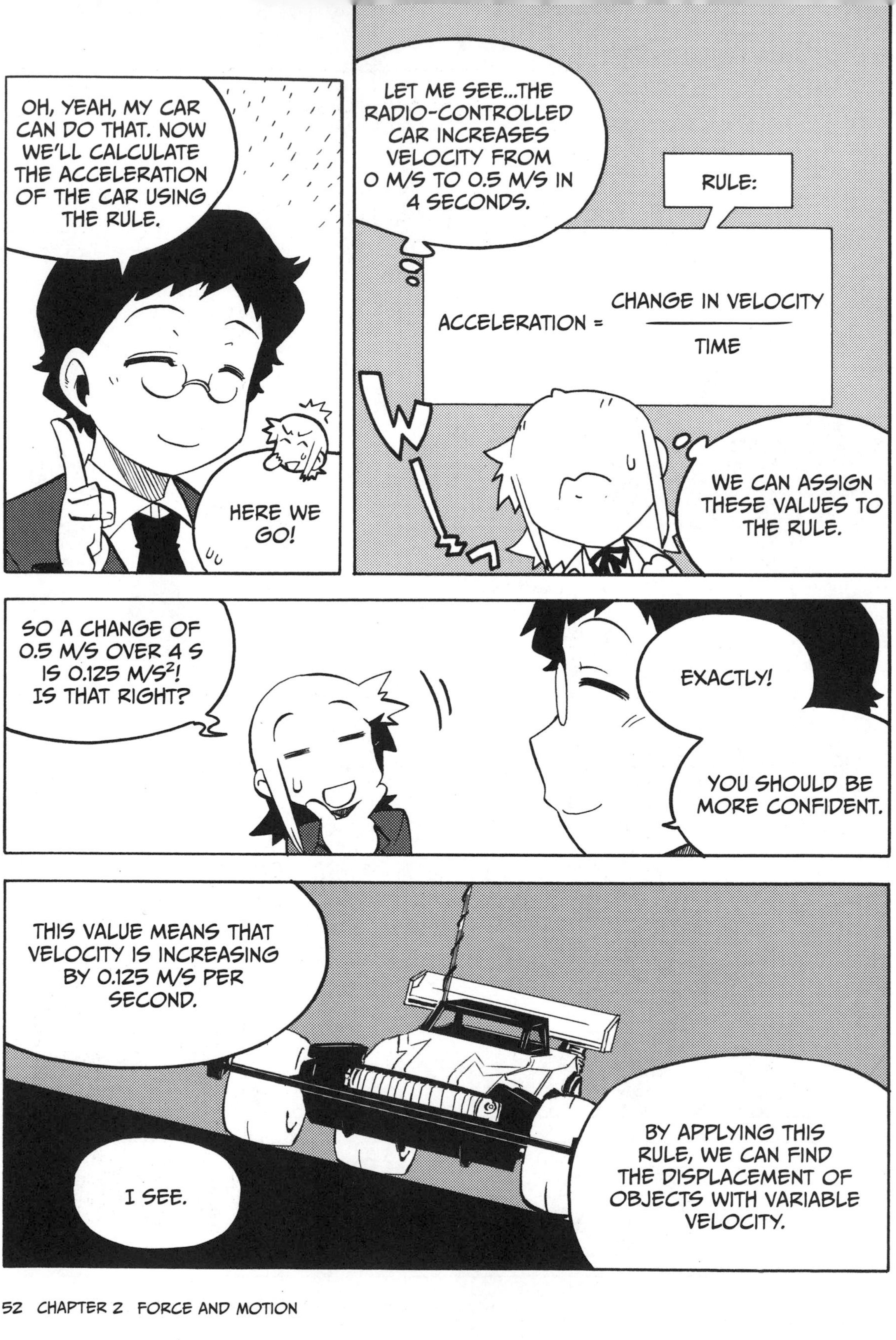
OH, YEAH, MY CAR CAN DO THAT. NOW WE'LL CALCULATE THE ACCELERATION OF THE CAR USING THE RULE.
HERE WE GO!
LET ME SEE...THE RADIO-CONTROLLED CAR INCREASES VELOCITY FROM 0 M/S TO 0.5 M/S IN 4 SECONDS.
RULE:
ACCELERATION = CHANGE IN VELOCITY / TIME
W
WE CAN ASSIGN THESE VALUES TO THE RULE.
SO A CHANGE OF 0.5 M/S OVER 4 S IS 0.125 M/S²! IS THAT RIGHT?
EXACTLY!
YOU SHOULD BE MORE CONFIDENT.
THIS VALUE MEANS THAT VELOCITY IS INCREASING BY 0.125 M/S PER SECOND.
I SEE.
BY APPLYING THIS RULE, WE CAN FIND THE DISPLACEMENT OF OBJECTS WITH VARIABLE VELOCITY.

LABORATORY

FINDING THE DISTANCE TRAVELED WHEN VELOCITY VARIES

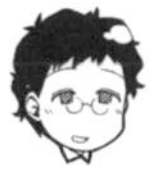

Let's change the setting so as to steadily increase the velocity up to 0.5 m/s. Here's a quiz for you. Given that velocity has attained 0.5 m/s in four seconds, how far has the radio-controlled car moved?

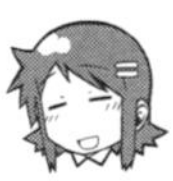

Hmm . . . starting at 0 m/s, the peak velocity is 0.5 m/s. So let me calculate, assuming the average speed, 0.25 m/s, for the velocity. 0.25 m/s × 4 s = 1 m!

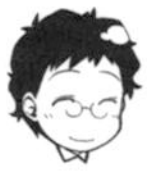

That's right! You are so sharp. But can you explain why you can get the right answer with that calculation?

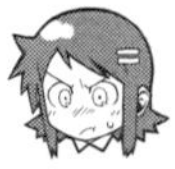

Uhm . . . remember, teaching me is *your* job, Nonomura-kun!

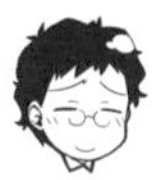

Ha ha, true enough. Before giving you a direct answer, let me explain how we can find the distance traveled when the velocity varies. When velocity is constant, we've learned that the distance traveled can be found by calculating the expression (speed × time). Now, given that d m (meters) represents the distance traveled in t s (seconds) and the constant velocity is v m/s, then distance = speed × time can be expressed in the following equation:

$$d = vt$$

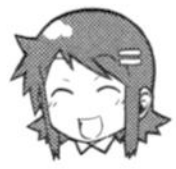

Well, duh!

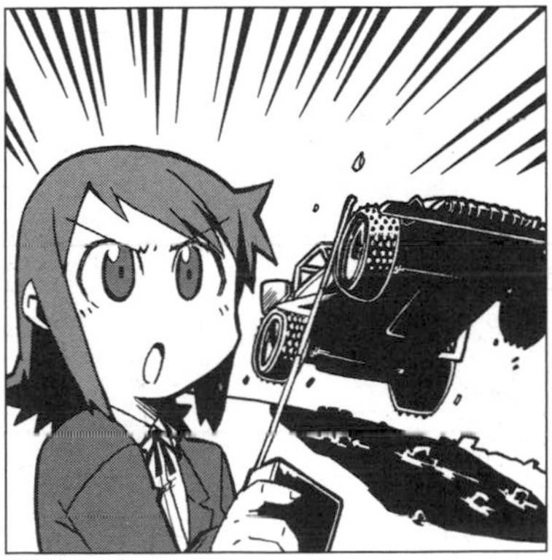

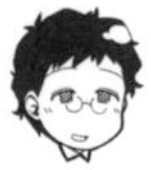

If you plot that relationship with velocity on the vertical axis and time on the horizontal axis, you get the following graph.

Velocity
v
Distance traveled
0
t
Time

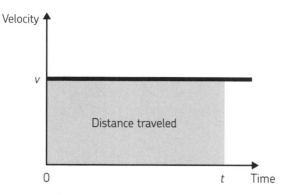

The shaded area represents the distance traveled. This chart is commonly referred to as a *v-t graph*, as it graphs velocity and time. That's the area of a rectangle having a horizontal length of t and a vertical length of v.

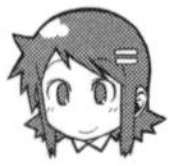

I see. It seems a little strange that an area represents a distance.

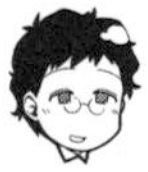

The area here is not a typical geometric area—this is a graph, like the ones you've seen in math class. The area of a geometric rectangle might be measured in square meters (m^2). But in our example, the units are time (seconds) for the horizontal axis and velocity (m/s) for the vertical axis. So the product of these two is equal to s × m/s = m. That's our unit for distance.

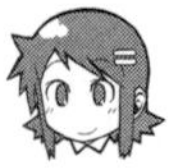

It's easy to find a distance when an object goes at a constant speed. But what about finding the distance when the speed is variable?

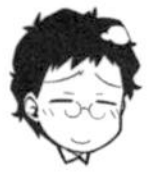

The only tool available to us is this equation:

distance = speed × time

So we can divide the time into segments to create a lot of "small rectangles" and then calculate distances respectively, assuming a constant velocity for each time segment.

What do you mean?

Look at the chart on the left below.

Velocity
Distances traveled in respective short time segments
v
Distance traveled (m)
0
t
Time
Velocity
v
Distance traveled (m)
0
t
Time

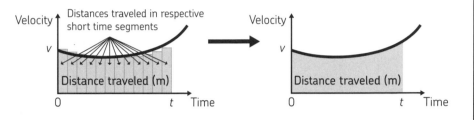

So we can find the area of each slender rectangle created by dividing time into short segments, and then adding up the areas to find the distance traveled.

It bothers me that those little rectangles won't exactly fit the graph. Wouldn't they bring about errors?

I see your concern. Then we can sub-divide the rectangles into smaller segments. By repeating division into even smaller segments until everything fits as shown in the chart on the right above, the distance we get becomes more and more precise.

Well, I guess so . . . if you could do that . . .

If we divided them into infinitely slender rectangles, we'd find exactly how far the object has moved. After all, the ultimate answer we get by dividing distance = speed × time into short time segments is the area created under a v-t graph. That's how we can find the distance traveled by finding the corresponding area. In summary,

distance traveled = area under a v-t graph

Just like that.*

---

* Students of calculus may notice that this process of finding an area under a graph is identical to *integration*.

Now, keeping in mind what we've learned so far, let's examine the reason why the distance you got intuitively is the right answer.

 All right!

 Your original calculation is the same as calculating an area on a velocity-time graph. The example with a radio-controlled car can be plotted into a chart like this one.

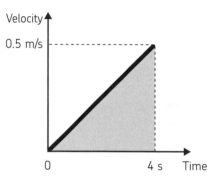

The area under the graph, as obtained from the rule for the area of a triangle, is as follows:

$$\tfrac{1}{2} \times \text{base (time)} \times \text{height (max velocity)} = \tfrac{1}{2} \times 4 \text{ s} \times 0.5 \text{ m/s} = 1 \text{ m}$$

This represents the distance traveled.

 We got 1 meter for the answer, just as we should.

 Let's find a general expression for the distance traveled, rather than using specific numeric values. Assuming velocity to be v and acceleration to be a, the relationship between the velocity and time for uniform accelerated motion is $v = at$.

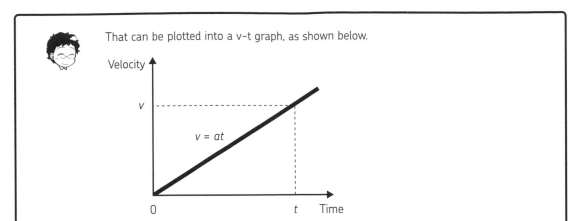

That can be plotted into a v-t graph, as shown below.

Velocity
v
v = at
0
t
Time

Let's assume *d* is the distance traveled in time *t*; its value should be equivalent to the area of a triangle with a base of *t* and height of *at* (which equals the final velocity of the object).

$$d = \frac{1}{2}at^2$$

Velocity
at
= ½(t × at)
Distance traveled (m)
0
t
Time

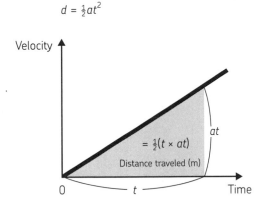

You see?

Ummmm . . . oh, I see how that works! The value we get by calculating $\frac{1}{2} \times 0.125\ m/s^2 \times (4\ s)^2 = 1\ m$. As it should be!

Now, Ninomiya-san, you can also calculate a distance traveled in uniform accelerated motion not by intuition but by the proper method.

NEWTON'S FIRST AND SECOND LAWS

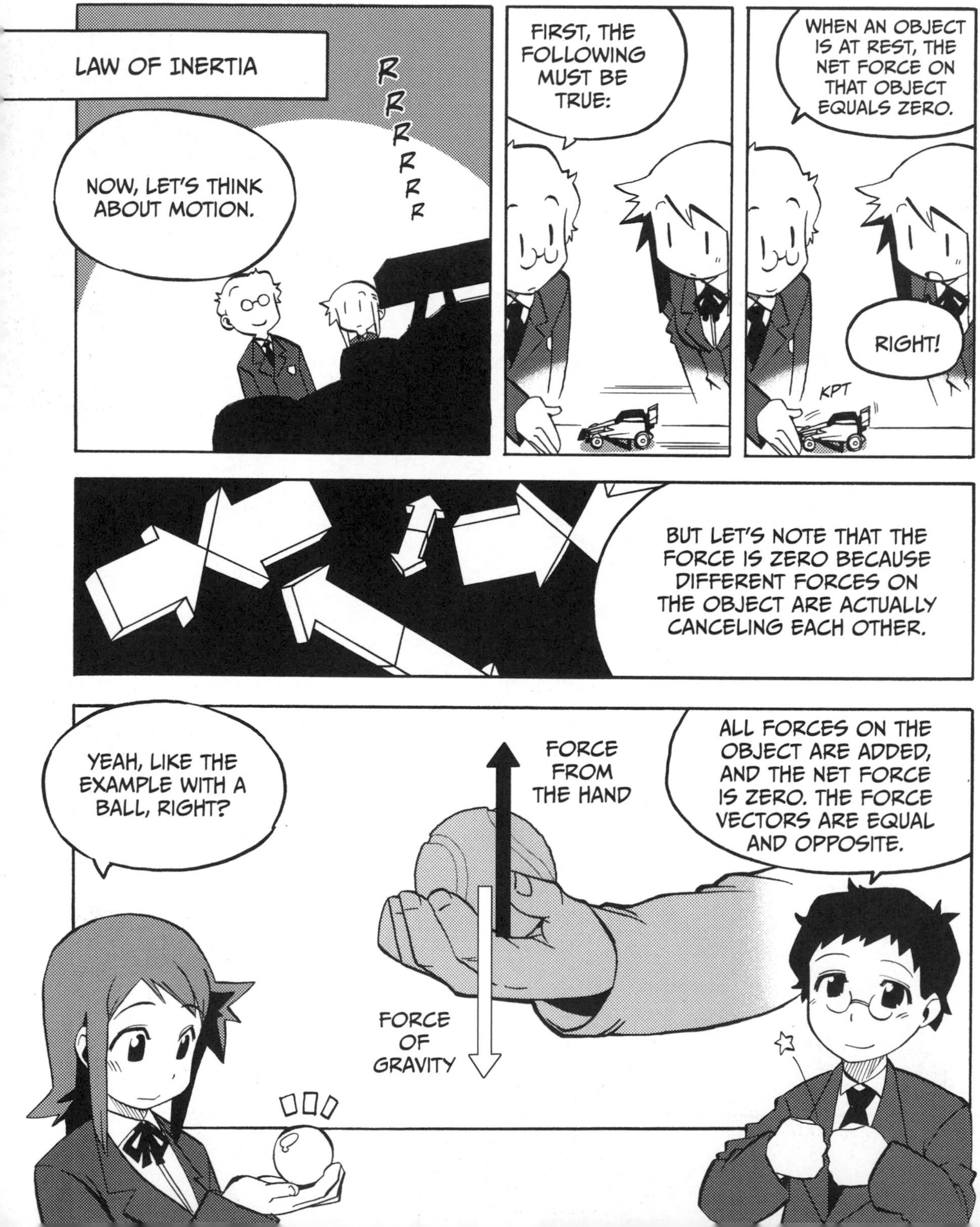
LAW OF INERTIA
NOW, LET'S THINK ABOUT MOTION.
RRRRRRR
FIRST, THE FOLLOWING MUST BE TRUE:
WHEN AN OBJECT IS AT REST, THE NET FORCE ON THAT OBJECT EQUALS ZERO.
RIGHT!
KPT
BUT LET'S NOTE THAT THE FORCE IS ZERO BECAUSE DIFFERENT FORCES ON THE OBJECT ARE ACTUALLY CANCELING EACH OTHER.
YEAH, LIKE THE EXAMPLE WITH A BALL, RIGHT?
FORCE FROM THE HAND
ALL FORCES ON THE OBJECT ARE ADDED, AND THE NET FORCE IS ZERO. THE FORCE VECTORS ARE EQUAL AND OPPOSITE.
FORCE OF GRAVITY

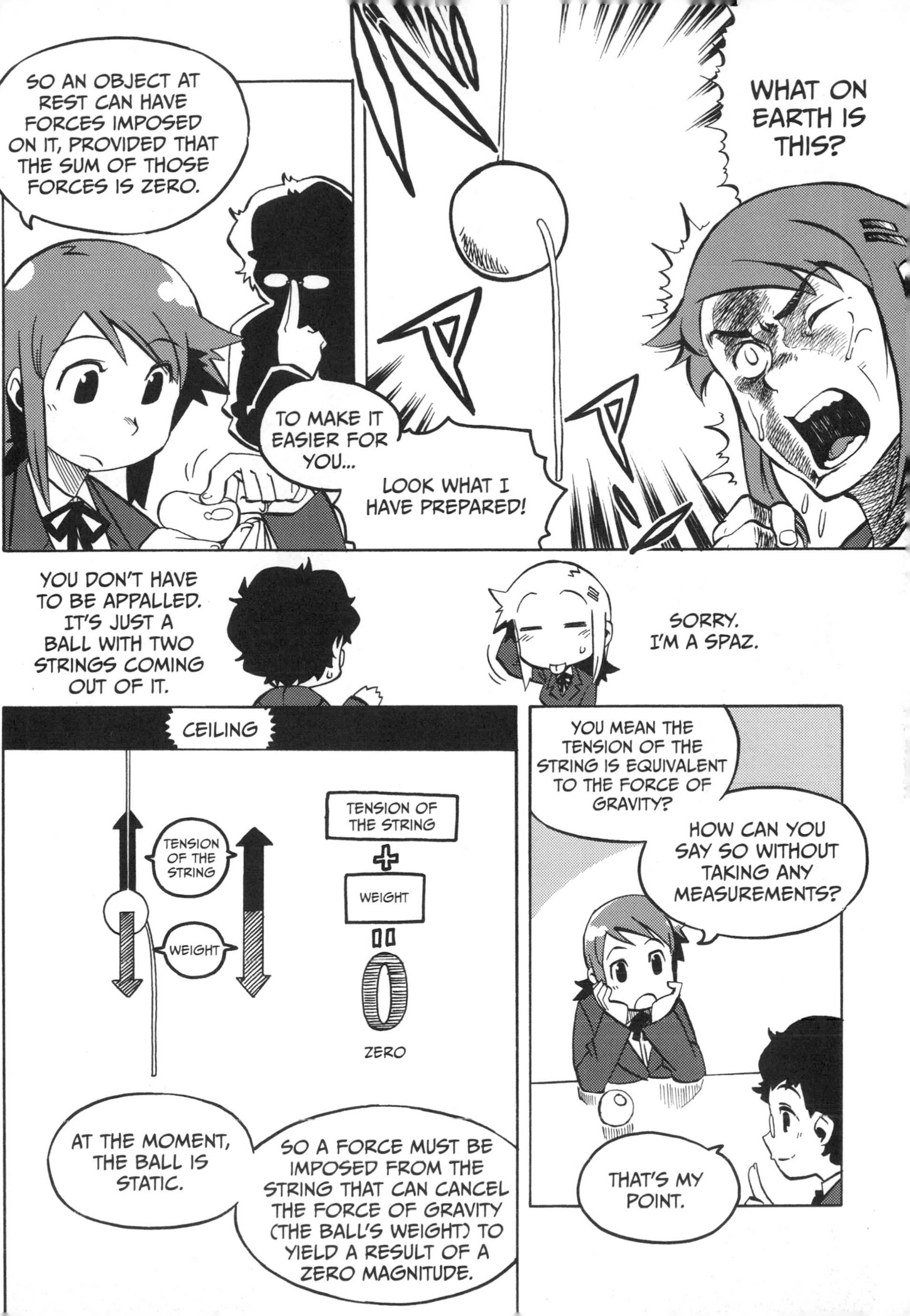
SO AN OBJECT AT REST CAN HAVE FORCES IMPOSED ON IT, PROVIDED THAT THE SUM OF THOSE FORCES IS ZERO.
TO MAKE IT EASIER FOR YOU...
LOOK WHAT I HAVE PREPARED!
WHAT ON EARTH IS THIS?
YOU DON'T HAVE TO BE APPALLED. IT'S JUST A BALL WITH TWO STRINGS COMING OUT OF IT.
SORRY. I'M A SPAZ.
CEILING
TENSION OF THE STRING
WEIGHT
TENSION OF THE STRING
+
WEIGHT
=
0
ZERO
AT THE MOMENT, THE BALL IS STATIC.
SO A FORCE MUST BE IMPOSED FROM THE STRING THAT CAN CANCEL THE FORCE OF GRAVITY (THE BALL'S WEIGHT) TO YIELD A RESULT OF A ZERO MAGNITUDE.
YOU MEAN THE TENSION OF THE STRING IS EQUIVALENT TO THE FORCE OF GRAVITY?
HOW CAN YOU SAY SO WITHOUT TAKING ANY MEASUREMENTS?
THAT'S MY POINT.

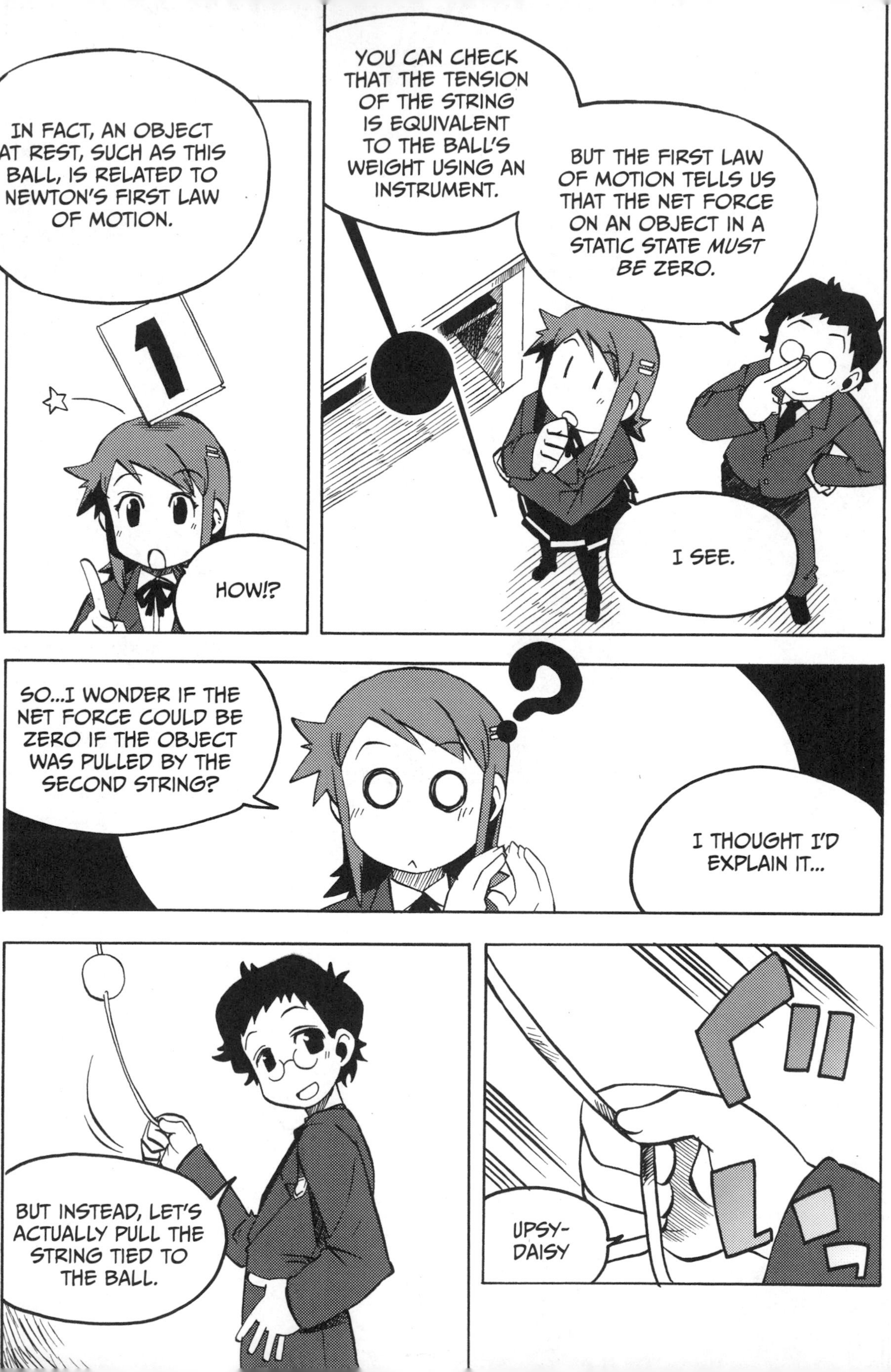
IN FACT, AN OBJECT AT REST, SUCH AS THIS BALL, IS RELATED TO NEWTON'S FIRST LAW OF MOTION.
1
HOW!?
YOU CAN CHECK THAT THE TENSION OF THE STRING IS EQUIVALENT TO THE BALL'S WEIGHT USING AN INSTRUMENT.
BUT THE FIRST LAW OF MOTION TELLS US THAT THE NET FORCE ON AN OBJECT IN A STATIC STATE *MUST* BE ZERO.
I SEE.
SO...I WONDER IF THE NET FORCE COULD BE ZERO IF THE OBJECT WAS PULLED BY THE SECOND STRING?
I THOUGHT I'D EXPLAIN IT...
BUT INSTEAD, LET'S ACTUALLY PULL THE STRING TIED TO THE BALL.
UPSY-DAISY

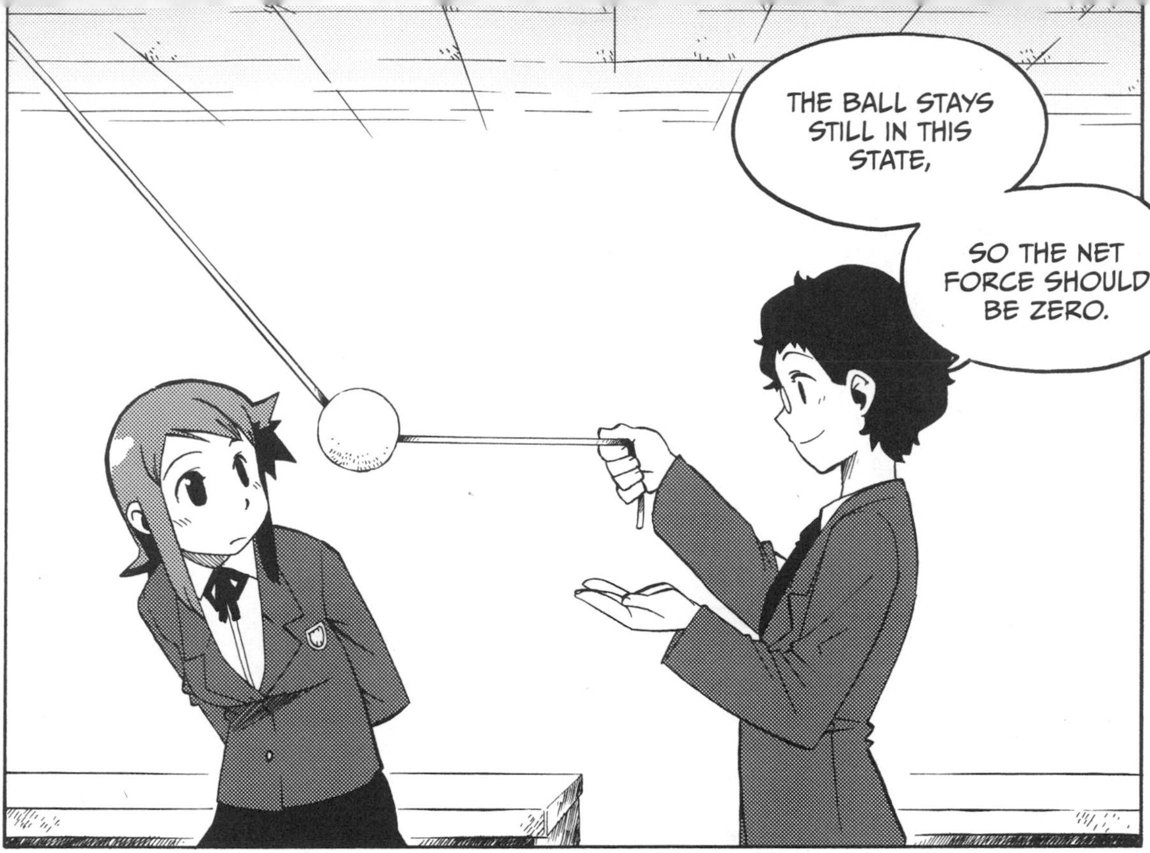
THE BALL STAYS STILL IN THIS STATE,
SO THE NET FORCE SHOULD BE ZERO.

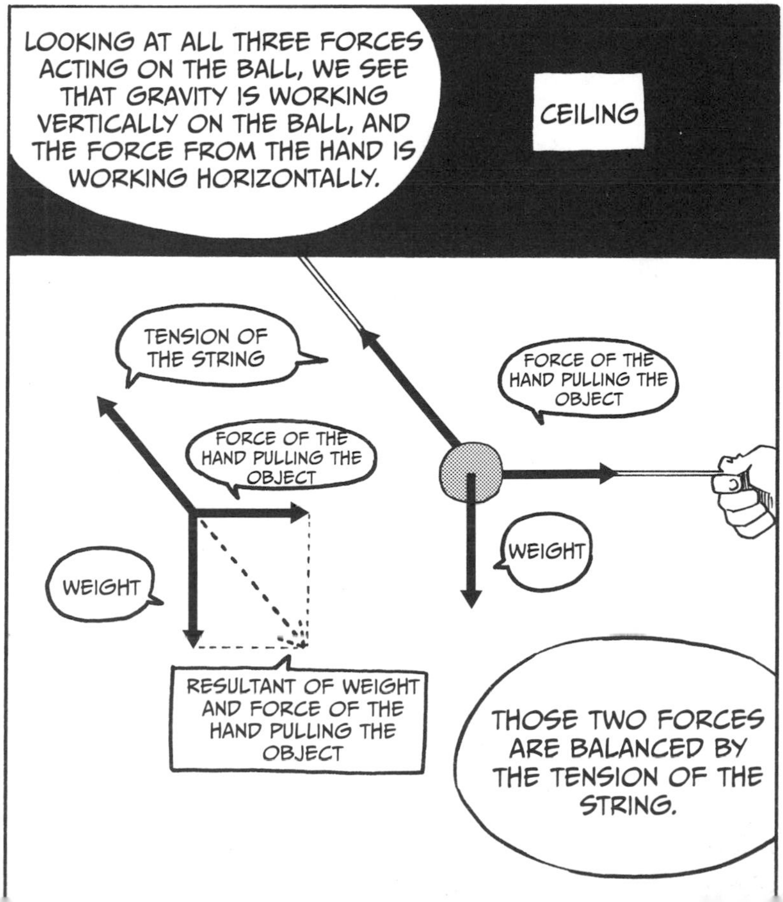
LOOKING AT ALL THREE FORCES ACTING ON THE BALL, WE SEE THAT GRAVITY IS WORKING VERTICALLY ON THE BALL, AND THE FORCE FROM THE HAND IS WORKING HORIZONTALLY.
CEILING
TENSION OF THE STRING
FORCE OF THE HAND PULLING THE OBJECT
FORCE OF THE HAND PULLING THE OBJECT
WEIGHT
WEIGHT
RESULTANT OF WEIGHT AND FORCE OF THE HAND PULLING THE OBJECT
THOSE TWO FORCES ARE BALANCED BY THE TENSION OF THE STRING.

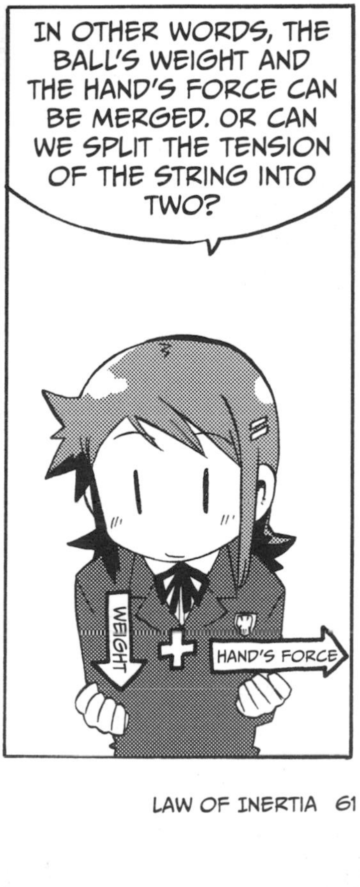
IN OTHER WORDS, THE BALL'S WEIGHT AND THE HAND'S FORCE CAN BE MERGED. OR CAN WE SPLIT THE TENSION OF THE STRING INTO TWO?
WEIGHT
HAND'S FORCE

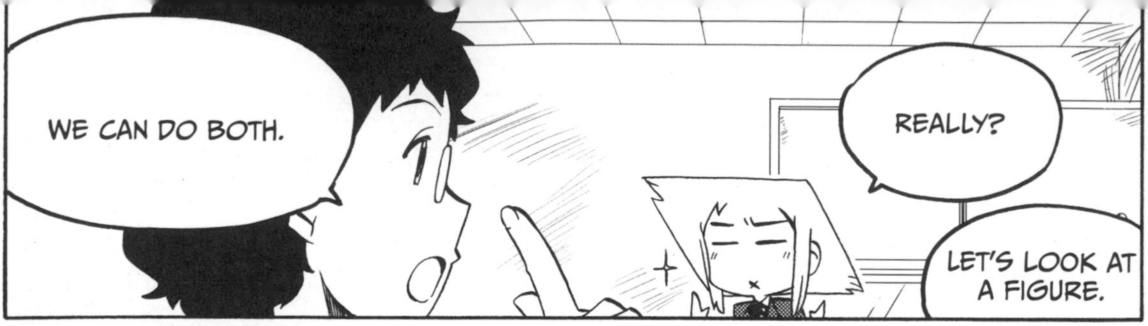
WE CAN DO BOTH.
REALLY?
LET'S LOOK AT A FIGURE.

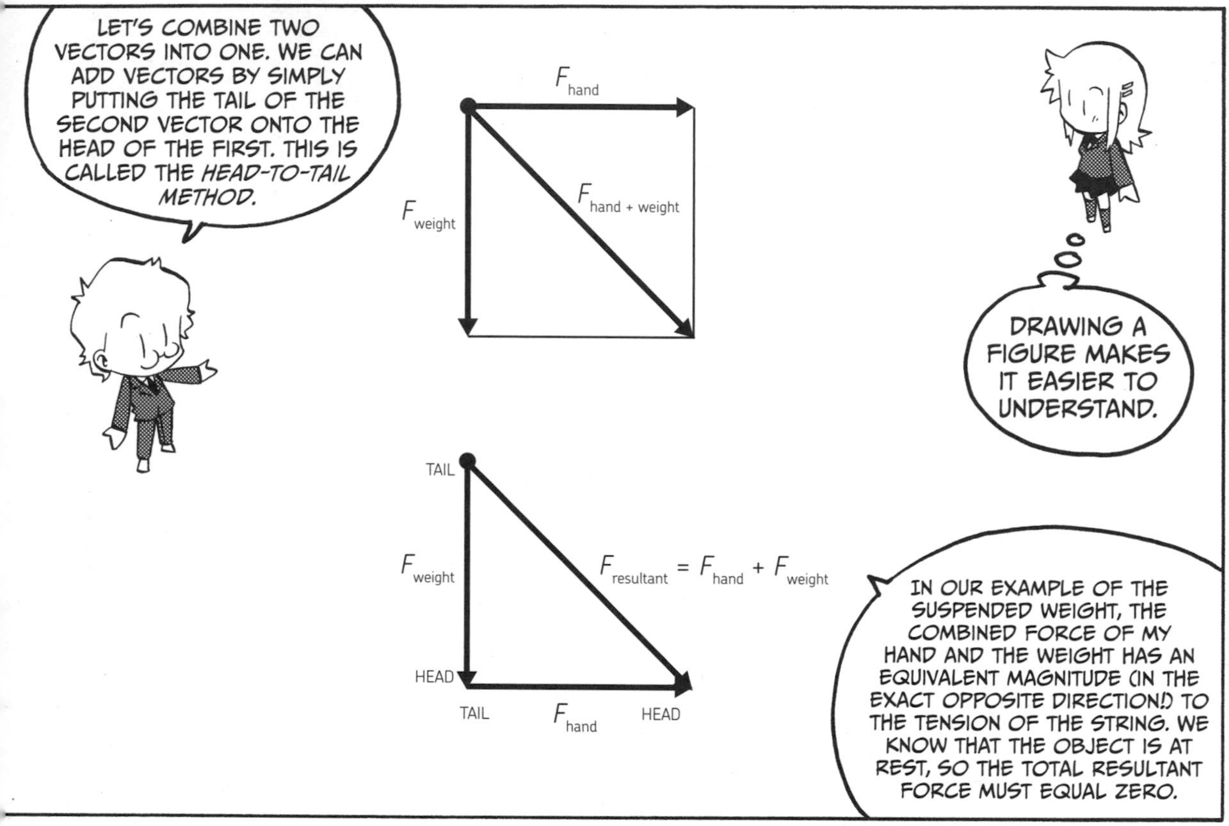
LET'S COMBINE TWO VECTORS INTO ONE. WE CAN ADD VECTORS BY SIMPLY PUTTING THE TAIL OF THE SECOND VECTOR ONTO THE HEAD OF THE FIRST. THIS IS CALLED THE *HEAD-TO-TAIL METHOD*.
F_{hand}
F_{weight}
$F_{hand + weight}$
DRAWING A FIGURE MAKES IT EASIER TO UNDERSTAND.
TAIL
F_{weight}
$F_{resultant} = F_{hand} + F_{weight}$
HEAD
TAIL
F_{hand}
HEAD
IN OUR EXAMPLE OF THE SUSPENDED WEIGHT, THE COMBINED FORCE OF MY HAND AND THE WEIGHT HAS AN EQUIVALENT MAGNITUDE (IN THE EXACT OPPOSITE DIRECTION!) TO THE TENSION OF THE STRING. WE KNOW THAT THE OBJECT IS AT REST, SO THE TOTAL RESULTANT FORCE MUST EQUAL ZERO.

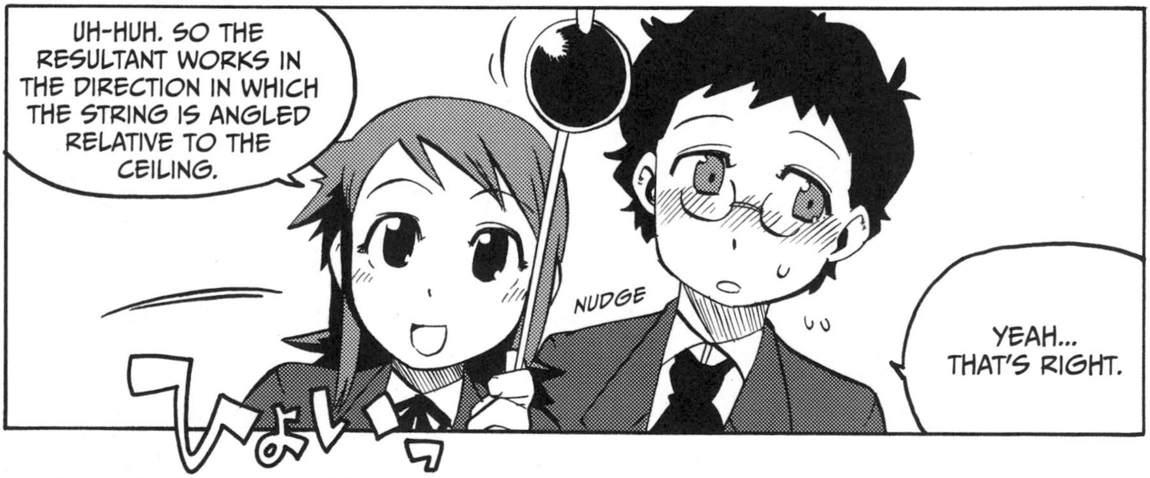
UH-HUH. SO THE RESULTANT WORKS IN THE DIRECTION IN WHICH THE STRING IS ANGLED RELATIVE TO THE CEILING.
NUDGE
ひょいっ
YEAH... THAT'S RIGHT.

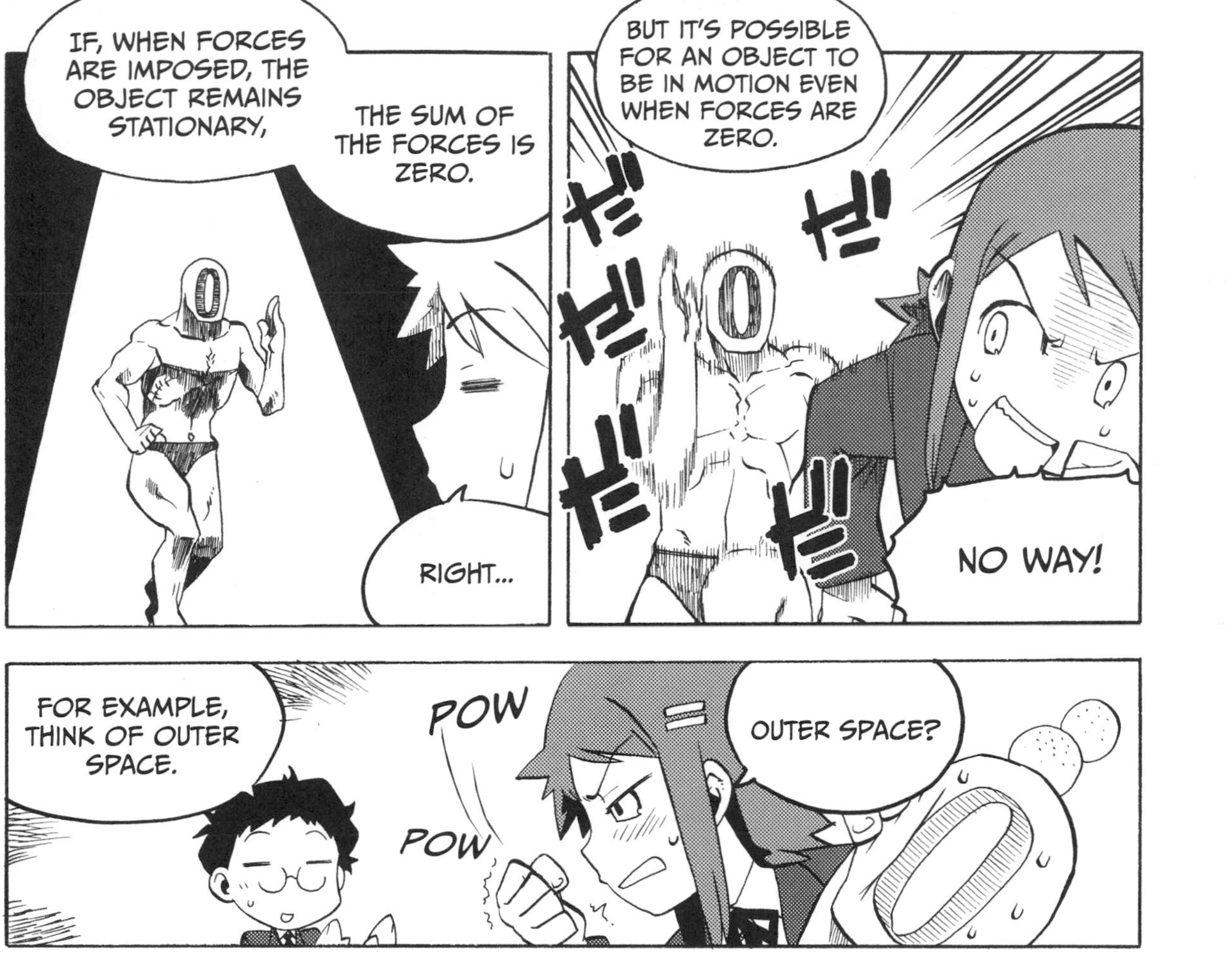
IF, WHEN FORCES ARE IMPOSED, THE OBJECT REMAINS STATIONARY,
THE SUM OF THE FORCES IS ZERO.
RIGHT...
BUT IT'S POSSIBLE FOR AN OBJECT TO BE IN MOTION EVEN WHEN FORCES ARE ZERO.
NO WAY!
FOR EXAMPLE, THINK OF OUTER SPACE.
POW
POW
OUTER SPACE?

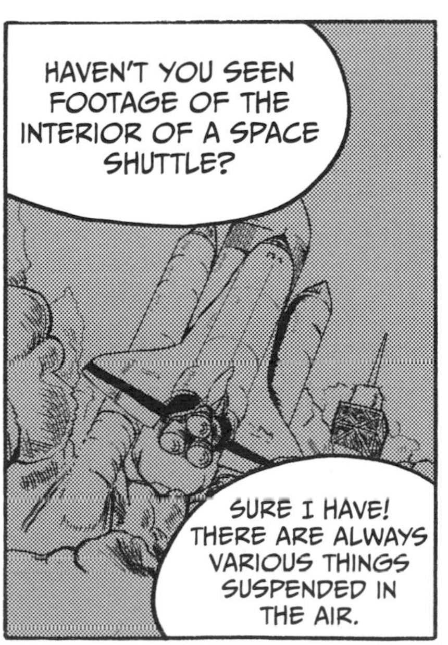
HAVEN'T YOU SEEN FOOTAGE OF THE INTERIOR OF A SPACE SHUTTLE?
SURE I HAVE! THERE ARE ALWAYS VARIOUS THINGS SUSPENDED IN THE AIR.

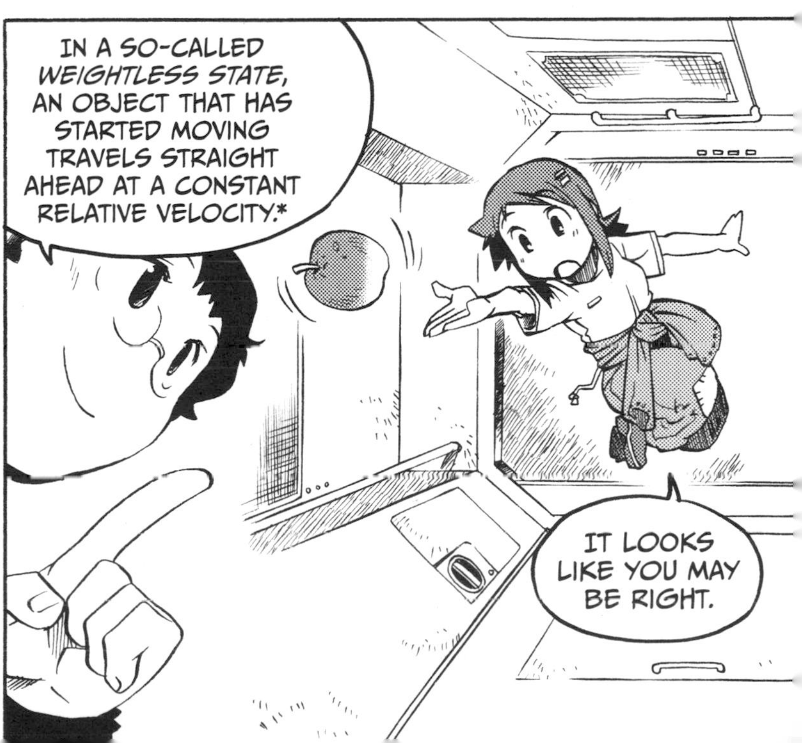
IN A SO-CALLED *WEIGHTLESS STATE*, AN OBJECT THAT HAS STARTED MOVING TRAVELS STRAIGHT AHEAD AT A CONSTANT RELATIVE VELOCITY.*
IT LOOKS LIKE YOU MAY BE RIGHT.

* IN ORBIT, OBJECTS ARE IN A STATE OF CONSTANT FREE FALL, MAKING THEIR APPARENT WEIGHT ZERO.

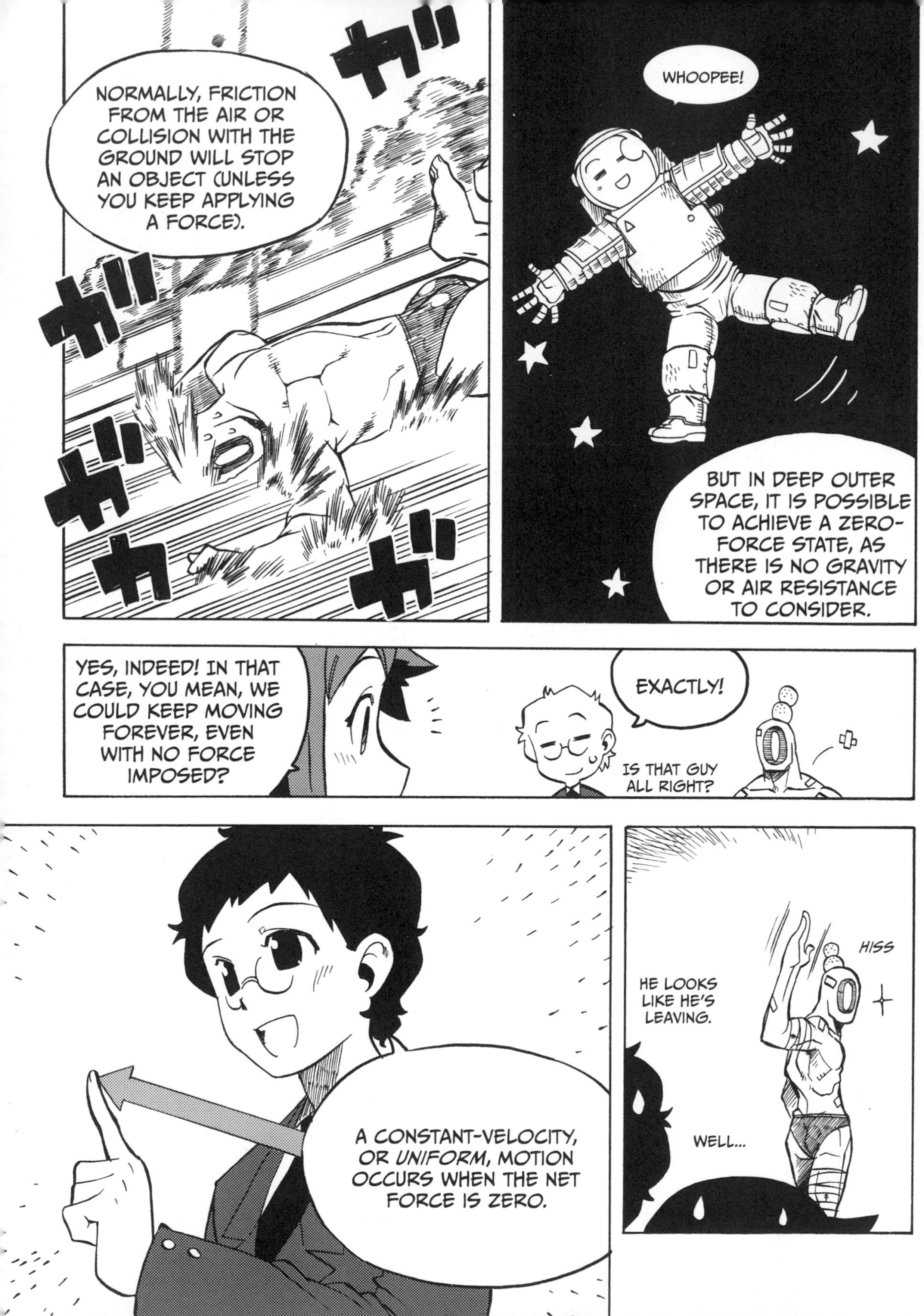
NORMALLY, FRICTION FROM THE AIR OR COLLISION WITH THE GROUND WILL STOP AN OBJECT (UNLESS YOU KEEP APPLYING A FORCE).
WHOOPEE!
BUT IN DEEP OUTER SPACE, IT IS POSSIBLE TO ACHIEVE A ZERO-FORCE STATE, AS THERE IS NO GRAVITY OR AIR RESISTANCE TO CONSIDER.
YES, INDEED! IN THAT CASE, YOU MEAN, WE COULD KEEP MOVING FOREVER, EVEN WITH NO FORCE IMPOSED?
EXACTLY!
IS THAT GUY ALL RIGHT?
A CONSTANT-VELOCITY, OR *UNIFORM*, MOTION OCCURS WHEN THE NET FORCE IS ZERO.
HISS
HE LOOKS LIKE HE'S LEAVING.
WELL...

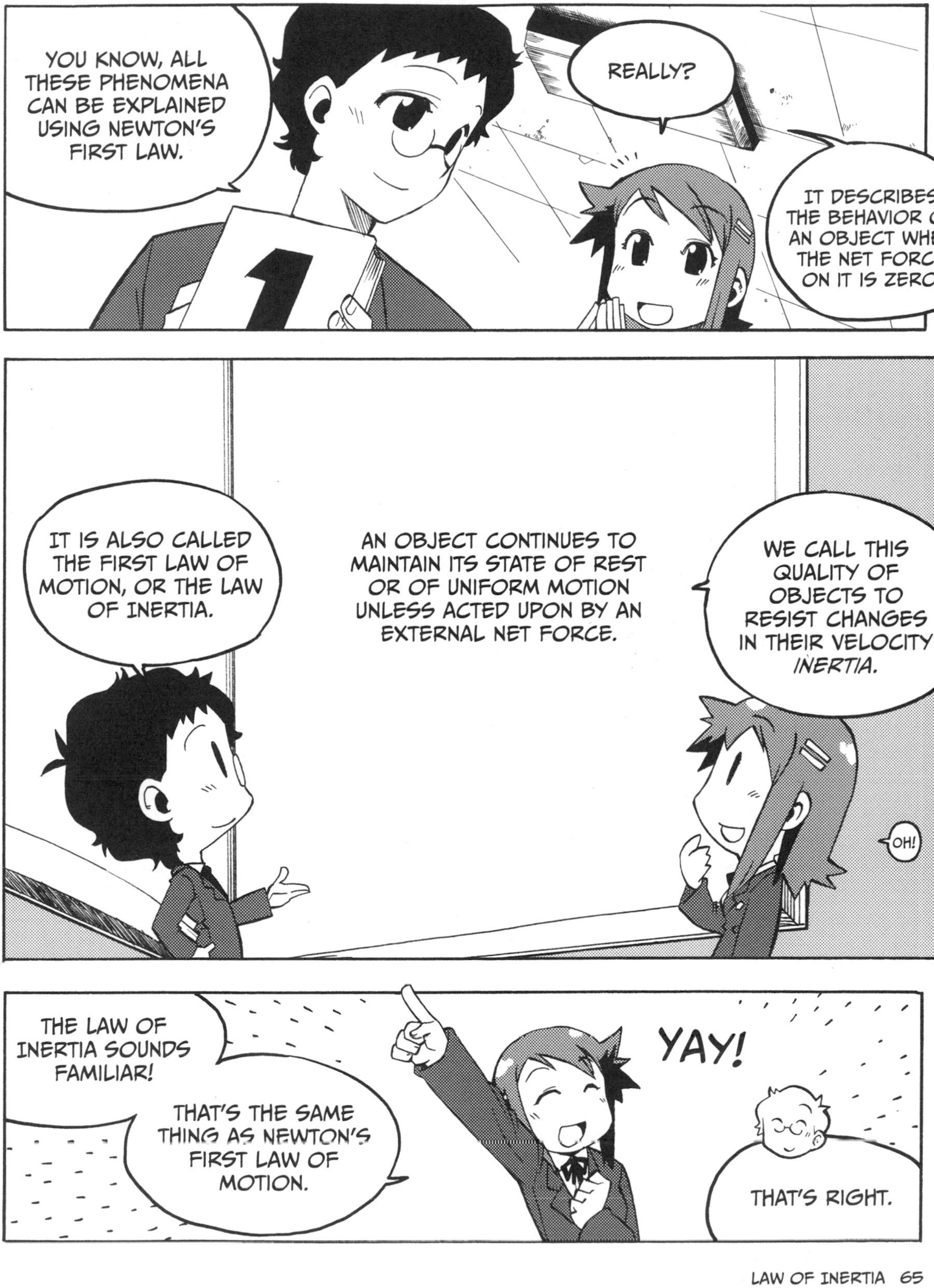
YOU KNOW, ALL THESE PHENOMENA CAN BE EXPLAINED USING NEWTON'S FIRST LAW.
REALLY?
IT DESCRIBES THE BEHAVIOR OF AN OBJECT WHEN THE NET FORCE ON IT IS ZERO.
IT IS ALSO CALLED THE FIRST LAW OF MOTION, OR THE LAW OF INERTIA.
AN OBJECT CONTINUES TO MAINTAIN ITS STATE OF REST OR OF UNIFORM MOTION UNLESS ACTED UPON BY AN EXTERNAL NET FORCE.
WE CALL THIS QUALITY OF OBJECTS TO RESIST CHANGES IN THEIR VELOCITY *INERTIA*.
OH!
THE LAW OF INERTIA SOUNDS FAMILIAR!
THAT'S THE SAME THING AS NEWTON'S FIRST LAW OF MOTION.
YAY!
THAT'S RIGHT.

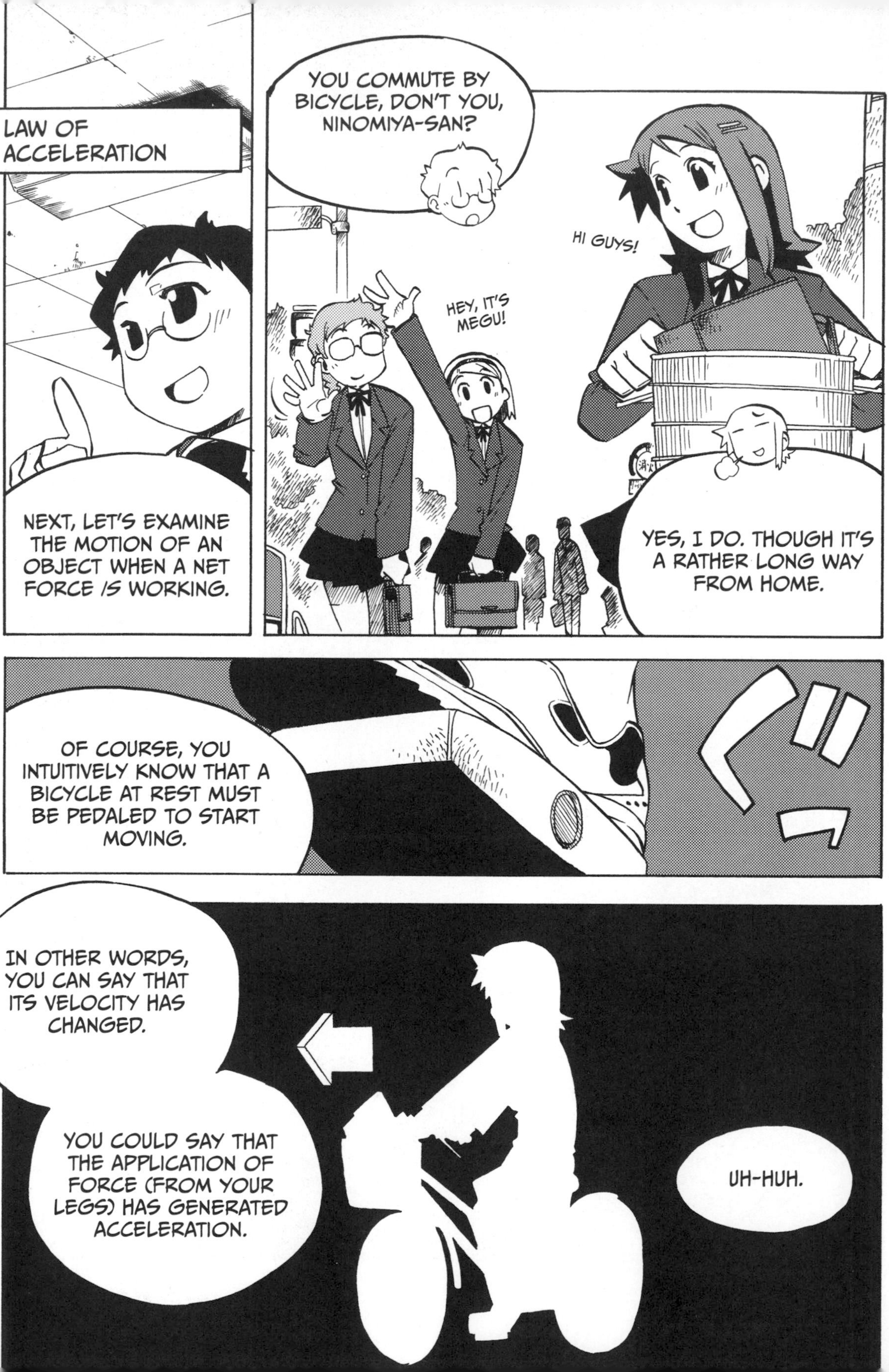
LAW OF ACCELERATION
NEXT, LET'S EXAMINE THE MOTION OF AN OBJECT WHEN A NET FORCE *IS* WORKING.
YOU COMMUTE BY BICYCLE, DON'T YOU, NINOMIYA-SAN?
HEY, IT'S MEGU!
HI GUYS!
YES, I DO. THOUGH IT'S A RATHER LONG WAY FROM HOME.
OF COURSE, YOU INTUITIVELY KNOW THAT A BICYCLE AT REST MUST BE PEDALED TO START MOVING.
IN OTHER WORDS, YOU CAN SAY THAT ITS VELOCITY HAS CHANGED.
YOU COULD SAY THAT THE APPLICATION OF FORCE (FROM YOUR LEGS) HAS GENERATED ACCELERATION.
UH-HUH.

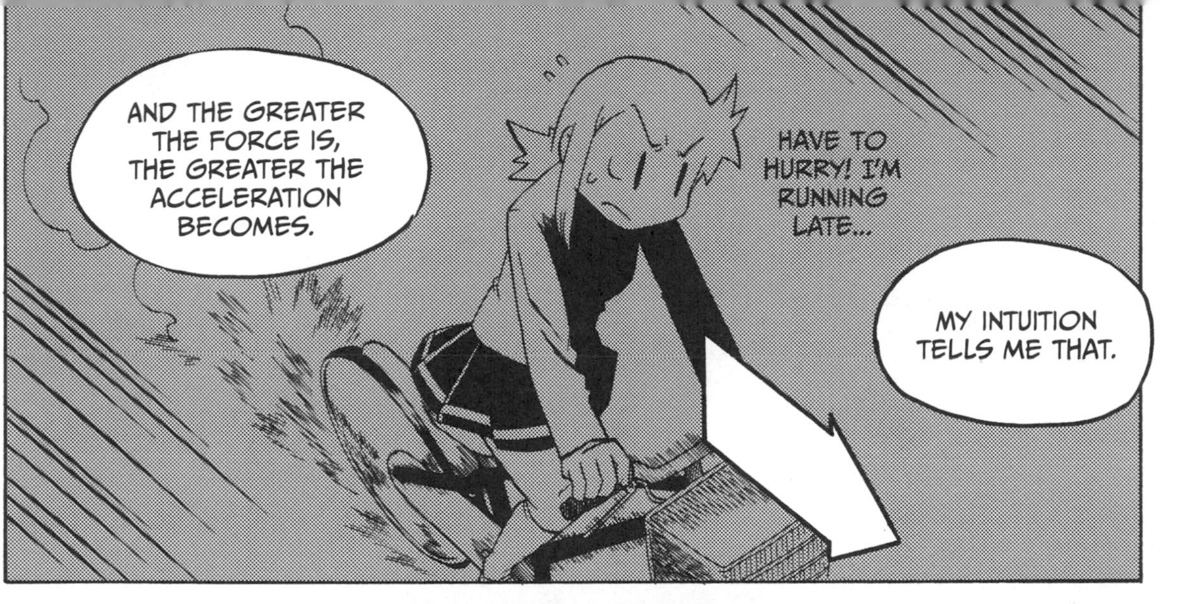
AND THE GREATER THE FORCE IS, THE GREATER THE ACCELERATION BECOMES.
HAVE TO HURRY! I'M RUNNING LATE...
MY INTUITION TELLS ME THAT.

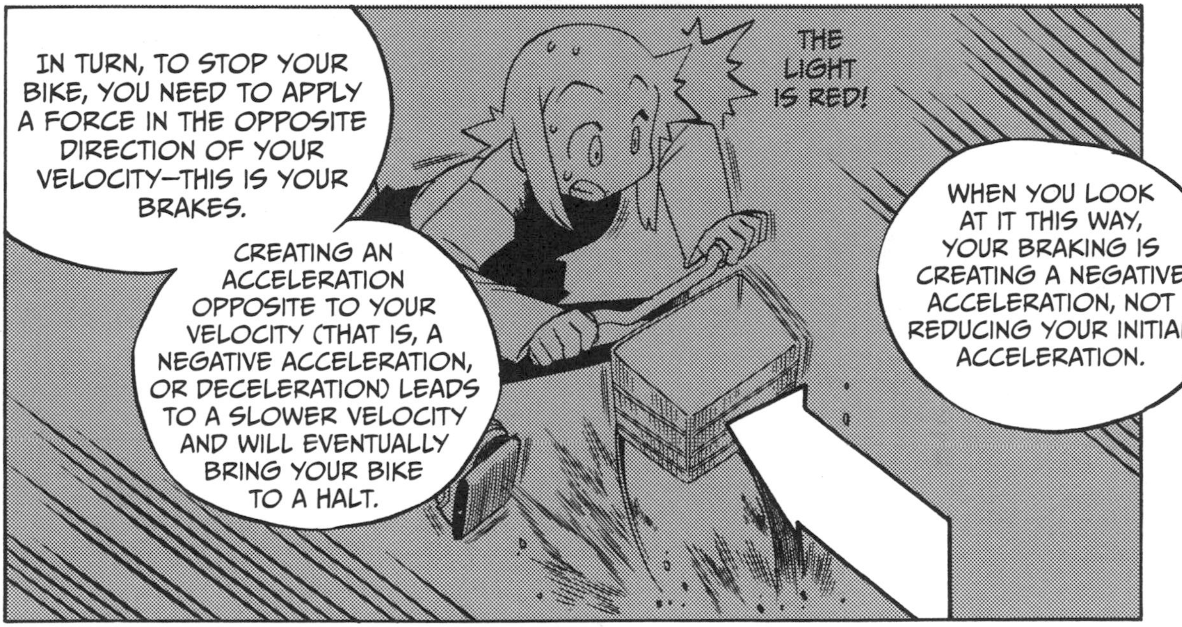
IN TURN, TO STOP YOUR BIKE, YOU NEED TO APPLY A FORCE IN THE OPPOSITE DIRECTION OF YOUR VELOCITY—THIS IS YOUR BRAKES.
THE LIGHT IS RED!
CREATING AN ACCELERATION OPPOSITE TO YOUR VELOCITY (THAT IS, A NEGATIVE ACCELERATION, OR DECELERATION) LEADS TO A SLOWER VELOCITY AND WILL EVENTUALLY BRING YOUR BIKE TO A HALT.
WHEN YOU LOOK AT IT THIS WAY, YOUR BRAKING IS CREATING A NEGATIVE ACCELERATION, NOT REDUCING YOUR INITIAL ACCELERATION.

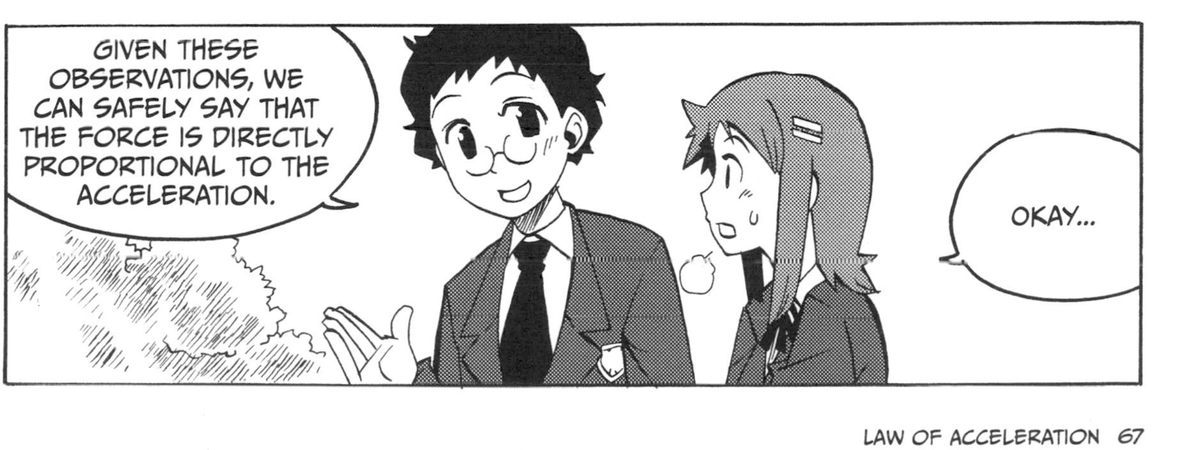
GIVEN THESE OBSERVATIONS, WE CAN SAFELY SAY THAT THE FORCE IS DIRECTLY PROPORTIONAL TO THE ACCELERATION.
OKAY...

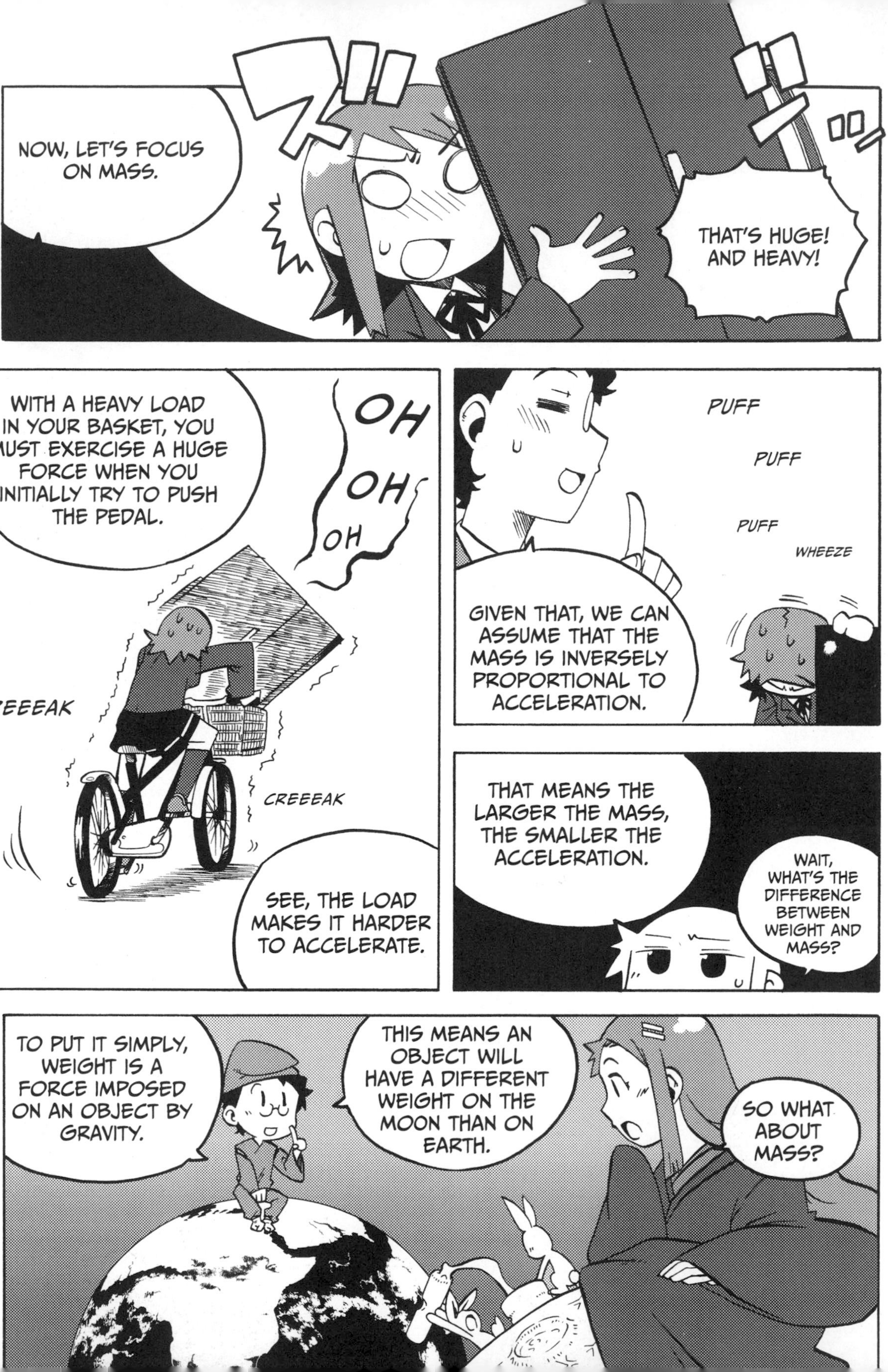
NOW, LET'S FOCUS ON MASS.
THAT'S HUGE! AND HEAVY!
WITH A HEAVY LOAD IN YOUR BASKET, YOU MUST EXERCISE A HUGE FORCE WHEN YOU INITIALLY TRY TO PUSH THE PEDAL.
OH OH OH
CREEEAK
CREEEAK
SEE, THE LOAD MAKES IT HARDER TO ACCELERATE.
PUFF
PUFF
PUFF
WHEEZE
GIVEN THAT, WE CAN ASSUME THAT THE MASS IS INVERSELY PROPORTIONAL TO ACCELERATION.
THAT MEANS THE LARGER THE MASS, THE SMALLER THE ACCELERATION.
WAIT, WHAT'S THE DIFFERENCE BETWEEN WEIGHT AND MASS?
TO PUT IT SIMPLY, WEIGHT IS A FORCE IMPOSED ON AN OBJECT BY GRAVITY.
THIS MEANS AN OBJECT WILL HAVE A DIFFERENT WEIGHT ON THE MOON THAN ON EARTH.
SO WHAT ABOUT MASS?

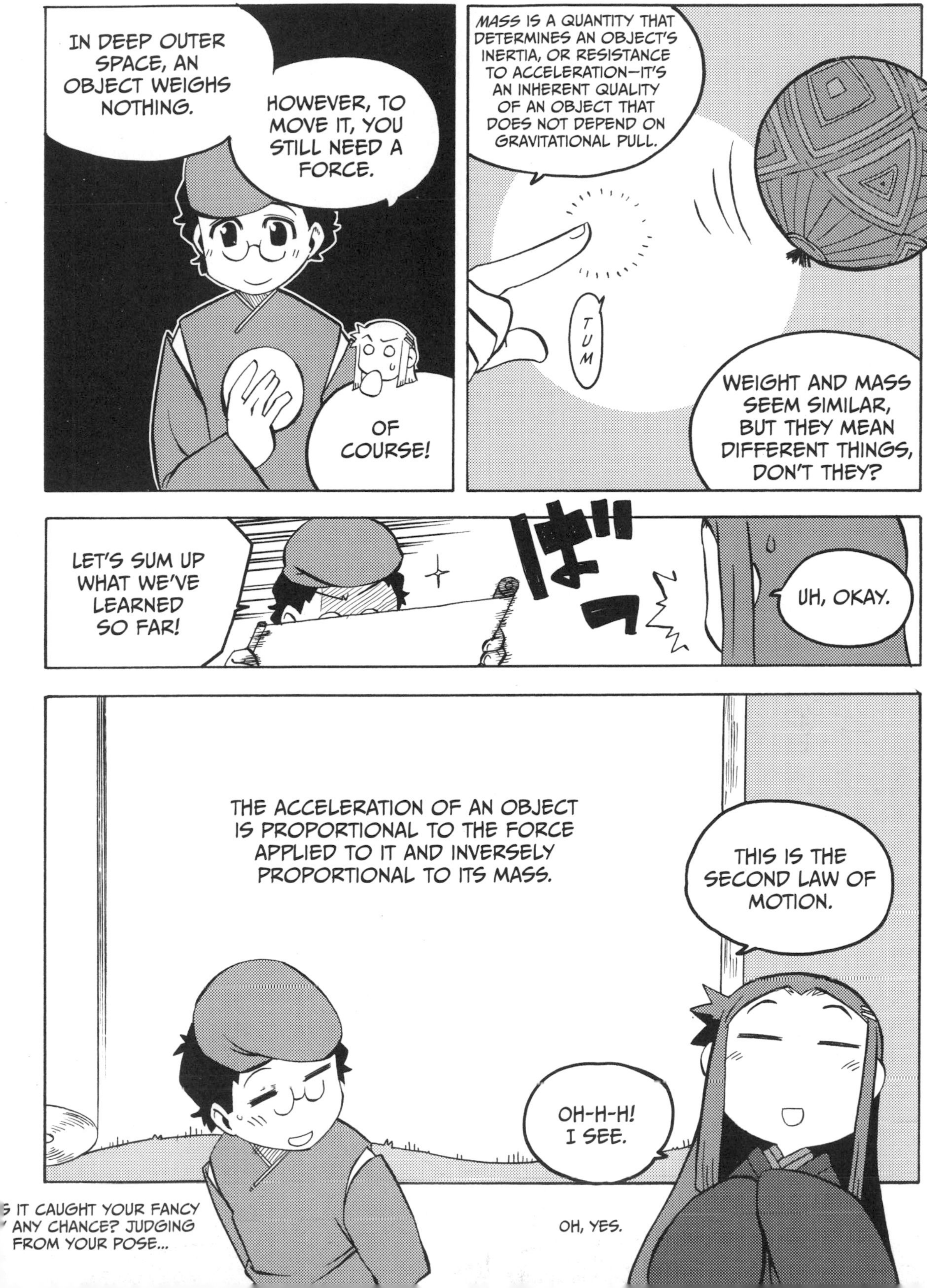

IN DEEP OUTER SPACE, AN OBJECT WEIGHS NOTHING.
HOWEVER, TO MOVE IT, YOU STILL NEED A FORCE.
OF COURSE!
MASS IS A QUANTITY THAT DETERMINES AN OBJECT'S INERTIA, OR RESISTANCE TO ACCELERATION—IT'S AN INHERENT QUALITY OF AN OBJECT THAT DOES NOT DEPEND ON GRAVITATIONAL PULL.
TUM
WEIGHT AND MASS SEEM SIMILAR, BUT THEY MEAN DIFFERENT THINGS, DON'T THEY?
LET'S SUM UP WHAT WE'VE LEARNED SO FAR!
ばっ
UH, OKAY.
THE ACCELERATION OF AN OBJECT IS PROPORTIONAL TO THE FORCE APPLIED TO IT AND INVERSELY PROPORTIONAL TO ITS MASS.
THIS IS THE SECOND LAW OF MOTION.
OH-H-H! I SEE.
S IT CAUGHT YOUR FANCY ANY CHANCE? JUDGING FROM YOUR POSE...
OH, YES.

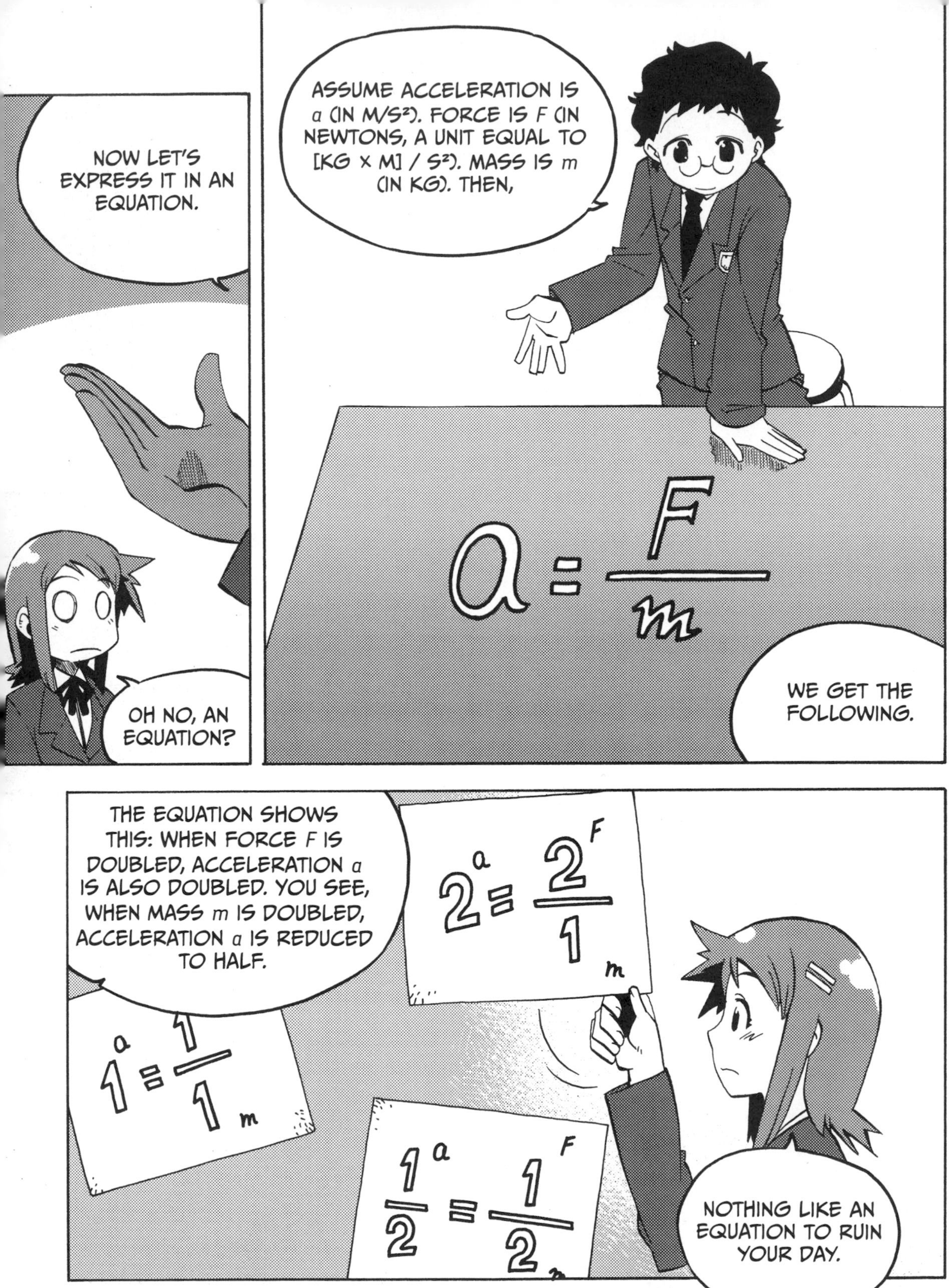
NOW LET'S EXPRESS IT IN AN EQUATION.
OH NO, AN EQUATION?
ASSUME ACCELERATION IS a (IN M/S²). FORCE IS F (IN NEWTONS, A UNIT EQUAL TO [KG × M] / S²). MASS IS m (IN KG). THEN,
$a = \frac{F}{m}$
WE GET THE FOLLOWING.
THE EQUATION SHOWS THIS: WHEN FORCE F IS DOUBLED, ACCELERATION a IS ALSO DOUBLED. YOU SEE, WHEN MASS m IS DOUBLED, ACCELERATION a IS REDUCED TO HALF.
$2^{a} = \frac{2^{F}}{1_{m}}$
$1^{a} = \frac{1}{1_{m}}$
$\frac{1}{2}^{a} = \frac{1^{F}}{2_{m}}$
NOTHING LIKE AN EQUATION TO RUIN YOUR DAY.

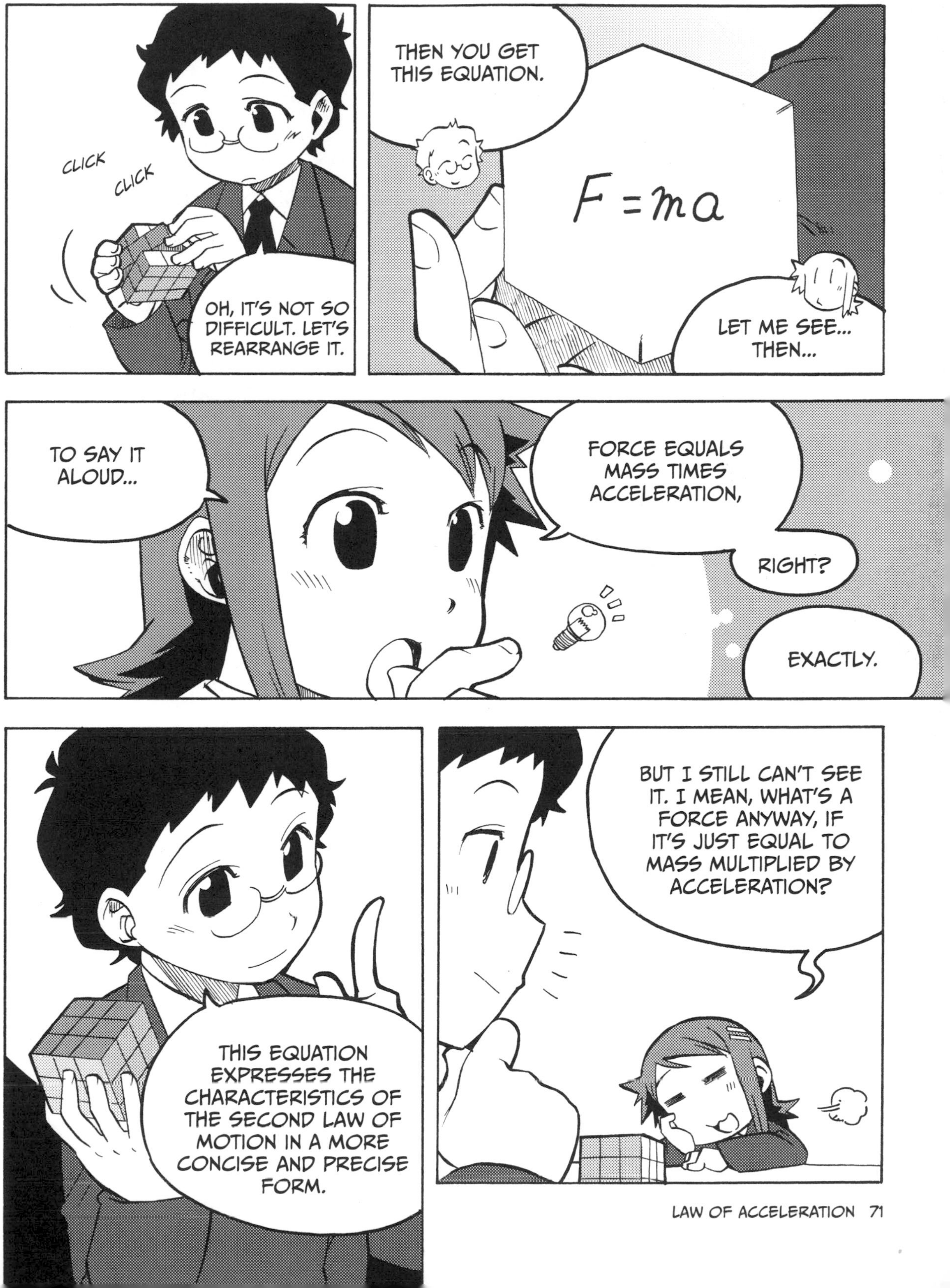
CLICK
CLICK
OH, IT'S NOT SO DIFFICULT. LET'S REARRANGE IT.
THEN YOU GET THIS EQUATION.
F = ma
LET ME SEE... THEN...
TO SAY IT ALOUD...
FORCE EQUALS MASS TIMES ACCELERATION,
RIGHT?
EXACTLY.
THIS EQUATION EXPRESSES THE CHARACTERISTICS OF THE SECOND LAW OF MOTION IN A MORE CONCISE AND PRECISE FORM.
BUT I STILL CAN'T SEE IT. I MEAN, WHAT'S A FORCE ANYWAY, IF IT'S JUST EQUAL TO MASS MULTIPLIED BY ACCELERATION?

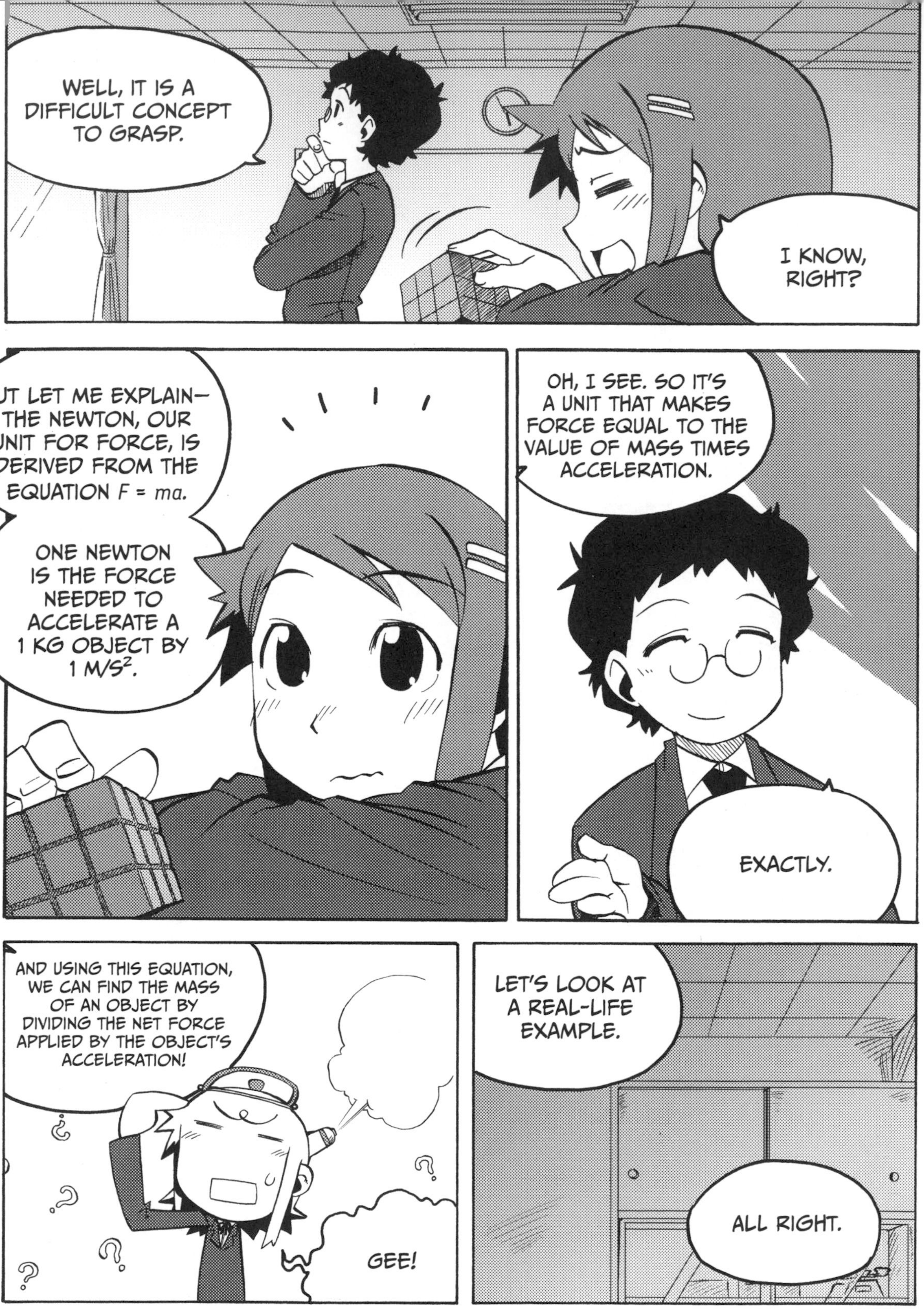
WELL, IT IS A DIFFICULT CONCEPT TO GRASP.
I KNOW, RIGHT?
BUT LET ME EXPLAIN—THE NEWTON, OUR UNIT FOR FORCE, IS DERIVED FROM THE EQUATION $F = ma$.
ONE NEWTON IS THE FORCE NEEDED TO ACCELERATE A 1 KG OBJECT BY 1 M/S^2.
OH, I SEE. SO IT'S A UNIT THAT MAKES FORCE EQUAL TO THE VALUE OF MASS TIMES ACCELERATION.
EXACTLY.
AND USING THIS EQUATION, WE CAN FIND THE MASS OF AN OBJECT BY DIVIDING THE NET FORCE APPLIED BY THE OBJECT'S ACCELERATION!
GEE!
LET'S LOOK AT A REAL-LIFE EXAMPLE.
ALL RIGHT.

LABORATORY

FINDING THE PRECISE VALUE OF A FORCE

 Earlier, we pushed each other while we were on roller blades. Let's say that I captured our motion on video.

 I didn't realize you were taping us!

 Oh, that's just the scenario I'm setting up.

 Jeez, don't scare me. How does that relate to the second law of motion?

 Suppose I have analyzed the video, and I've created a v–t graph of your motion.

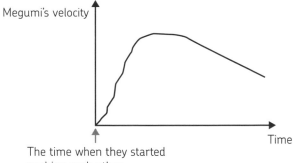

Megumi's velocity

Time

The time when they started
pushing each other

 We can see that velocity increases sharply from zero, which must be when I'm at rest, and then drops gradually after that. But the initial increase in velocity is wobbly.

In a case like this, it may be a good idea to draw a line segment that represents the average increase in velocity. In other words, we'll simplify the scenario to assume this is a case of uniform acceleration.

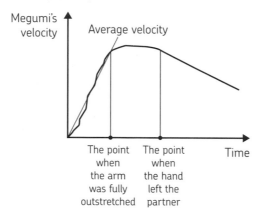

Megumi's velocity

Average velocity

The point when the arm was fully outstretched

The point when the hand left the partner

Time

 I see.

 You can find acceleration by calculating the change in velocity over time—acceleration = change in velocity / time. In this case, let's assume that your acceleration is equal to 0.6 m/s^2. To find the force I applied to your hands, let's also assume your mass is 40 kg, Ninomiya-san.

$$F = ma = 40 \text{ kg} \times 0.6 \text{ m/s}^2 = 24 \text{ kg} \times \text{m/s}^2, \text{ or 24N}$$

 We've found the precise value of the force! So, this is important! We can measure the exact force on an object by measuring its acceleration and its mass.

 Now, if you know that I have a mass of 60 kg, can you predict my acceleration, due to the application of an equal and opposite 24N of force?

 Oh, I see. We're combining the second and third laws of motion. F_{Megumi} must equal F_{Ryota}. Since $F = ma$, we know that $F / m = a$. In your case, that's 24N / 60 kg, or 0.4 m/s^2. So we can use these laws to predict the movement of objects. Neat!

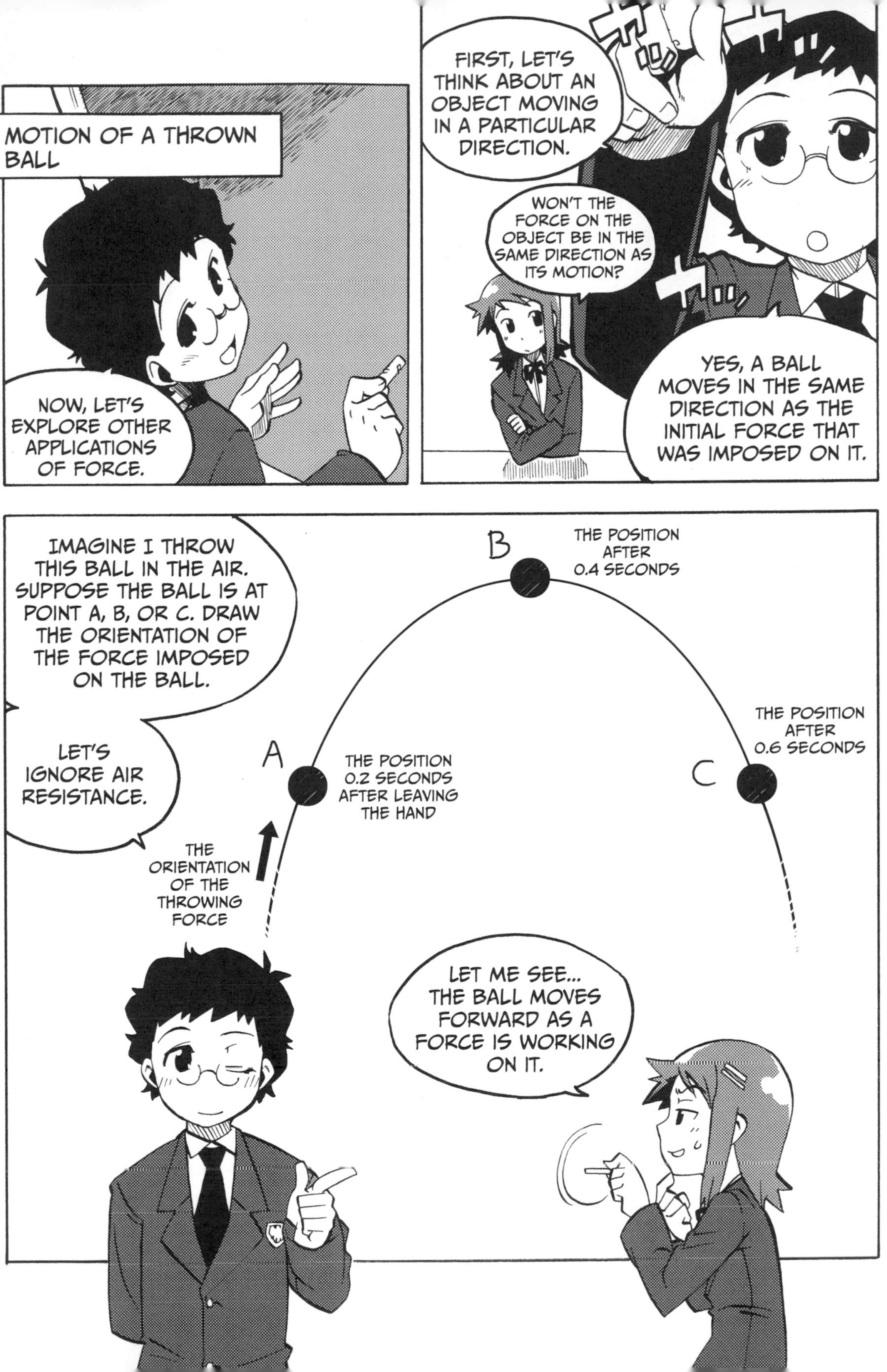
MOTION OF A THROWN BALL
NOW, LET'S EXPLORE OTHER APPLICATIONS OF FORCE.
FIRST, LET'S THINK ABOUT AN OBJECT MOVING IN A PARTICULAR DIRECTION.
WON'T THE FORCE ON THE OBJECT BE IN THE SAME DIRECTION AS ITS MOTION?
YES, A BALL MOVES IN THE SAME DIRECTION AS THE INITIAL FORCE THAT WAS IMPOSED ON IT.
IMAGINE I THROW THIS BALL IN THE AIR. SUPPOSE THE BALL IS AT POINT A, B, OR C. DRAW THE ORIENTATION OF THE FORCE IMPOSED ON THE BALL.
LET'S IGNORE AIR RESISTANCE.
B
THE POSITION AFTER 0.4 SECONDS
A
THE POSITION 0.2 SECONDS AFTER LEAVING THE HAND
C
THE POSITION AFTER 0.6 SECONDS
THE ORIENTATION OF THE THROWING FORCE
LET ME SEE... THE BALL MOVES FORWARD AS A FORCE IS WORKING ON IT.

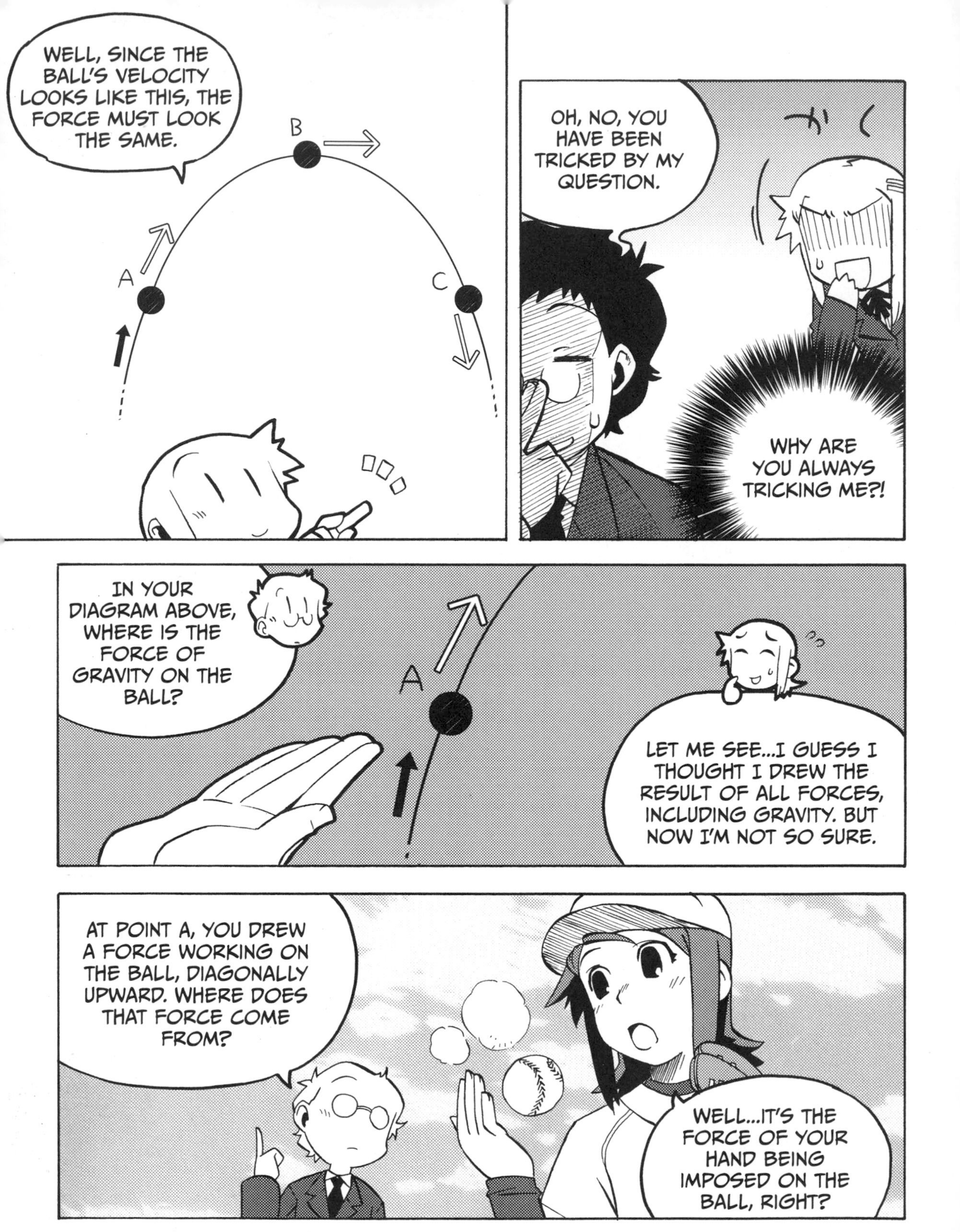

WELL, SINCE THE BALL'S VELOCITY LOOKS LIKE THIS, THE FORCE MUST LOOK THE SAME.
B
A
C
OH, NO, YOU HAVE BEEN TRICKED BY MY QUESTION.
WHY ARE YOU ALWAYS TRICKING ME?!
IN YOUR DIAGRAM ABOVE, WHERE IS THE FORCE OF GRAVITY ON THE BALL?
A
LET ME SEE...I GUESS I THOUGHT I DREW THE RESULT OF ALL FORCES, INCLUDING GRAVITY. BUT NOW I'M NOT SO SURE.
AT POINT A, YOU DREW A FORCE WORKING ON THE BALL, DIAGONALLY UPWARD. WHERE DOES THAT FORCE COME FROM?
WELL...IT'S THE FORCE OF YOUR HAND BEING IMPOSED ON THE BALL, RIGHT?

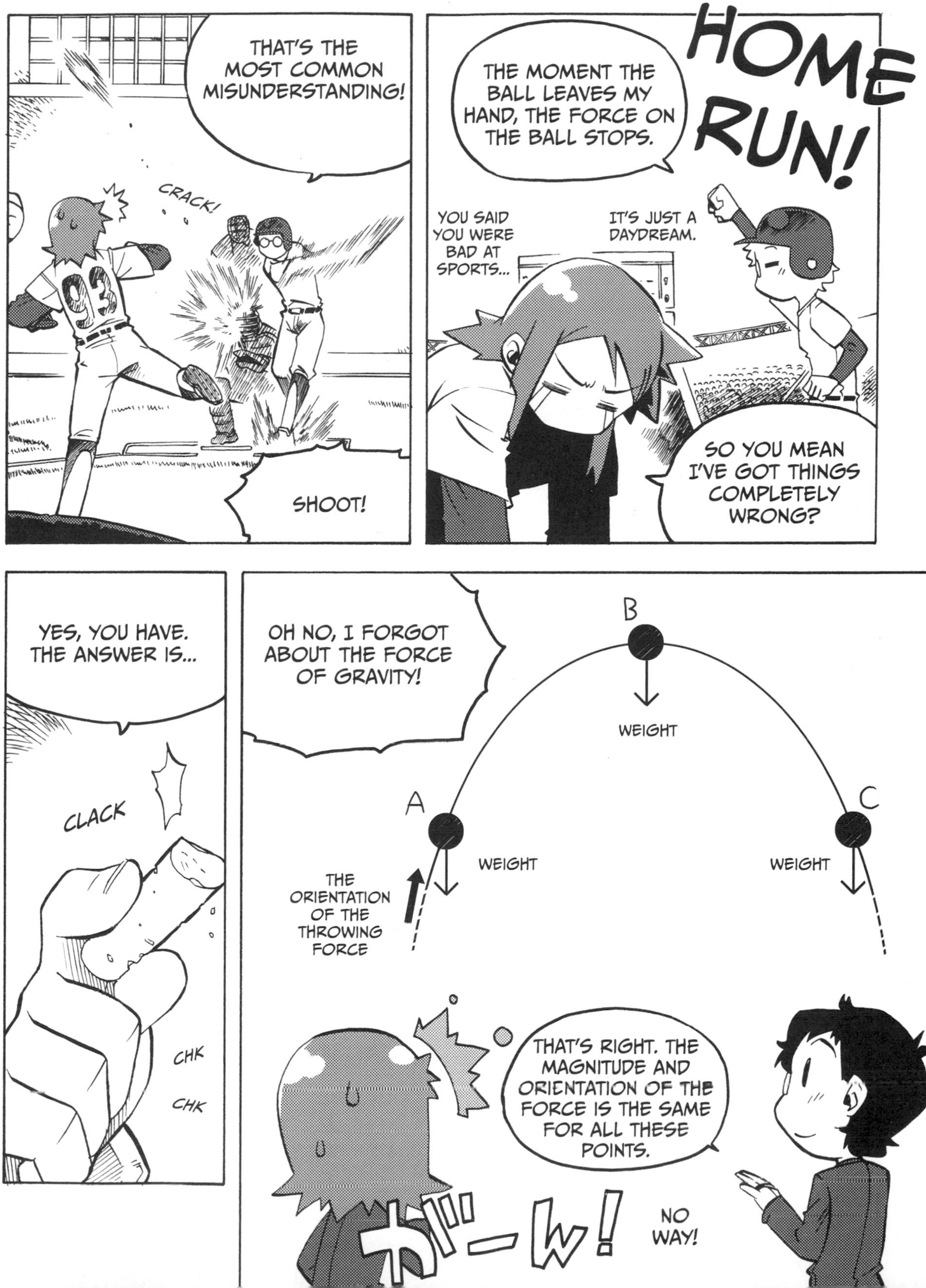
THAT'S THE MOST COMMON MISUNDERSTANDING!
CRACK!
SHOOT!
THE MOMENT THE BALL LEAVES MY HAND, THE FORCE ON THE BALL STOPS.
HOME RUN!
YOU SAID YOU WERE BAD AT SPORTS...
IT'S JUST A DAYDREAM.
SO YOU MEAN I'VE GOT THINGS COMPLETELY WRONG?
YES, YOU HAVE. THE ANSWER IS...
CLACK
CHK
CHK
OH NO, I FORGOT ABOUT THE FORCE OF GRAVITY!
B
WEIGHT
A
WEIGHT
THE ORIENTATION OF THE THROWING FORCE
C
WEIGHT
THAT'S RIGHT. THE MAGNITUDE AND ORIENTATION OF THE FORCE IS THE SAME FOR ALL THESE POINTS.
NO WAY!

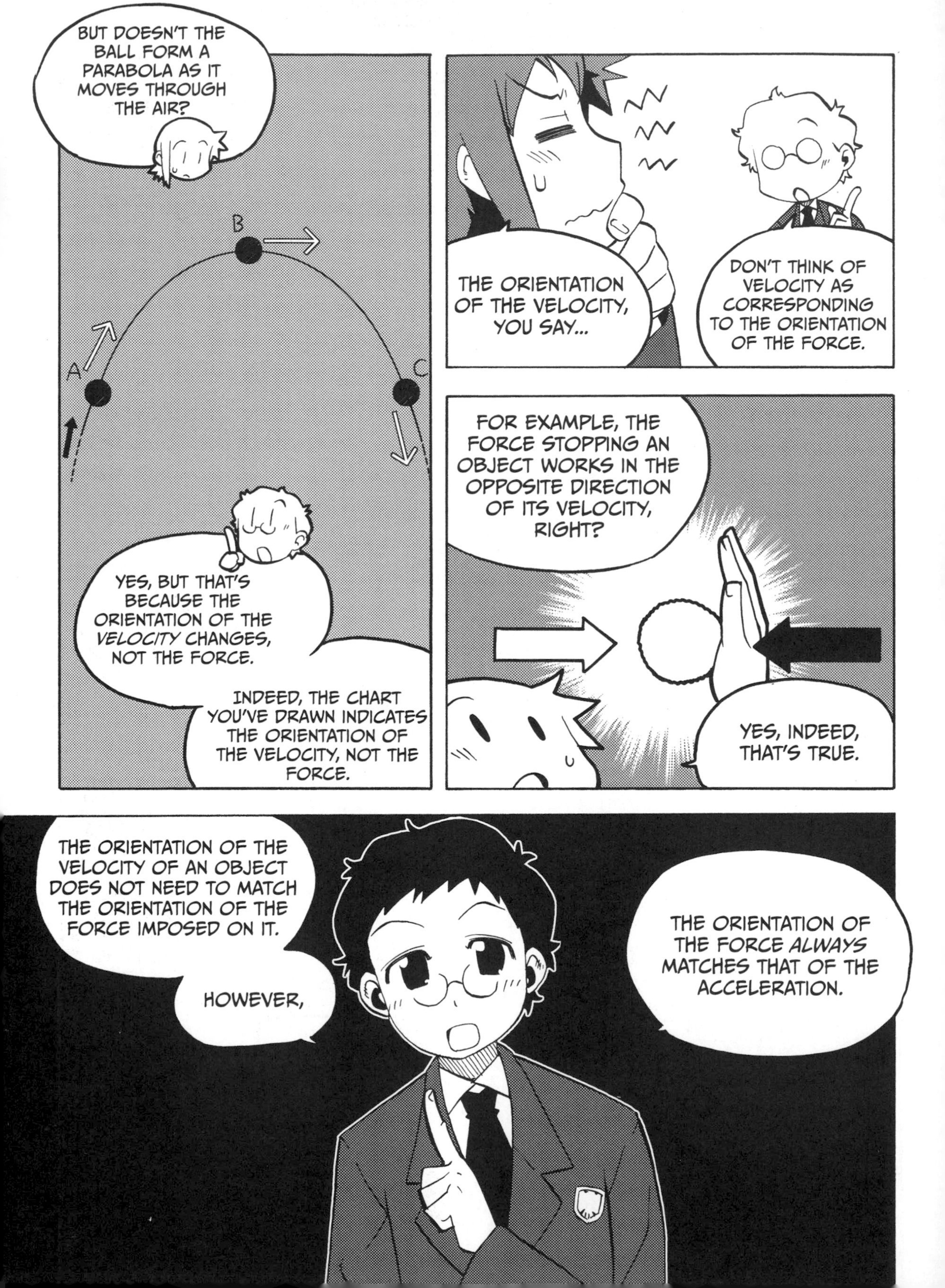
BUT DOESN'T THE BALL FORM A PARABOLA AS IT MOVES THROUGH THE AIR?
B
A
C
YES, BUT THAT'S BECAUSE THE ORIENTATION OF THE *VELOCITY* CHANGES, NOT THE FORCE.
INDEED, THE CHART YOU'VE DRAWN INDICATES THE ORIENTATION OF THE VELOCITY, NOT THE FORCE.
THE ORIENTATION OF THE VELOCITY, YOU SAY...
DON'T THINK OF VELOCITY AS CORRESPONDING TO THE ORIENTATION OF THE FORCE.
FOR EXAMPLE, THE FORCE STOPPING AN OBJECT WORKS IN THE OPPOSITE DIRECTION OF ITS VELOCITY, RIGHT?
YES, INDEED, THAT'S TRUE.
THE ORIENTATION OF THE VELOCITY OF AN OBJECT DOES NOT NEED TO MATCH THE ORIENTATION OF THE FORCE IMPOSED ON IT.
HOWEVER,
THE ORIENTATION OF THE FORCE *ALWAYS* MATCHES THAT OF THE ACCELERATION.

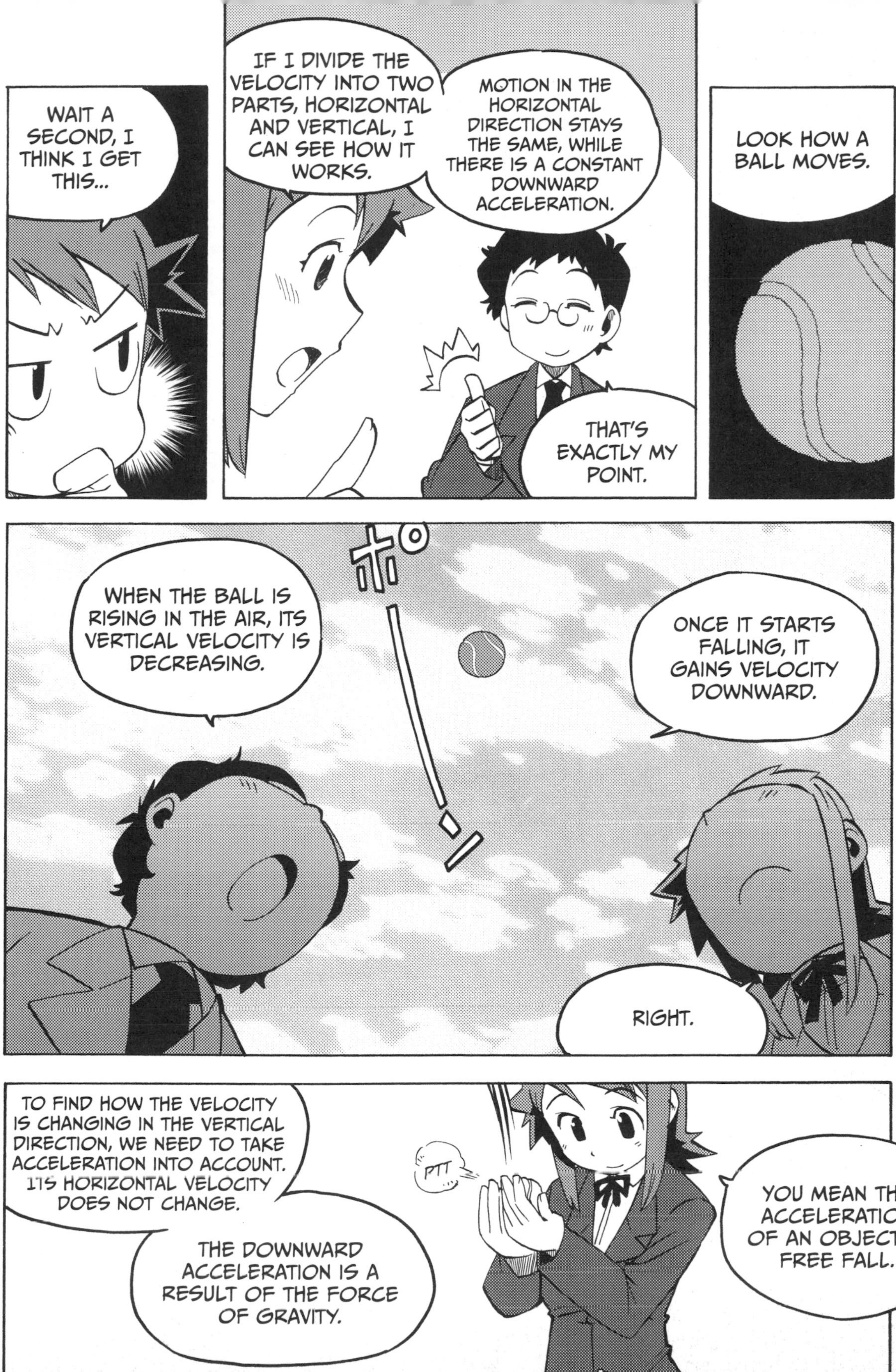

WAIT A SECOND, I THINK I GET THIS...
IF I DIVIDE THE VELOCITY INTO TWO PARTS, HORIZONTAL AND VERTICAL, I CAN SEE HOW IT WORKS.
MOTION IN THE HORIZONTAL DIRECTION STAYS THE SAME, WHILE THERE IS A CONSTANT DOWNWARD ACCELERATION.
THAT'S EXACTLY MY POINT.
LOOK HOW A BALL MOVES.
ポーン
WHEN THE BALL IS RISING IN THE AIR, ITS VERTICAL VELOCITY IS DECREASING.
ONCE IT STARTS FALLING, IT GAINS VELOCITY DOWNWARD.
RIGHT.
TO FIND HOW THE VELOCITY IS CHANGING IN THE VERTICAL DIRECTION, WE NEED TO TAKE ACCELERATION INTO ACCOUNT. ITS HORIZONTAL VELOCITY DOES NOT CHANGE.
THE DOWNWARD ACCELERATION IS A RESULT OF THE FORCE OF GRAVITY.
PTT
YOU MEAN THE ACCELERATION OF AN OBJECT IN FREE FALL.

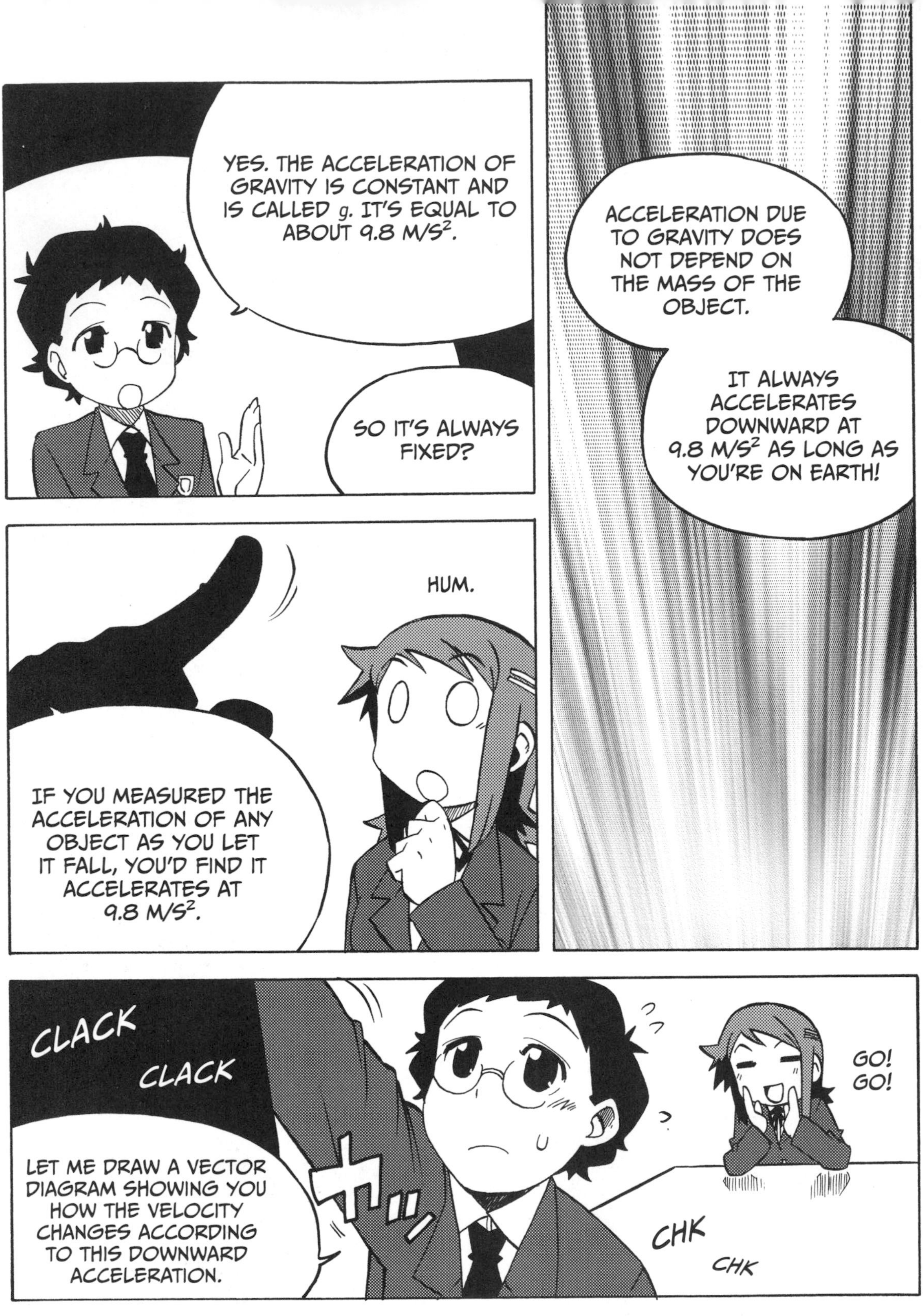
YES. THE ACCELERATION OF GRAVITY IS CONSTANT AND IS CALLED *g*. IT'S EQUAL TO ABOUT 9.8 M/S^2.
SO IT'S ALWAYS FIXED?
HUM.
IF YOU MEASURED THE ACCELERATION OF ANY OBJECT AS YOU LET IT FALL, YOU'D FIND IT ACCELERATES AT 9.8 M/S^2.
ACCELERATION DUE TO GRAVITY DOES NOT DEPEND ON THE MASS OF THE OBJECT.
IT ALWAYS ACCELERATES DOWNWARD AT 9.8 M/S^2 AS LONG AS YOU'RE ON EARTH!
CLACK
CLACK
LET ME DRAW A VECTOR DIAGRAM SHOWING YOU HOW THE VELOCITY CHANGES ACCORDING TO THIS DOWNWARD ACCELERATION.
GO! GO!
CHK
CHK

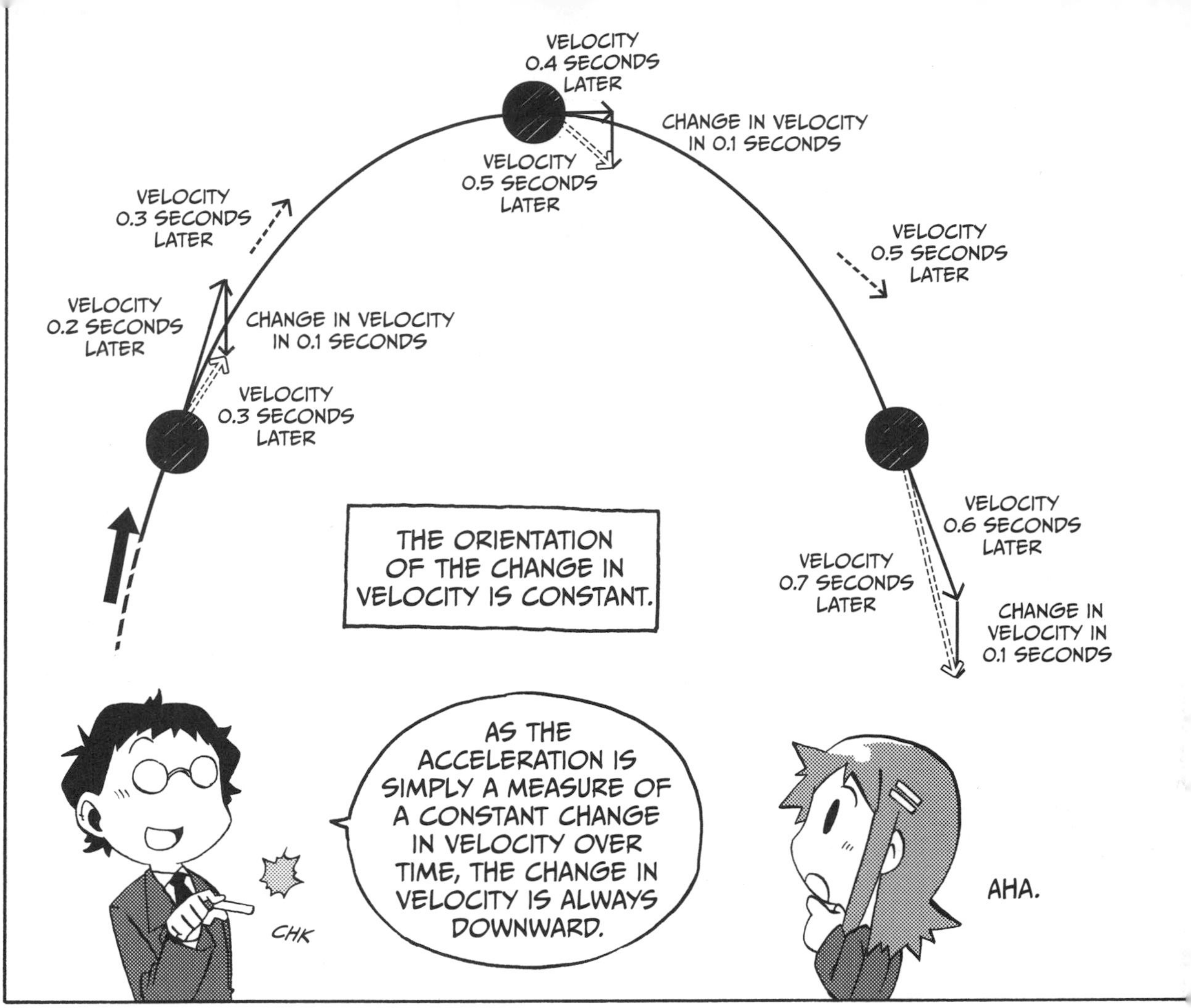
VELOCITY 0.4 SECONDS LATER
CHANGE IN VELOCITY IN 0.1 SECONDS
VELOCITY 0.5 SECONDS LATER
VELOCITY 0.3 SECONDS LATER
VELOCITY 0.5 SECONDS LATER
VELOCITY 0.2 SECONDS LATER
CHANGE IN VELOCITY IN 0.1 SECONDS
VELOCITY 0.3 SECONDS LATER
VELOCITY 0.6 SECONDS LATER
THE ORIENTATION OF THE CHANGE IN VELOCITY IS CONSTANT.
VELOCITY 0.7 SECONDS LATER
CHANGE IN VELOCITY IN 0.1 SECONDS
AS THE ACCELERATION IS SIMPLY A MEASURE OF A CONSTANT CHANGE IN VELOCITY OVER TIME, THE CHANGE IN VELOCITY IS ALWAYS DOWNWARD.
CHK
AHA.

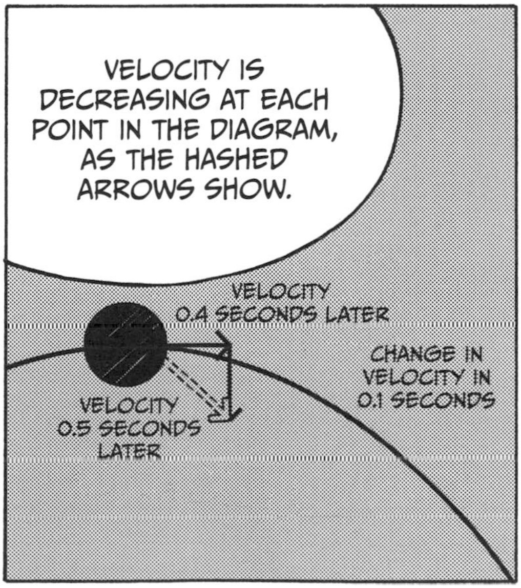
VELOCITY IS DECREASING AT EACH POINT IN THE DIAGRAM, AS THE HASHED ARROWS SHOW.
VELOCITY 0.4 SECONDS LATER
CHANGE IN VELOCITY IN 0.1 SECONDS
VELOCITY 0.5 SECONDS LATER

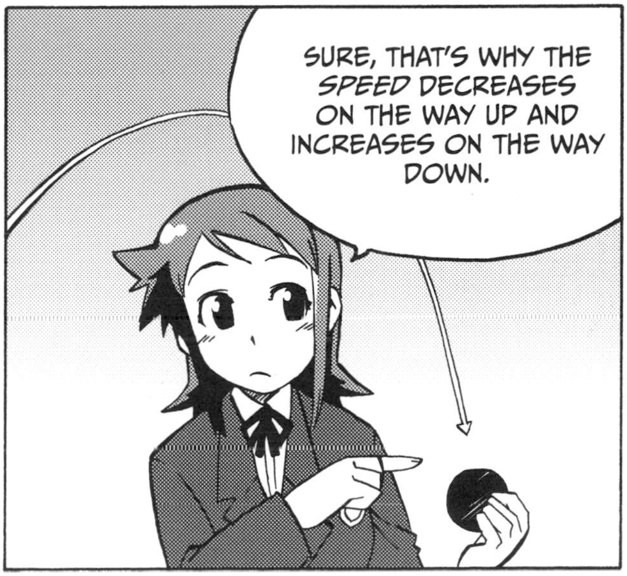
SURE, THAT'S WHY THE *SPEED* DECREASES ON THE WAY UP AND INCREASES ON THE WAY DOWN.

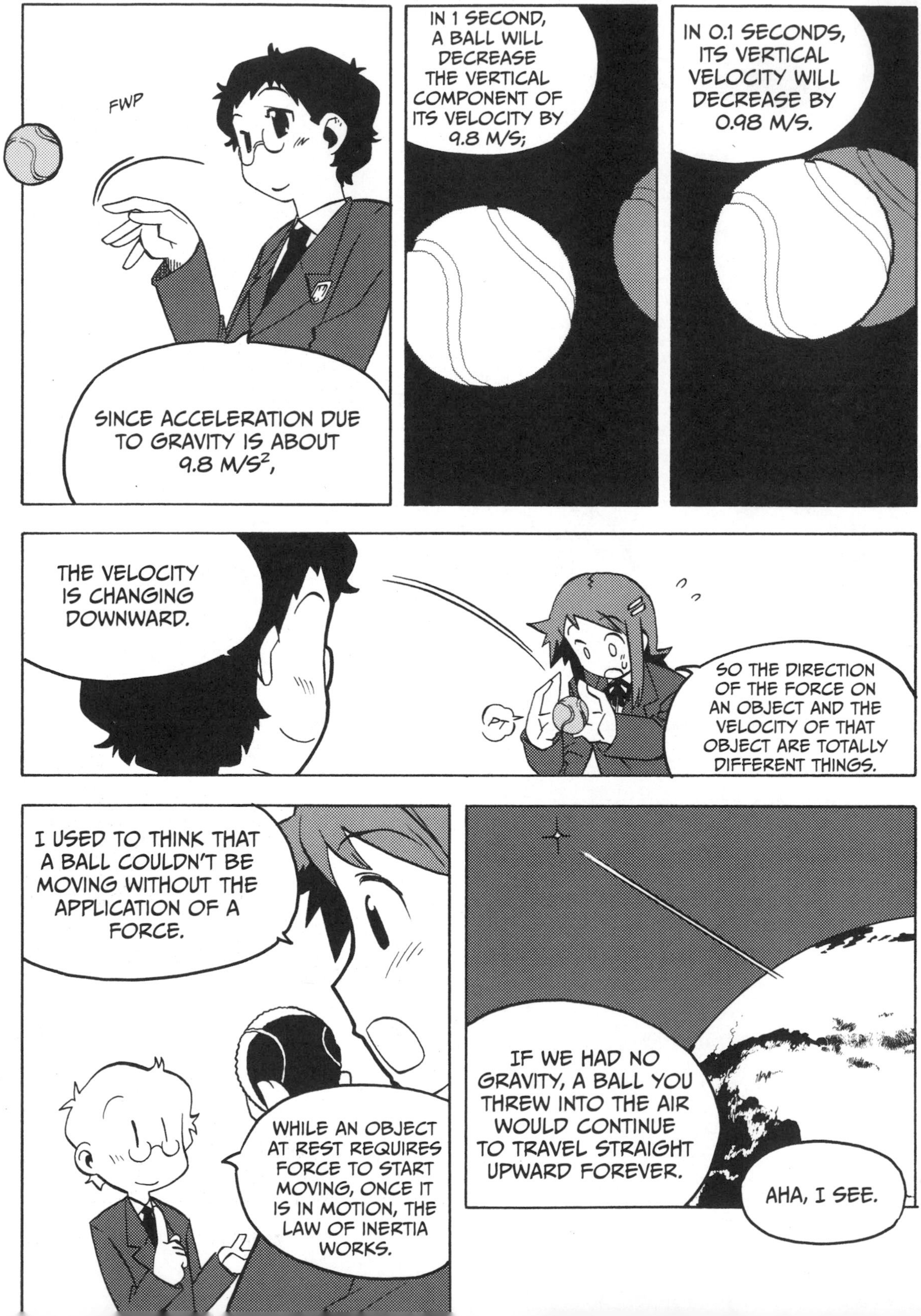
FWP
SINCE ACCELERATION DUE TO GRAVITY IS ABOUT 9.8 M/S²,
IN 1 SECOND, A BALL WILL DECREASE THE VERTICAL COMPONENT OF ITS VELOCITY BY 9.8 M/S;
IN 0.1 SECONDS, ITS VERTICAL VELOCITY WILL DECREASE BY 0.98 M/S.
THE VELOCITY IS CHANGING DOWNWARD.
PT
SO THE DIRECTION OF THE FORCE ON AN OBJECT AND THE VELOCITY OF THAT OBJECT ARE TOTALLY DIFFERENT THINGS.
I USED TO THINK THAT A BALL COULDN'T BE MOVING WITHOUT THE APPLICATION OF A FORCE.
WHILE AN OBJECT AT REST REQUIRES FORCE TO START MOVING, ONCE IT IS IN MOTION, THE LAW OF INERTIA WORKS.
IF WE HAD NO GRAVITY, A BALL YOU THREW INTO THE AIR WOULD CONTINUE TO TRAVEL STRAIGHT UPWARD FOREVER.
AHA, I SEE.

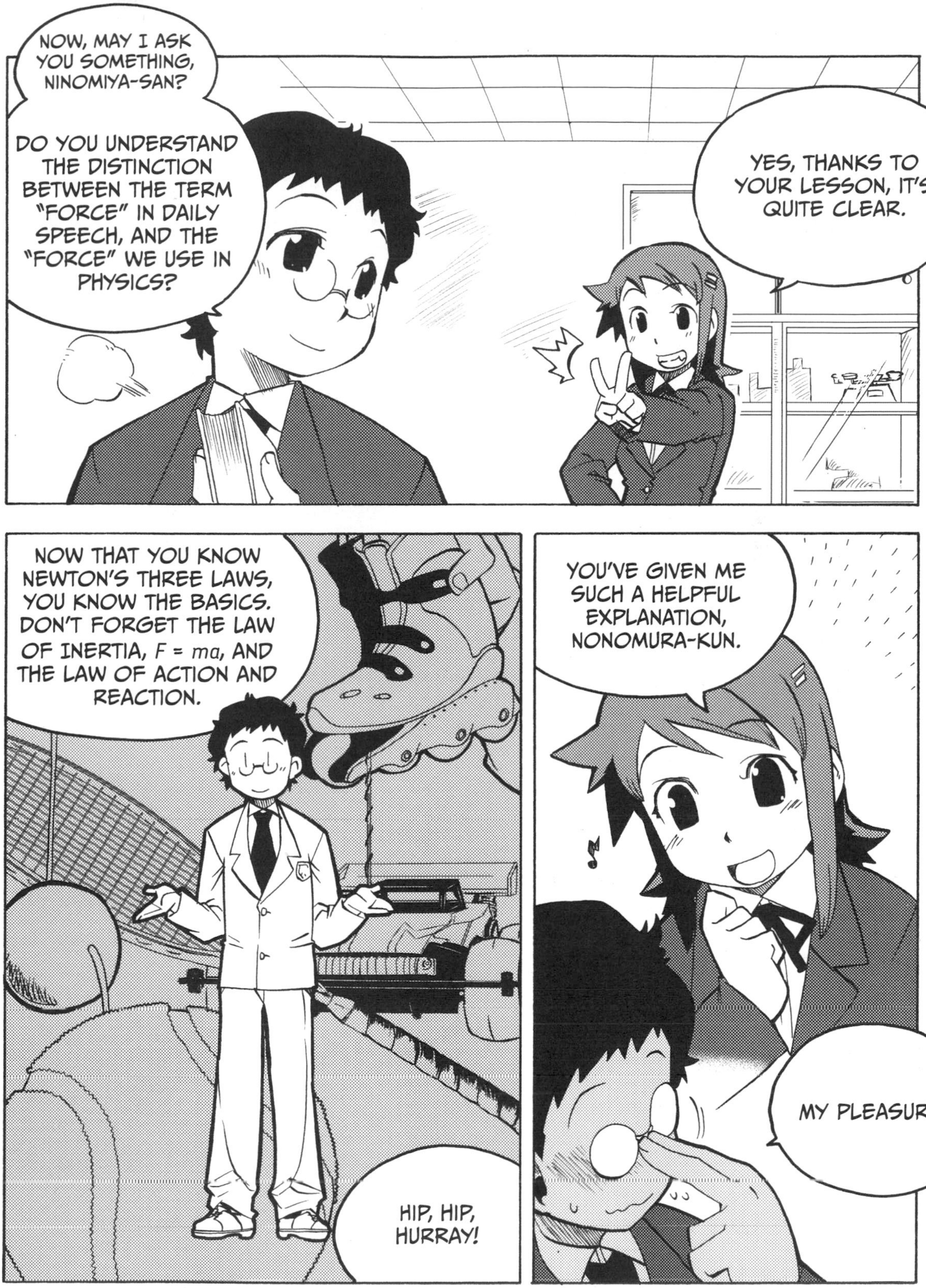

NOW, MAY I ASK YOU SOMETHING, NINOMIYA-SAN?
DO YOU UNDERSTAND THE DISTINCTION BETWEEN THE TERM "FORCE" IN DAILY SPEECH, AND THE "FORCE" WE USE IN PHYSICS?
YES, THANKS TO YOUR LESSON, IT'S QUITE CLEAR.
NOW THAT YOU KNOW NEWTON'S THREE LAWS, YOU KNOW THE BASICS. DON'T FORGET THE LAW OF INERTIA, $F = ma$, AND THE LAW OF ACTION AND REACTION.
HIP, HIP, HURRAY!
YOU'VE GIVEN ME SUCH A HELPFUL EXPLANATION, NONOMURA-KUN.
MY PLEASURE.

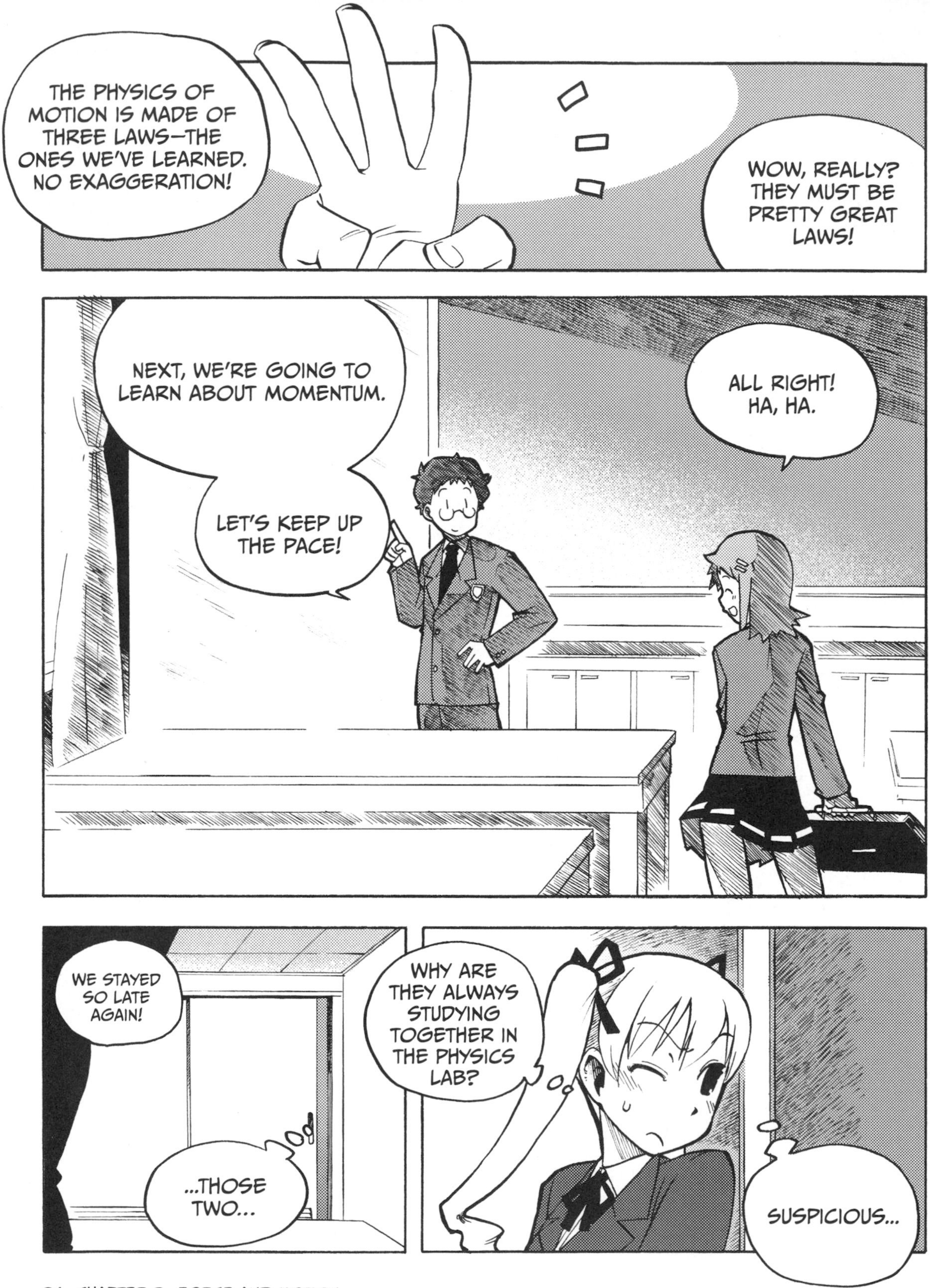
THE PHYSICS OF MOTION IS MADE OF THREE LAWS—THE ONES WE'VE LEARNED. NO EXAGGERATION!
WOW, REALLY? THEY MUST BE PRETTY GREAT LAWS!
NEXT, WE'RE GOING TO LEARN ABOUT MOMENTUM.
LET'S KEEP UP THE PACE!
ALL RIGHT! HA, HA.
WE STAYED SO LATE AGAIN!
...THOSE TWO...
WHY ARE THEY ALWAYS STUDYING TOGETHER IN THE PHYSICS LAB?
SUSPICIOUS...

THREE RULES OF UNIFORM ACCELERATED MOTION

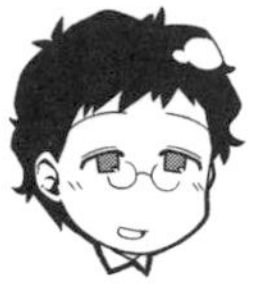

Let's look at the uniform accelerated motion of an object traveling in a straight line. Assuming the initial velocity of the object is v_1, the velocity after time t is v_2, the distance traveled in time t is d, and the uniform acceleration of the object is a, the following three rules are true:

❶ $v_2 = at + v_1$

❷ $d = v_1 t + \frac{1}{2}at^2$

❸ $v_2^2 - v_1^2 = 2ad$

Let's derive these rules. First, let's look at rule ❶. If the acceleration is constant, the following is true:

change in velocity = acceleration × time

Since the change in velocity is equal to $v_2 - v_1$, acceleration is a, and time is t, we can derive the following equation to get rule ❶:

$$v_2 = at + v_1$$

Next, let's derive rule ❷. On page 54, we learned that the distance an object travels can be expressed as the area under a v-t graph. According to rule ❶, the v-t graph should look like the following figure.

Velocity
$v_2 = at + v_1$
Area = $\frac{1}{2}at^2$
v_1
Area = $v_1 t$
0
t Time

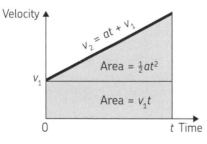

The area of this v-t graph is equal to the distance the object travels.

Since the area of the rectangular portion in the lower section of the v-t graph is $v_1 t$, and the area of the triangular portion above is $\frac{1}{2}at^2$, we get the following equation:

$$d = v_1 t + \frac{1}{2}at^2$$

NOTE *Technically,* d *in this equation represents displacement. What's the distinction between distance and displacement?* Distance *is a scalar quantity, and* displacement *is a vector quantity, as it has a specific direction. We'll occasionally use the term* distance *informally to refer to distance with a particular direction (displacement) in this book.*

Rule ❸ can be derived by removing t from rules ❶ and ❷. First, let's solve equation ❶ for t:

$$\frac{(v_2 - v_1)}{a} = t$$

If we substitute this value into rule ❷, the following equation will be the result:

$$d = v_1\left(\frac{v_2 - v_1}{a}\right) + \frac{1}{2}a\left(\frac{v_2 - v_1}{a}\right)^2$$

$$d = \frac{v_1v_2 - v_1^2}{a} + \frac{1}{2}a\left(\frac{v_2^2 - 2v_1v_2 + v_1^2}{a_2}\right)$$

$$d = \frac{2v_1v_2 - 2v_1^2 + v_2^2 - 2v_1v_2 + v_1^2}{2a}$$

$$d = \frac{v_2^2 - v_1^2}{2a}$$

Voilà! Simply multiply both sides by $2a$, and you've just derived rule ❸!

ADDING VECTORS: THE HEAD-TO-TAIL METHOD

Because force is a vector, we need to make calculations according to the rules of vectors I explained in Chapter 1. If two vectors are parallel, adding them is simple—you can either add their magnitudes or subtract one from the other (if the two vectors are in opposite directions).

However, in the real world, we'll have to add vectors pointing in all different directions. To do this, we'll use the head-to-tail method. To illustrate, let's assume that an object is receiving two forces, $\vec{F}_A$ and $\vec{F}_B$, as seen below.

$\vec{F}_B$
Object
$\vec{F}_A$
$\vec{F}_A + \vec{F}_B$

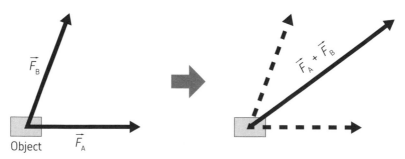

The total force on the object is equal to one combined force, represented by the arrow shown on the right. This arrow is the sum of the forces $\vec{F}_A + \vec{F}_B$, and we'll call it $\vec{F}_{resultant}$. But how can we find its exact magnitude and direction?

$\vec{F}_B$
$\vec{F}_A + \vec{F}_B$
$\vec{F}_A$

Head
$\vec{F}_{resultant} = \vec{F}_A + \vec{F}_B$
$\vec{F}_B$
Tail
Tail
$\vec{F}_A$
Head

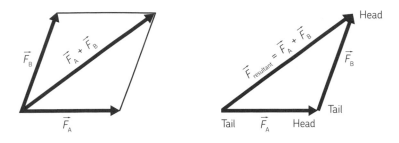

To determine the magnitude and the direction of a resultant force, you can simply place the head of one vector onto the tail of the second. The resultant force connects from the tail of the first vector to the head of the second. The resultant vector $\vec{F}_{resultant}$ forms a triangle with $\vec{F}_A$ and $\vec{F}_B$, as you can see on the right. You can use the head-to-tail method for any vector, not just forces, and you can find the resultant force from three or more forces by repeatedly applying the head-to-tail method.

THE COMPOSITION AND DECOMPOSITION OF FORCES

To make forces easier to understand and analyze, we'll often split them into their horizontal and vertical constituent parts. That's because the head-to-tail method also works in reverse. That is, we can split a single diagonal force into the addition of its horizontal and vertical parts. Let's look at an example.

θ
$\vec{F}_{tension}$
$\vec{F}_{hand}$
$\vec{F}_{gravity}$
$\vec{F}_{tension}$
θ
$\vec{F}_{hand}$
$\vec{F}_{gravity}$
$\vec{F}_{gravity} + \vec{F}_{hand}$

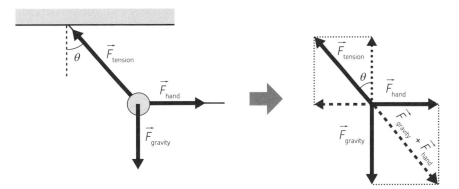

Let's look at the balance of forces when a weight hanging from the ceiling is pulled horizontally (see page 61). As shown on the right in the figure above, let's assume gravity is $\vec{F}_{gravity}$, the force from the hand pulling horizontally is $\vec{F}_{hand}$, and the tension of the string is $\vec{F}_{tension}$. When the weight is stationary, the three forces are balanced. Thus, adding the three forces as vectors yields zero:

$$\vec{F}_{gravity} + \vec{F}_{hand} + \vec{F}_{tension} = 0$$

You can rewrite this equation as the following:

$$\vec{F}_{gravity} + \vec{F}_{hand} = -\vec{F}_{tension}$$

With this in mind, let's revisit our diagram, thinking in terms of horizontal and vertical forces. Because the object is at rest, forces in the horizontal direction must equal zero. In the same way, the sum of the forces in the vertical direction must equal zero.

What are the horizontal forces in play? F_{hand} and the horizontal component of the tension of the string, $F_{tension}$. They are acting in opposite directions, and the object is at rest, so these two forces must be equal:

$$F_{hand} = \text{horizontal component of } F_{tension}$$

What are the vertical forces acting on the object? The force of gravity downward and the vertical portion of the tension of the string, $F_{tension}$. They are acting in opposite directions, and the object is at rest, so these forces must also be equal:

$$F_{gravity} = \text{vertical component of } F_{tension}$$

So, how can we actually "decompose" the force of the tension into its horizontal and vertical parts? We'll use concepts from *trigonometry*, the study of triangles.

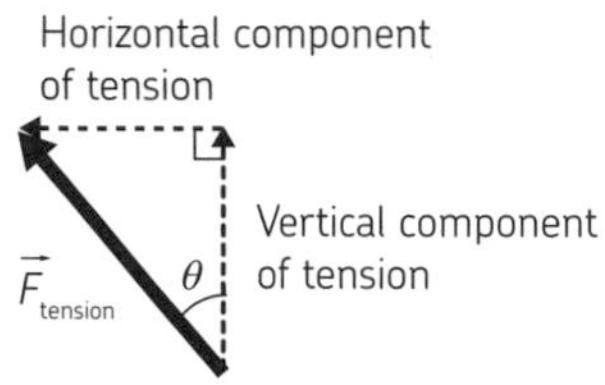

Horizontal component of tension
Vertical component of tension
$\vec{F}_{tension}$
θ

Remember the head-to-tail method of addition of vectors? Here we'll decompose our diagonal force, $F_{tension}$, into its horizontal and vertical parts, forming a right triangle. If the angle of this triangle is represented by θ, we can represent the horizontal and vertical constituent parts in terms of this angle! Recalling the previous two equations, we get the following:

❶ $F_{hand} = \sin\theta \times F_{tension}$

❷ $F_{gravity} = \cos\theta \times F_{tension}$

Now, if we simply divide equation ❶ by equation ❷, we'll be able to discount the force of the tension:

$$\frac{\sin\theta}{\cos\theta} = \frac{F_{hand}}{F_{gravity}}$$

This is equal to the following:

$$\tan\theta = \frac{F_{hand}}{F_{gravity}}$$

That means we can then represent the force of the hand in terms of the force of gravity and the angle of the string!

$$F_{hand} = \tan\theta \times F_{gravity}$$

WAIT A SECOND,
WHAT'S ALL THIS SINE AND COSINE STUFF?

If you've never studied trigonometry, don't worry—it's not too difficult to understand. *Trigonometry* is simply the study of the relationship between the length of a triangle's sides and its angles, especially right triangles. Because we often split forces and velocities into their horizontal and vertical parts, we'll use trigonometry frequently.

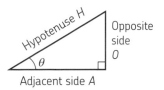

Let's look at the example below. Consider a right triangle with an angle of θ.

Sine, cosine, and tangent (the three main trigonometric functions) are simply representations of the ratios of the three sides of this triangle.

The sine of the angle theta (sin θ) is equal to the ratio of the opposite side (O) to the hypotenuse (H). In an equation, it looks like this:

$$\sin \theta = \frac{O}{H}$$

The other trigonometric functions are simply representations of different ratios! For example, the cosine of theta (cos θ) is equal to the ratio of the adjacent side (A) to the hypotenuse, and the tangent of theta (tan θ) is the ratio of the opposite side to the adjacent side. The equations look like this:

$$\cos \theta = \frac{A}{H}$$

$$\tan \theta = \frac{O}{A}$$

If you have trouble remembering these different ratios and what they mean, try using the mnemonic device *SOHCAHTOA*.

$$\sin = \boldsymbol{O} / \boldsymbol{H}, \cos = \boldsymbol{A} / \boldsymbol{H}, \tan = \boldsymbol{O} / \boldsymbol{A}$$

Whenever you're confused about whether to use sine, cosine, or tangent, just think about SOHCAHTOA, the magic triangular island of trigonometry.

The magical island of

SOHCAHTOA!

NEWTON'S FIRST LAW OF MOTION

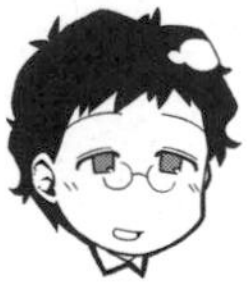

Newton's first law of motion states, "An object continues to maintain its state of rest or of uniform motion unless acted upon by an external net force." An isolated object in outer space, where no gravity is being exerted, will eternally stay at rest or travel with uniform velocity unless another force is imposed on it. A stationary object can have forces acting on it—however, the sum of these forces must be equal to zero. For example, a stationary object sitting on a desk is subject to the downward pull of gravity. The object remains stationary because it receives a vertical upward force from the desk, yielding a resultant force of zero.

Now that we understand the forces acting on a stationary object, we can move on to understanding what happens when the net force on an object is *not* zero.

NEWTON'S SECOND LAW OF MOTION

When a force is imposed on an object, that object starts moving with a uniform acceleration proportional to the net force applied and inversely proportional to its mass. Assuming the vector of a force imposed on the object is F, the acceleration of the object is a, and the mass of the object is m, the second law of motion yields the following equation:

$$F = ma$$

Mass is a quantity that has only a magnitude, so it is a scalar quantity. However, recall that force and acceleration are vectors—so pay special attention to the acceleration of the object and the orientation of the force. They will be in the same direction!

The radio-controlled car you saw on page 49 moves in a square and attains a uniform velocity while it travels in a straight line. At this time, the net force on the car is zero. However, when the car turns around a corner, a force must be exerted to change the direction of its velocity. This is an important distinction: Acceleration does not have to change the *magnitude* of a velocity! It can simply change the *direction* of a velocity!

THE ORIENTATION OF VELOCITY, ACCELERATION, AND FORCE

According to the second law of motion, the orientation of acceleration always matches the orientation of the force. However, the orientation of velocity does not directly correspond to the orientation of either the force or the acceleration. From the relationship between acceleration and velocity (explained on page 52) comes the following equation:

$$\text{change in velocity} = \text{acceleration} \times \text{time}$$

This means that the orientation of the change in velocity matches the orientation of acceleration! It's a subtle distinction, but an important one.

Let's look at an example. Suppose there is an object in motion at constant velocity v. When no force is acting on the object, it moves in a straight line at velocity v_1, according to the first law of motion. If a vertical force is imposed on the object for time t, how would the object's velocity change? Assuming that the acceleration created by the force is a and the velocity after the force is imposed is v_2, you can derive the following equation:

$$v_2 - v_1 = at$$

or

$$v_2 = v_1 + at$$

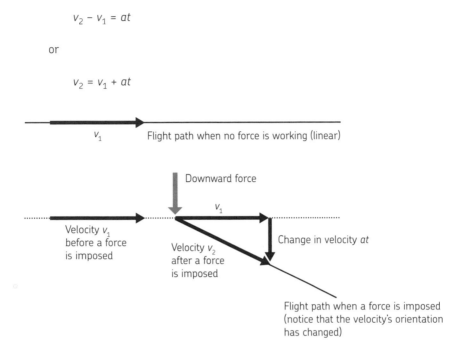

v_1

Flight path when no force is working (linear)

Downward force

v_1

Velocity v_1 before a force is imposed

Velocity v_2 after a force is imposed

Change in velocity at

Flight path when a force is imposed (notice that the velocity's orientation has changed)

Thus, the addition of a force changes the direction of an object's motion. We can easily predict this object's motion by splitting v_2 into its constituent horizontal and vertical parts. Its horizontal velocity must be equal to v_1, as there has been no force in the horizontal direction. The change in the object's vertical velocity is simply at!

In the example of throwing a ball on page 75, the force of gravity continues to act on the ball, even when the ball is moving upward. When the ball is rising in the air, its vertical velocity is decreasing due to the force of gravity. Once it starts falling, it gains velocity downward. The ball's horizontal velocity does not change; only its vertical velocity varies. The ball's motion follows the shape of a parabola, as shown in the following figure.

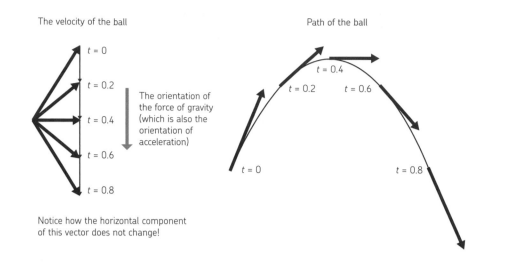

The velocity of the ball

t = 0
t = 0.2
t = 0.4
t = 0.6
t = 0.8
The orientation of the force of gravity (which is also the orientation of acceleration)

Notice how the horizontal component of this vector does not change!

Path of the ball

t = 0
t = 0.2
t = 0.4
t = 0.6
t = 0.8

AN OBJECT DOES NOT HAVE ITS OWN FORCE

Those who have not studied physics tend to think, "An object in motion has a force." This is a common but incorrect notion. As we learned in Chapter 1, force is generated between paired elements whose movement affects each other. An object in motion does not have an internal force that causes it to stay in motion—it's simply the result of the first law of motion.

Let's look at the example of a ball being thrown up in the air. The ball receives a force from the hand until the moment it leaves the hand. (In response, due to the law of action and reaction, the hand receives a force from the ball—but this force has nothing to do with the ball's motion.) Once the ball leaves the hand, it only receives the force of gravity from the earth. The force on the ball from the hand does not remain after the ball leaves the hand.

THE UNIT FOR FORCE

Newton's second law gives us the unit for force:

$$\text{force} = \text{mass} \times \text{acceleration}$$

In this equation, the unit for mass is kilograms (kg), while the unit for acceleration is meters per second squared (m/s^2). Therefore, the unit for force is equal to kg × m/s^2. To represent this more easily, we can use a unit called a *newton (N)*:

$$1 \text{ newton} = 1 (\text{kg} \times \text{m/s}^2)$$

You can use newtons to represent forces. As you can probably guess, this unit is named after the great Isaac Newton, who established the foundations of physics. A force of 1N is equivalent to the force required for accelerating an object with a mass of 1 kg by 1 m/s^2.

MEASURING MASS AND FORCE

Weight
m_2
Gravitational mass

How can we determine the mass of an object? Mass can be measured with a scale, which takes into account the fact that the force of gravity working on an object (that is, its weight) is proportional to its mass. Mass that is measured based on gravity is referred to as *gravitational mass*.

However, mass that is calculated using Newton's second law represents a measurement of the resistance of an object against acceleration; this mass has no direct relation to gravity. Mass as calculated by Newton's second law (mass = force / acceleration) is referred to as *inertial mass*.

a_1
a_2
1
2
m_1
m_2
$m_1 \left| \frac{a_1}{a_2} \right| = m_2$
Inertial mass

Inertial mass can be measured by combining Newton's second law and the law of action and reaction. First, we need an object with a known mass (we'll call it the *reference object* and label it m_1 in our diagram). Then, we'll arrange the object whose mass we want to measure (we'll call it the *measurement object* and label it m_2 in our diagram) and the reference object so that their forces work on each other through a collision. In this collision, there are no external forces working on the objects.

At this time, the forces of the reference object and the measurement object working on each other are subject to the law of action and reaction. That is, they must be equal:

If $F_1 = m_1a_1$ and $F_2 = m_2a_2$, we know that $F_1 = F_2$, due to the law of action and reaction. Therefore, we can express that relationship like so:

$$m_1a_1 = m_2a_2$$

Since we're trying to solve for m_2, our measurement object, we'll rearrange that equation as follows:

$$m_2 = \frac{m_1a_1}{a_2}$$

Of course, these accelerations are actually in opposite directions, so we'll consider their magnitude alone.

The acceleration of an object can be found by measuring the distance the object travels and the time it takes to travel that distance. If you have these measurements, you can find the inertial mass of the measurement object.

Although experiments have shown that gravitational mass is the same as inertial mass, Newton's Laws don't say that this *has* to be the case. Our understanding of this relationship comes from Einstein, who founded general relativity on the *equivalence principle*—the idea that inertial and gravitational mass are the same. This is still an active area of research.

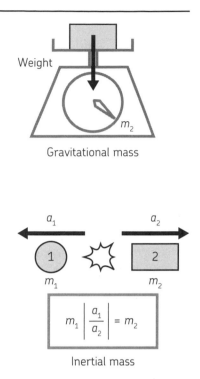

Once we've determined the mass of the objects in the collision, we can determine the force they've applied to each other. As the force causes the object to accelerate, we can measure this acceleration. We can then substitute this value into the following equation to determine an exact value of the force we've applied:

mass × acceleration = force

DETERMINING WEIGHT

The force of gravity from the earth acting on an object with mass m is expressed as follows:

❶ $F = mg$

In this equation, g is the magnitude of gravitational acceleration—about 9.8 m/s^2 when measured near the ground surface. This relationship is derived from the equation for universal gravitation.

h
m
R
Center of the earth!
M
R + h
m
M

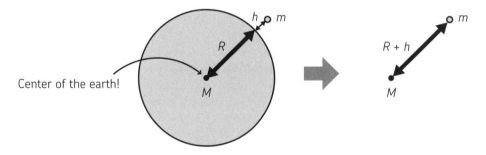

Consider an object with mass m located at an altitude h above the earth.

Assume that the earth is a perfect sphere with radius R, mass M, and a uniform density. Doing this, we can also assume that the gravity generated outside the surface of the earth by the entire globe is equivalent to the gravity of a point with mass equivalent to M. Using the equation that describes universal gravitation, which we saw on page 43, we can calculate the forces and acceleration due to the earth's gravitational pull.

Therefore, the magnitude of gravity from the earth acting on an object is equal to the value expressed below:

$$F = G\frac{Mm}{(R+h)^2}$$

Also note that the force of gravity on an object near the earth's surface (where $h = 0$) is as follows:

$$F = G\frac{Mm}{R^2} \quad \text{where} \quad G\frac{M}{R^2} = g$$

Since we know that force here is also equal to mass times acceleration, we can set this equation equal to ❶.

NOTE *Remember from page 43 that G is the universal gravitational constant.*

$$mg = G\frac{Mm}{R^2}$$

$$g = G\frac{M}{R^2}$$

The radius of the earth is about 6.38×10^6 m, and its mass is about 5.98×10^{24} kg. Using these values, you can calculate the value of g, the acceleration of an object due to gravity:

$$g = G\frac{M}{R^2} = 6.67 \times 10^{-11} \times \frac{5.98 \times 10^{24}}{(6.38 \times 10^6)^2} \approx 9.8 \text{ m/s}^2$$

This is gravitational acceleration—notice that it does not depend on the mass of the smaller object (m). Strictly speaking, because the earth is not a perfect sphere, gravitational acceleration close to the earth's surface varies slightly, depending on the location. Even so, you can safely approximate this value as 9.8 m/s².

Now try to find the magnitude of gravitational acceleration at a point in the orbit of a space shuttle going around the earth. A space shuttle travels about 300 to 500 km from the earth's surface.

Assume $h = 500$ km above the earth's surface. $R + h = (6.38 \times 10^6 \text{ m}) + (0.5 \times 10^6 \text{ m}) = 6.88 \times 10^6$ m. Using this calculation, you can find the acceleration due to gravity at this altitude:

$$g = G\frac{M}{(R+h)^2} = 6.67 \times 10^{-11} \times \frac{5.98 \times 10^{24}}{(6.88 \times 10^6)^2} \approx 8.4 \text{ m/s}^2$$

In other words, a space shuttle is affected by gravity that is about 86 percent (8.4 / 9.8 = 0.86) the strength of the gravity working on the earth's surface. Since the distance from the earth to a traveling space shuttle is about one-tenth of the radius of the earth, it is only reasonable to assume that the space shuttle is still heavily affected by the earth's gravity.

Then why does it feel like there is no gravity inside a space shuttle? It is because the shuttle is always "falling" as it is pulled by the earth's gravity. Einstein theorized that if the cable holding an elevator breaks, a person inside the falling elevator would find himself in a weightless environment much like outer space. Just like an elevator with a broken cable, the space shuttle's acceleration is oriented toward the center of the earth due to gravity. However, it always falls with a velocity oriented perpendicular to the direction of gravity; it does not move directly downward.

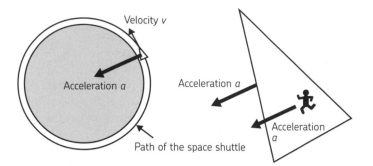

Velocity *v*

Acceleration *a*

Acceleration *a*

Acceleration *a*

Path of the space shuttle

For this reason, the space shuttle travels around the earth along a circular (or, more specifically, elliptical) path. The feeling of so-called zero-gravity is created because the space shuttle and everything inside it, including the astronauts, are "falling" at the same gravitational acceleration.

UNDERSTANDING PARABOLIC MOTION

We examined a ball in flight on page 75—that ball's motion is something we call *parabolic motion*. Here, let's take a more in-depth look at the ball's flight using some real numbers and equations.

In the figure below, the distance in the horizontal direction is expressed as *x*, the vertical direction as *y*, and the ball's mass as *m*. The force of gravity on the ball works downward along the y-axis, with a magnitude of *mg*. Represented in terms of its constituent parts, the force vector on the ball is expressed as follows:

force in x direction

$$F = (0, -mg)$$

force in y direction

Similarly, we can represent acceleration in terms of component elements as $a = (a_x, a_y)$. We know the following:

Acceleration in the x direction is $a_x = 0$.

Acceleration in the y direction is $a_y = -g$.

In short, the ball has a uniform velocity in the x direction, and uniform accelerated motion occurs in the y direction.

Given that we know these values, we can find the velocity of the ball at any time. When the ball is released, $t = 0$ and the velocity for throwing it is $v_1 = (v_{1_x}, v_{1_y})$, from rule ❶, you get:

$$v_{2_x} = v_{1_x}$$

$$v_{2_y} = v_{1_y} - gt$$

These equations indicate that velocity does not change in the x direction, but it does change downward in the y direction by −9.8 m/s in one second ($gt = -9.8\ \text{m/s}^2 \times 1\ \text{s} = -9.8\ \text{m/s}$).

$\vec{v} = (v_x, v_y)$
v_y
v_x
$\vec{a} = (0, -g)$
y direction
$\vec{F} = (0, -mg)$
x direction

B
$\vec{v}$
y
v_y
A
C
v_x
v_{1_y}
Initial velocity $\vec{v_1}$
θ
v_{1_x}
x

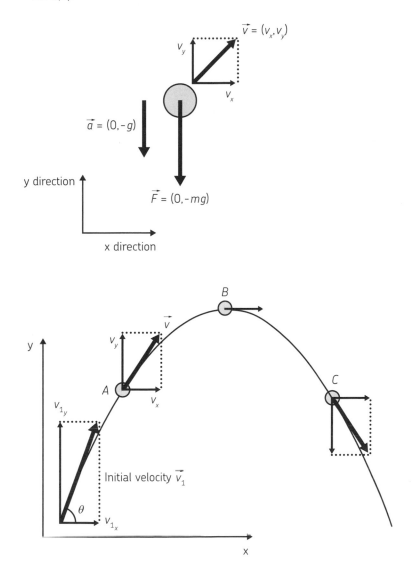

Next, let's find the location of the ball. Let's split it into constituent parts in the x and y directions:

$$x = v_{1_x} t$$

$$y = v_{1_y} t - \tfrac{1}{2} g t^2$$

Now, if there were only some way to eliminate the time variable in this second equation. Perhaps we should rearrange that first equation!

$$t = \frac{x}{v_{1_x}}$$

Substituting that into the second equation, we get the following:

$$y = v_{1_y} \left(\frac{x}{v_{1_x}} \right) - \tfrac{1}{2} g \left(\frac{x}{v_{1_x}} \right)^2$$

This is actually a quadratic function, and it will show a parabola when plotted. The origin is at the point where the ball leaves the hand.

From this equation, you can tell where the thrown ball will land. Actually, we can take the term $\left(\frac{x}{v_{1_x}} \right)$ out of this equation like so:

$$y = \frac{x}{v_{1_x}} \left(v_{1_y} - \tfrac{1}{2} g \left(\frac{x}{v_{1_x}} \right) \right)$$

And given that we know that the ball's landing point should be where $y = 0$ and $x \neq 0$, let's set y equal to 0:

$$0 = \frac{x}{v_{1_x}} \left(v_{1_y} - \tfrac{1}{2} g \left(\frac{x}{v_{1_x}} \right) \right)$$

$$v_{1_y} = \tfrac{1}{2} g \left(\frac{x}{v_{1_x}} \right)$$

Given this equation, we can solve for x, the distance that the ball travels!

❶ $$x = \frac{2 v_{1_x} v_{1_y}}{g}$$

By rewriting the expression and assigning θ to the angle of the throw, you can find the angle that would enable the ball to reach the farthest point for a given velocity. The initial velocity can be expressed as follows:

$$v_1 = (v_{1_x}, v_{1_y}) = (v_1 \cos \theta, v_1 \sin \theta)$$

You can use this to re-define the landing point in equation ❶:

$$x = \frac{2 \times v_1 \cos\theta \times v_1 \sin\theta}{g} = \frac{v_1^2 \sin 2\theta}{g}$$

This value reaches a maximum when $\sin(2\theta) = 1$.* Therefore, when throwing a ball at this fixed velocity, the ball reaches the farthest distance when thrown at an angle of 45 degrees.

USING CALCULUS TO FIND ACCELERATION AND VELOCITY

WARNING: CALCULUS AHEAD!

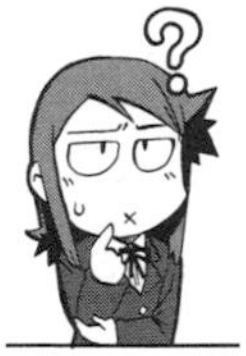

Normally, the velocity of an object changes over time. In this example, let's say Δt is a short amount of time during which we can assume the velocity is constant. Then we get the following approximation, where Δx represents the displacement created in time Δt:

$$v = \frac{\Delta x}{\Delta t}$$

In this equation, the smaller the value you assign to Δt, the more precise approximation you can get for the velocity. In an experiment, Δt can only have a finite value. Thus, we can only find the velocity to be an average value. But mathematically, we can assume a case where Δt infinitely approaches zero. In other words, we can define the velocity for a given moment as follows:

❶ $$v = \lim_{\Delta t \to 0} \frac{\Delta x}{\Delta t} = \frac{dx}{dt}$$ This is the very definition of a derivative.

The same is true with acceleration. Let's assign Δv to a short amount of time Δt, during which the velocity can be assumed to be virtually constant. Then acceleration a is expressed as follows:

$$a = \frac{\Delta v}{\Delta t}$$

When acceleration is not uniform, we can make the change in t infinitely small:

$$a = \lim_{\Delta t \to 0} \frac{\Delta v}{\Delta t} = \frac{dv}{dt}$$

This represents acceleration for a given instant. Also note that by substituting expression ❶ in this expression, we get the following:

$$a = \frac{d}{dt}\left(\frac{dx}{dt}\right) = \frac{d^2x}{dt^2}$$

* If you were confused by the math in these equations, remember that $\sin(2\theta) = 2\sin\theta\cos\theta$.

Thus, acceleration can be expressed as the second derivative of the displacement. Newton's second law ($F = ma$) can be expressed in differential calculus as follows:

$$m \frac{dv}{dt} = F \quad \text{or} \quad m \frac{d^2x}{dt^2} = F$$

USING THE AREA OF A V-T GRAPH TO FIND THE DISTANCE TRAVELED BY AN OBJECT

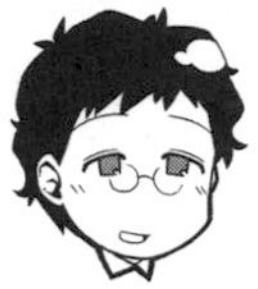

Next, let's examine how we can find the distance an object travels if we already know its velocity (see page 54). When the velocity is uniform, we know that the following equation holds true, and we can find the distance traveled (Δx) over a change in time (Δt).

v_i
Δx_i
Δt
Velocity
0
t_i
t Time
Velocity
$\int_0^t v dt$
0
t Time

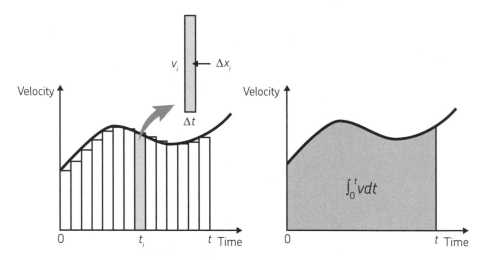

For a velocity that's changing magnitude, we can find an approximation by summing up the distances traveled in Δt-long time segments. In other words, we divide the time interval between point 0 and point t into n segments, assign t_i to the ith point in time, and assign v_i to the velocity at that moment.

Express the time as Δt, the velocity as v_i and the distance as Δx_i to yield the following equation:

$$x = v_1\Delta t + v_2\Delta t + \ldots + v_i\Delta t + \ldots + v_n\Delta t$$

The distance x traveled between point 0 and point t can be found using the following approximation:

$$x = \sum_{i=1}^{n} v_i\Delta t$$

When the rectangle is divided into infinitely small segments to allow Δt to infinitely approach zero (when n or the number of segments approaches infinity), the results will be much more precise:

$$x = \lim_{\Delta t \to 0} \sum_{i=1}^{n} v_i \Delta t = \int_0^t v\, dt$$

This is the very definition of integration. This equation shows that you can find the distance traveled using integral calculus representing the area under the v-t graph.

Now, given uniform accelerated motion with acceleration a, velocity v_1 at time $t = 0$, and velocity v_2 at time t, we know the following:

$$a = \frac{v_2 - v_1}{t}$$

From this equation, we can immediately tell that $v_2 = v_1 + at$, or rule ❶ on page 85. Now that we have an equation for final velocity as a function of time, we may substitute it into the integral equation to calculate displacement:

$$x = \int_0^t (v_1 + at)\, dt$$

Since v_1 and a are constants, this is a relatively simple integral to evaluate:

$$x = \left[v_1 t + \tfrac{1}{2} at^2 \right]_0^t$$

The lower limit of $t = 0$ makes evaluating this equation quite simple:

$$x = v_1 t + \tfrac{1}{2} at^2$$

We have just derived a rule that should look very familiar to you!

3
MOMENTUM

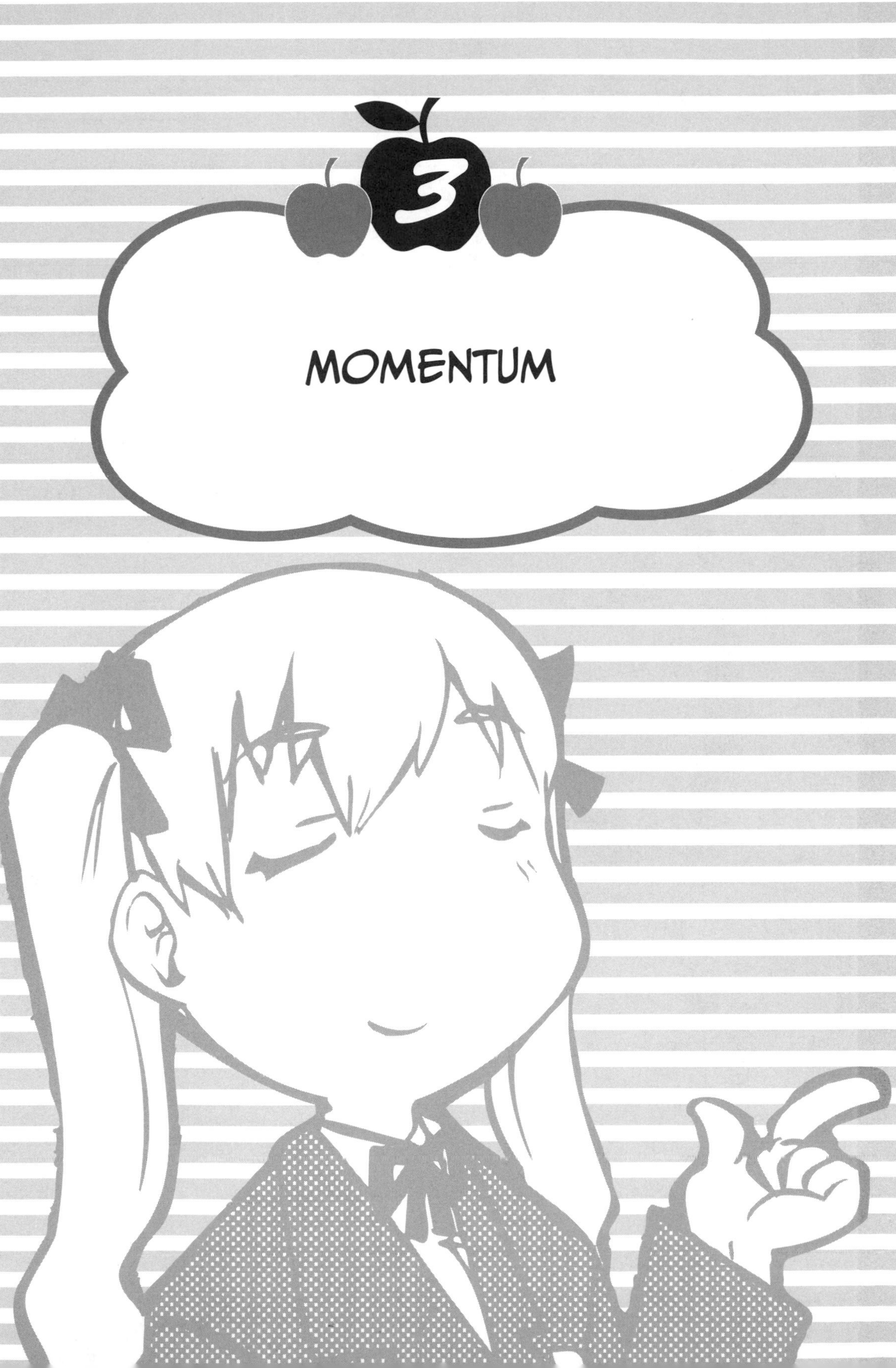

MOMENTUM AND IMPULSE

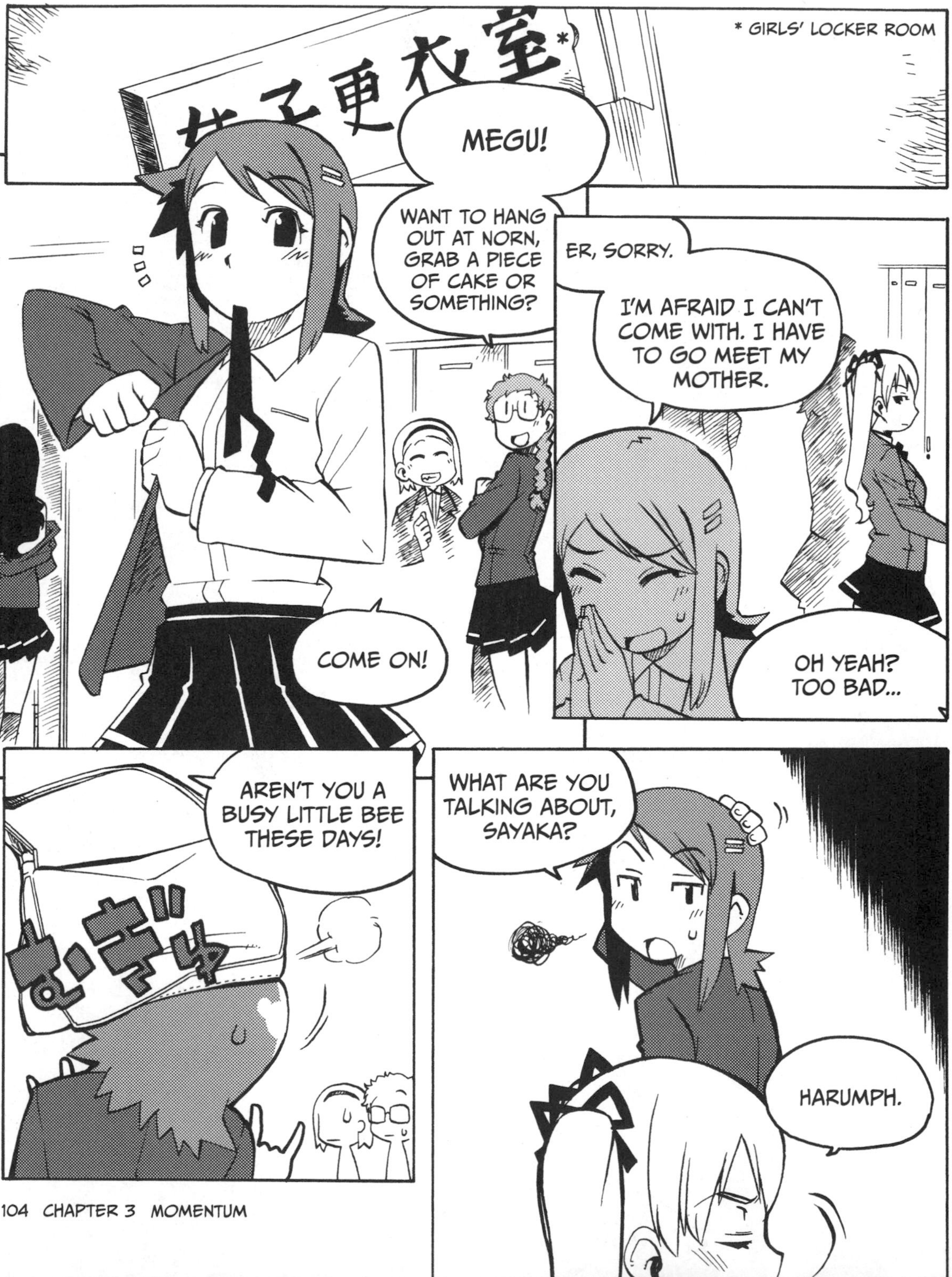

女子更衣室*
* GIRLS' LOCKER ROOM
MEGU!
WANT TO HANG OUT AT NORN, GRAB A PIECE OF CAKE OR SOMETHING?
COME ON!
ER, SORRY.
I'M AFRAID I CAN'T COME WITH. I HAVE TO GO MEET MY MOTHER.
OH YEAH? TOO BAD...
AREN'T YOU A BUSY LITTLE BEE THESE DAYS!
むぎゅ
WHAT ARE YOU TALKING ABOUT, SAYAKA?
HARUMPH.

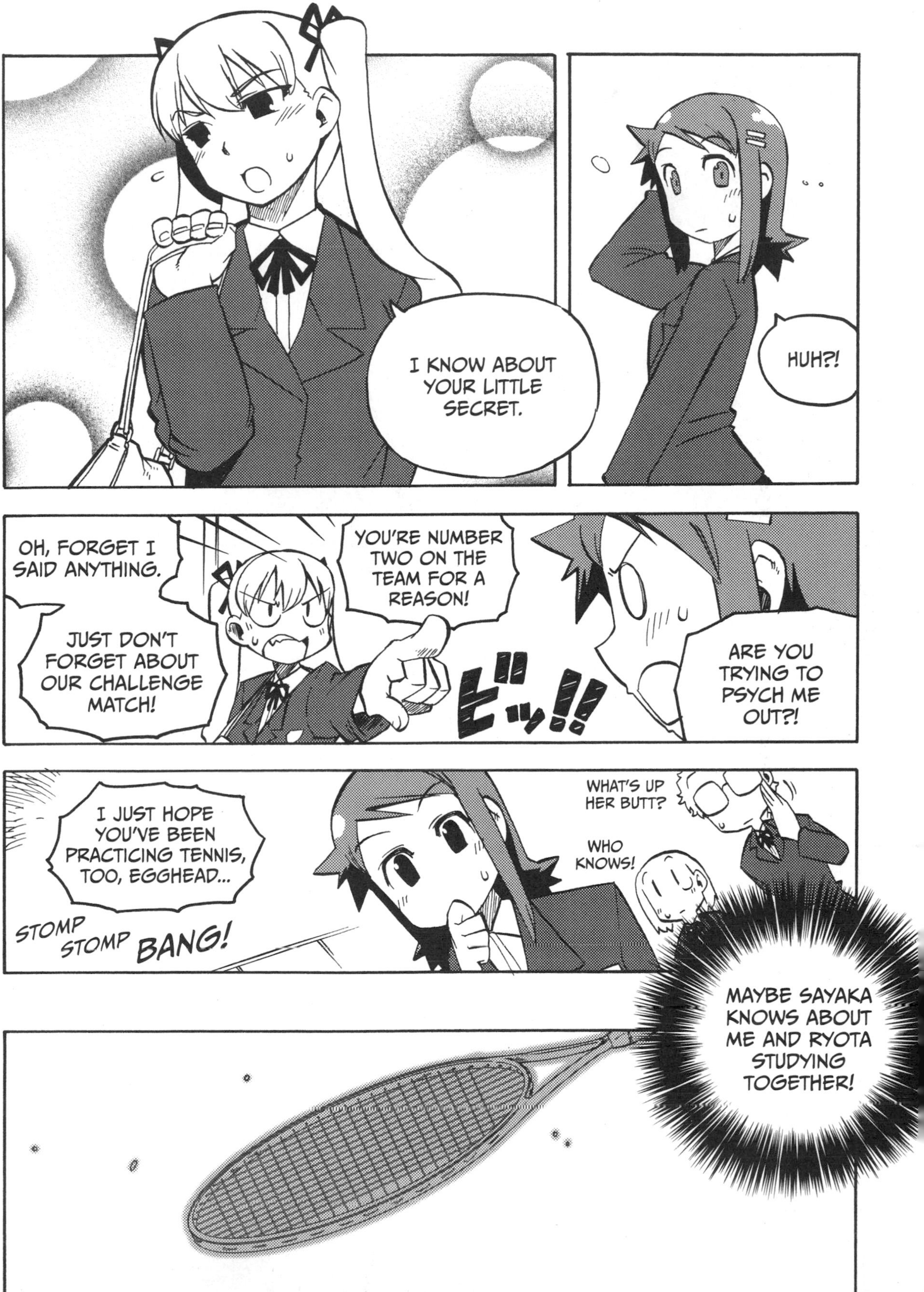

I KNOW ABOUT YOUR LITTLE SECRET.
HUH?!
OH, FORGET I SAID ANYTHING.
JUST DON'T FORGET ABOUT OUR CHALLENGE MATCH!
YOU'RE NUMBER TWO ON THE TEAM FOR A REASON!
ビッ!!
ARE YOU TRYING TO PSYCH ME OUT?!
I JUST HOPE YOU'VE BEEN PRACTICING TENNIS, TOO, EGGHEAD...
STOMP
STOMP
BANG!
WHAT'S UP HER BUTT?
WHO KNOWS!
MAYBE SAYAKA KNOWS ABOUT ME AND RYOTA STUDYING TOGETHER!

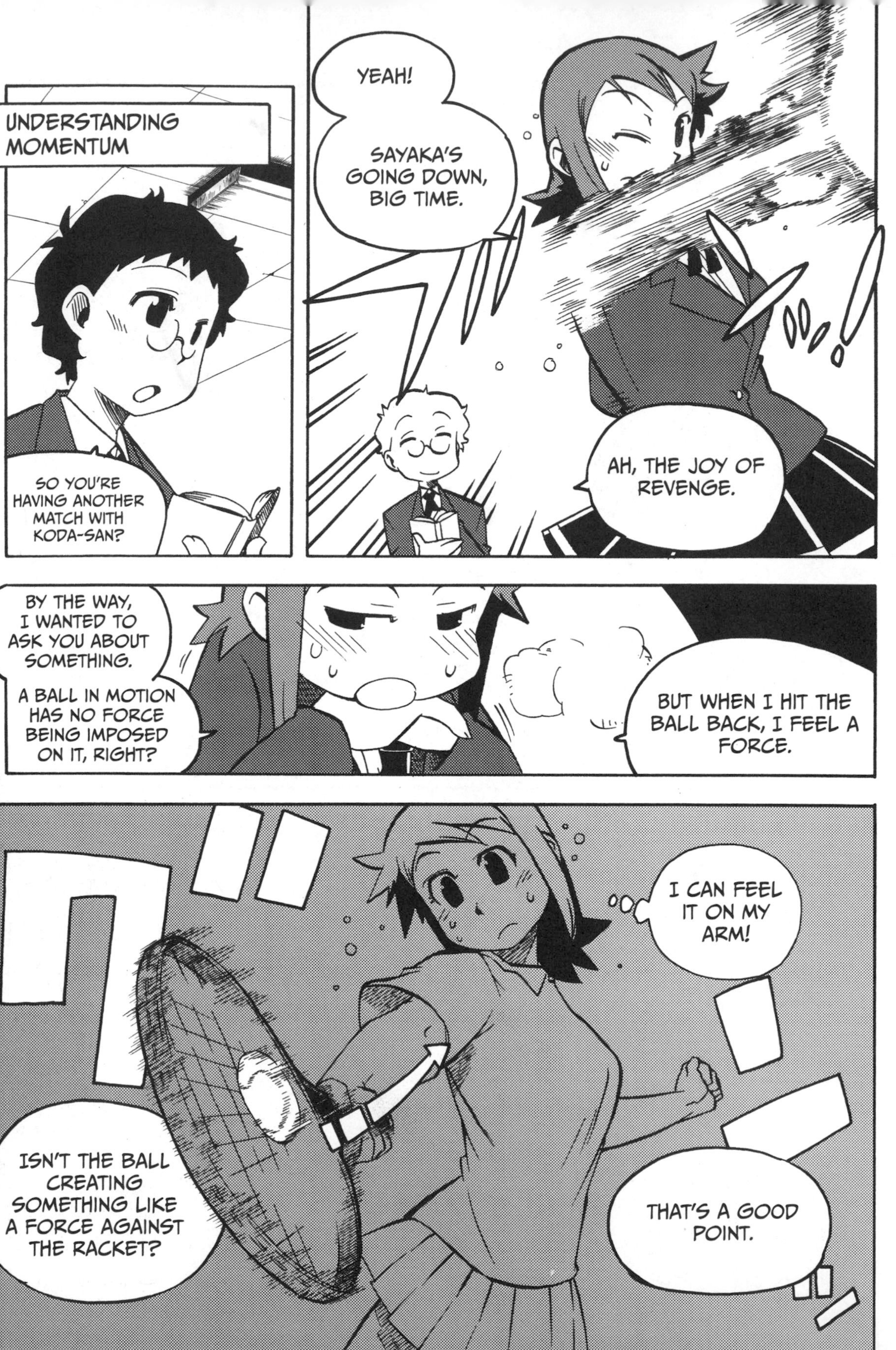
UNDERSTANDING MOMENTUM
SO YOU'RE HAVING ANOTHER MATCH WITH KODA-SAN?
YEAH!
SAYAKA'S GOING DOWN, BIG TIME.
AH, THE JOY OF REVENGE.
BY THE WAY, I WANTED TO ASK YOU ABOUT SOMETHING.
A BALL IN MOTION HAS NO FORCE BEING IMPOSED ON IT, RIGHT?
BUT WHEN I HIT THE BALL BACK, I FEEL A FORCE.
I CAN FEEL IT ON MY ARM!
ISN'T THE BALL CREATING SOMETHING LIKE A FORCE AGAINST THE RACKET?
THAT'S A GOOD POINT.

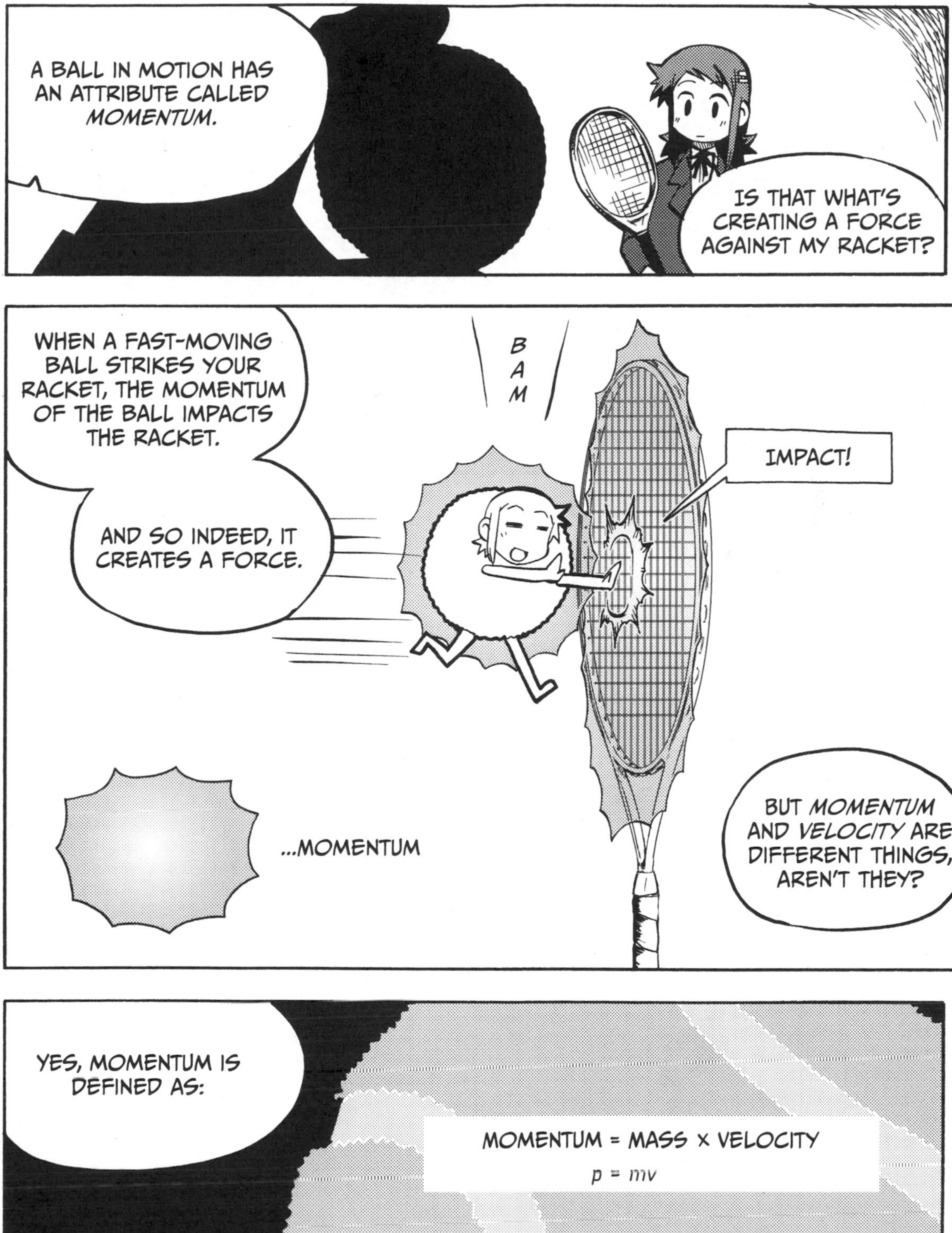
A BALL IN MOTION HAS AN ATTRIBUTE CALLED *MOMENTUM.*
IS THAT WHAT'S CREATING A FORCE AGAINST MY RACKET?
WHEN A FAST-MOVING BALL STRIKES YOUR RACKET, THE MOMENTUM OF THE BALL IMPACTS THE RACKET.
BAM
IMPACT!
AND SO INDEED, IT CREATES A FORCE.
...MOMENTUM
BUT *MOMENTUM* AND *VELOCITY* ARE DIFFERENT THINGS, AREN'T THEY?
YES, MOMENTUM IS DEFINED AS:
MOMENTUM = MASS × VELOCITY
$p = mv$

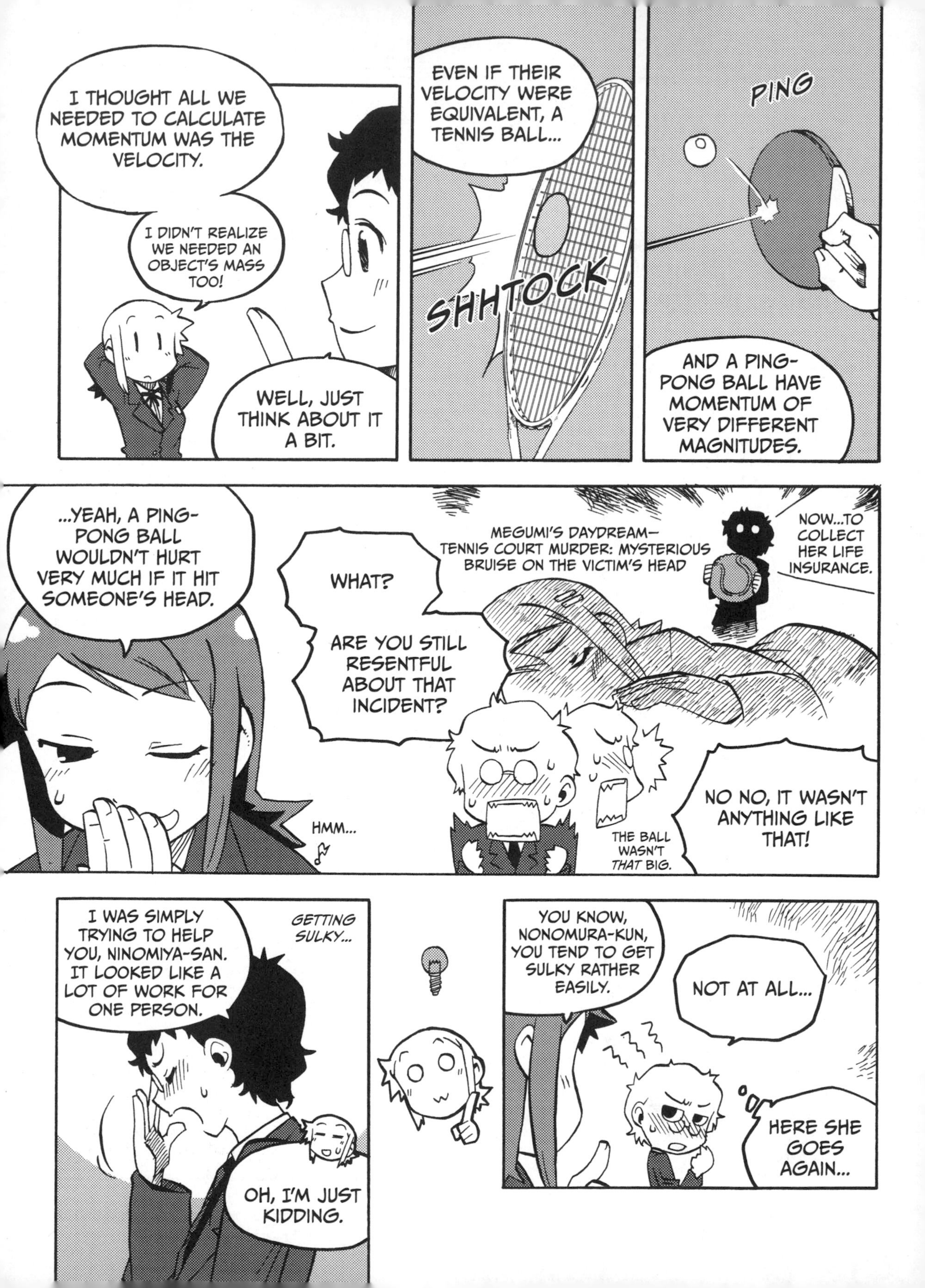
I THOUGHT ALL WE NEEDED TO CALCULATE MOMENTUM WAS THE VELOCITY.
I DIDN'T REALIZE WE NEEDED AN OBJECT'S MASS TOO!
WELL, JUST THINK ABOUT IT A BIT.
EVEN IF THEIR VELOCITY WERE EQUIVALENT, A TENNIS BALL...
SHHTOCK
PING
AND A PING-PONG BALL HAVE MOMENTUM OF VERY DIFFERENT MAGNITUDES.
...YEAH, A PING-PONG BALL WOULDN'T HURT VERY MUCH IF IT HIT SOMEONE'S HEAD.
WHAT?
ARE YOU STILL RESENTFUL ABOUT THAT INCIDENT?
HMM...
MEGUMI'S DAYDREAM—TENNIS COURT MURDER: MYSTERIOUS BRUISE ON THE VICTIM'S HEAD
NOW...TO COLLECT HER LIFE INSURANCE.
NO NO, IT WASN'T ANYTHING LIKE THAT!
THE BALL WASN'T *THAT* BIG.
I WAS SIMPLY TRYING TO HELP YOU, NINOMIYA-SAN. IT LOOKED LIKE A LOT OF WORK FOR ONE PERSON.
GETTING SULKY...
OH, I'M JUST KIDDING.
YOU KNOW, NONOMURA-KUN, YOU TEND TO GET SULKY RATHER EASILY.
NOT AT ALL...
HERE SHE GOES AGAIN...

LABORATORY

DIFFERENCE IN MOMENTUM DUE TO A DIFFERENCE IN MASS

 To help you understand how momentum works, I've brought in a softball and a tennis ball.

Let's examine the momentum of a softball traveling slowly and a tennis ball traveling quickly.

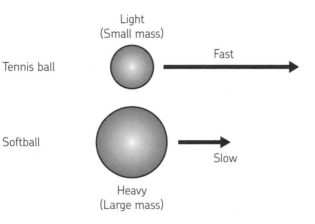

Light
(Small mass)

Tennis ball

Fast

Softball

Slow

Heavy
(Large mass)

 Let me see, the softball is much heavier than the tennis ball, right?

 Yes, of course. We know the following about the two balls:

$$m_{softball} > m_{tennis\ ball}$$

$$v_{softball} < v_{tennis\ ball}$$

However, we can't tell which ball has the greater momentum. Recall that momentum can be calculated as mass multiplied by velocity ($p = mv$). We'd need to know numerical values to determine the difference precisely.

Well, I know that a tennis ball has a mass of about 60 g.

And a softball is about 180 g.

So we're almost there. It's 60 g versus 180 g—the mass of a softball is about three times as great as that of a tennis ball.

Given these new facts and the relationship $p = mv$, to have an equivalent momentum, the tennis ball must have a velocity three times as great as the softball.

Oh, I see.

Softball
3m
v⃗
3mv⃗
Momentum
Tennis ball
m
3v⃗
3mv⃗

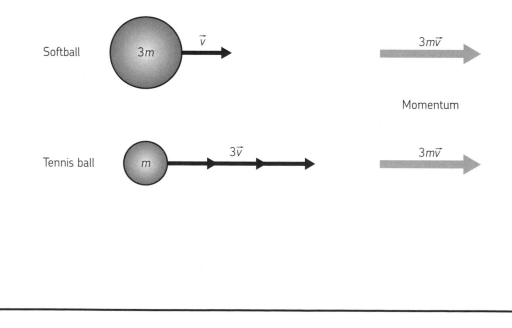

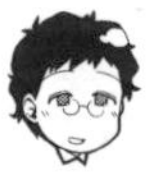

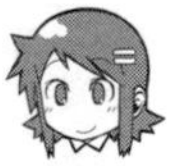

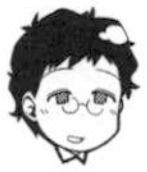

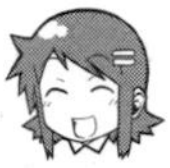

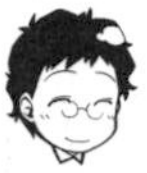

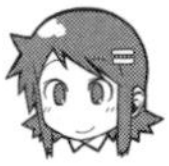

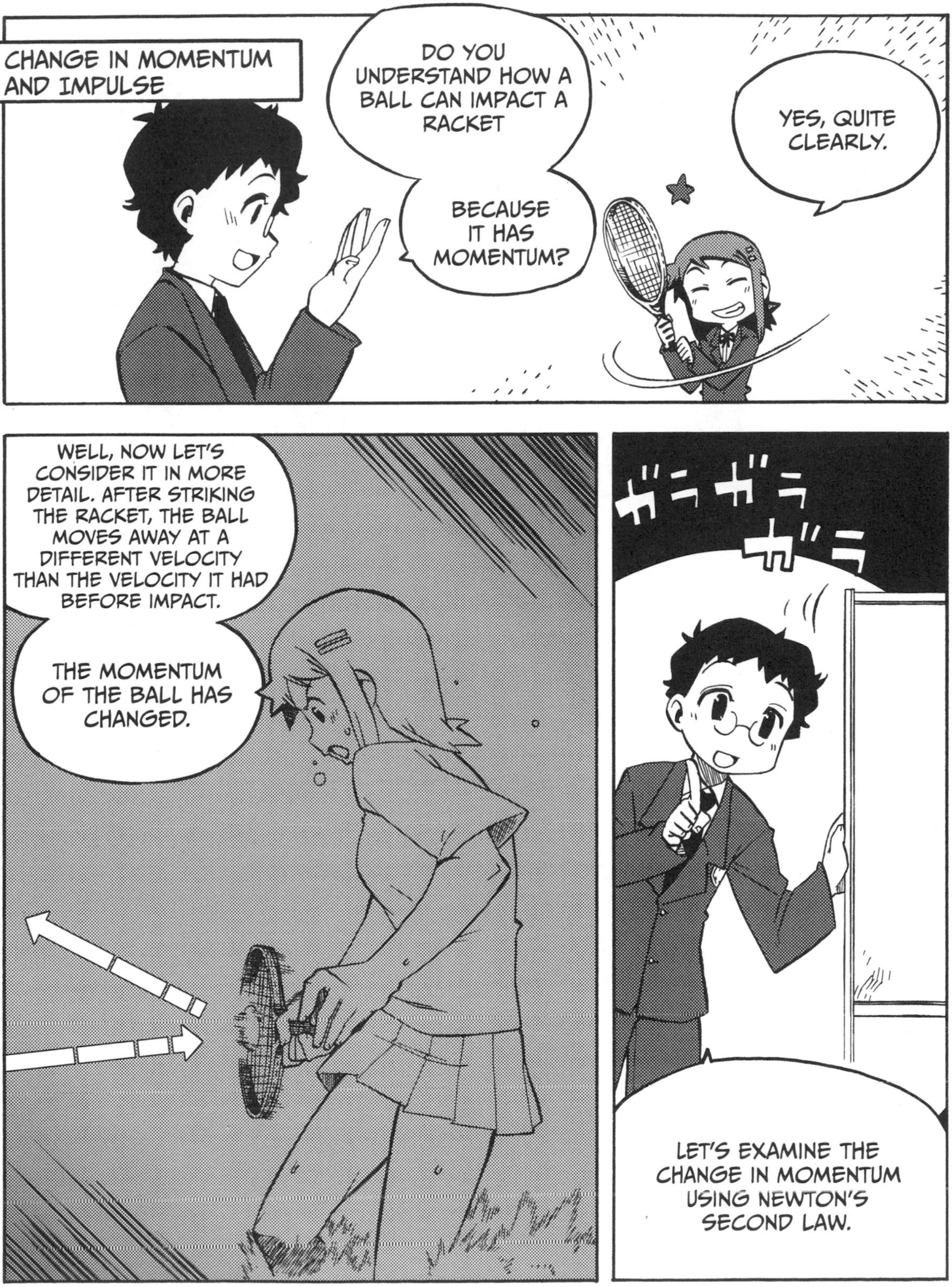
CHANGE IN MOMENTUM AND IMPULSE
DO YOU UNDERSTAND HOW A BALL CAN IMPACT A RACKET
BECAUSE IT HAS MOMENTUM?
YES, QUITE CLEARLY.
WELL, NOW LET'S CONSIDER IT IN MORE DETAIL. AFTER STRIKING THE RACKET, THE BALL MOVES AWAY AT A DIFFERENT VELOCITY THAN THE VELOCITY IT HAD BEFORE IMPACT.
THE MOMENTUM OF THE BALL HAS CHANGED.
ガラガラ
ガラ
LET'S EXAMINE THE CHANGE IN MOMENTUM USING NEWTON'S SECOND LAW.

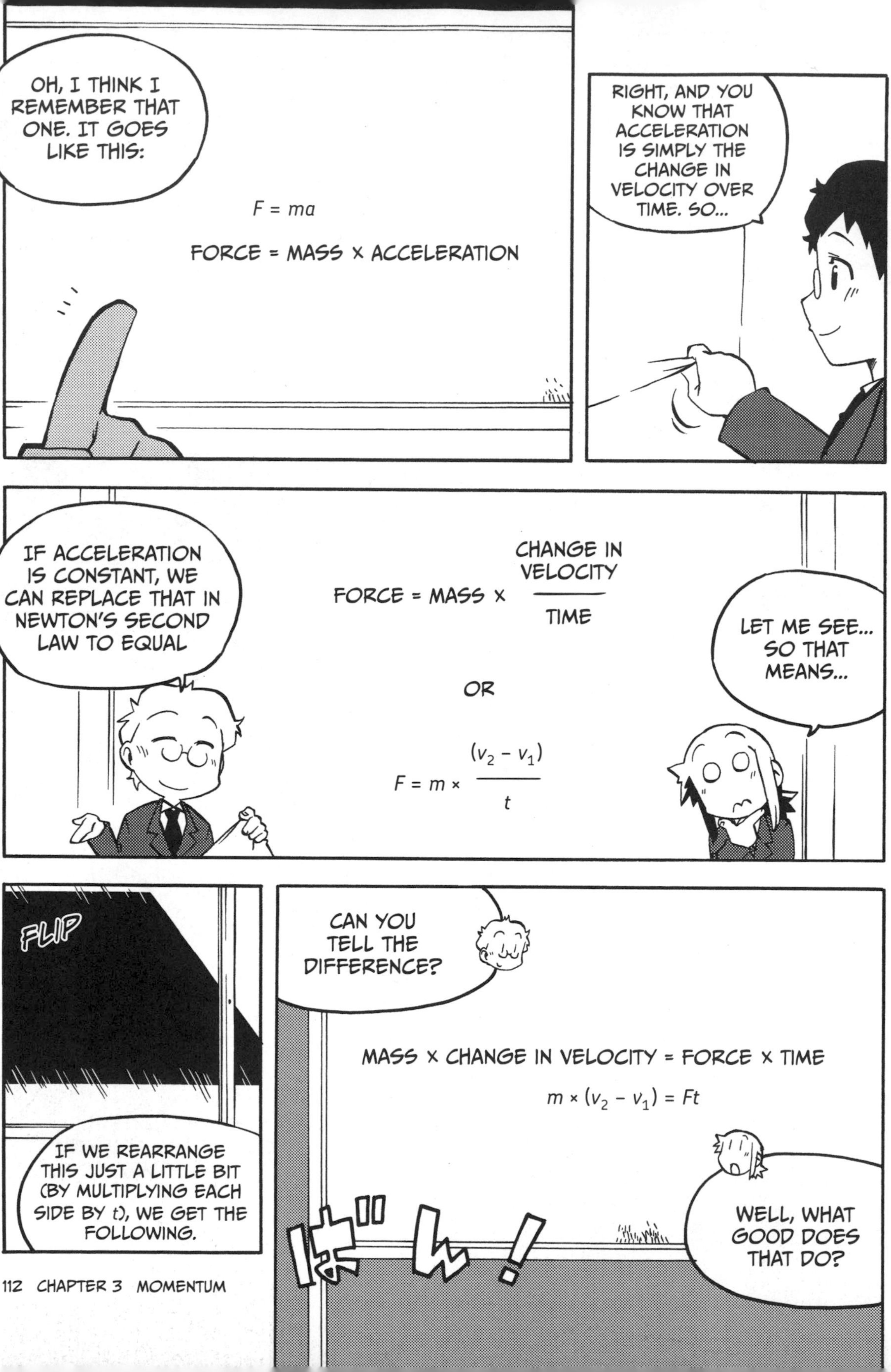
OH, I THINK I REMEMBER THAT ONE. IT GOES LIKE THIS:
$F = ma$
FORCE = MASS × ACCELERATION
RIGHT, AND YOU KNOW THAT ACCELERATION IS SIMPLY THE CHANGE IN VELOCITY OVER TIME. SO...
IF ACCELERATION IS CONSTANT, WE CAN REPLACE THAT IN NEWTON'S SECOND LAW TO EQUAL
FORCE = MASS × $\frac{\text{CHANGE IN VELOCITY}}{\text{TIME}}$
OR
$F = m \times \frac{(v_2 - v_1)}{t}$
LET ME SEE... SO THAT MEANS...
FLIP
IF WE REARRANGE THIS JUST A LITTLE BIT (BY MULTIPLYING EACH SIDE BY t), WE GET THE FOLLOWING.
CAN YOU TELL THE DIFFERENCE?
MASS × CHANGE IN VELOCITY = FORCE × TIME
$m \times (v_2 - v_1) = Ft$
ばん!
WELL, WHAT GOOD DOES THAT DO?

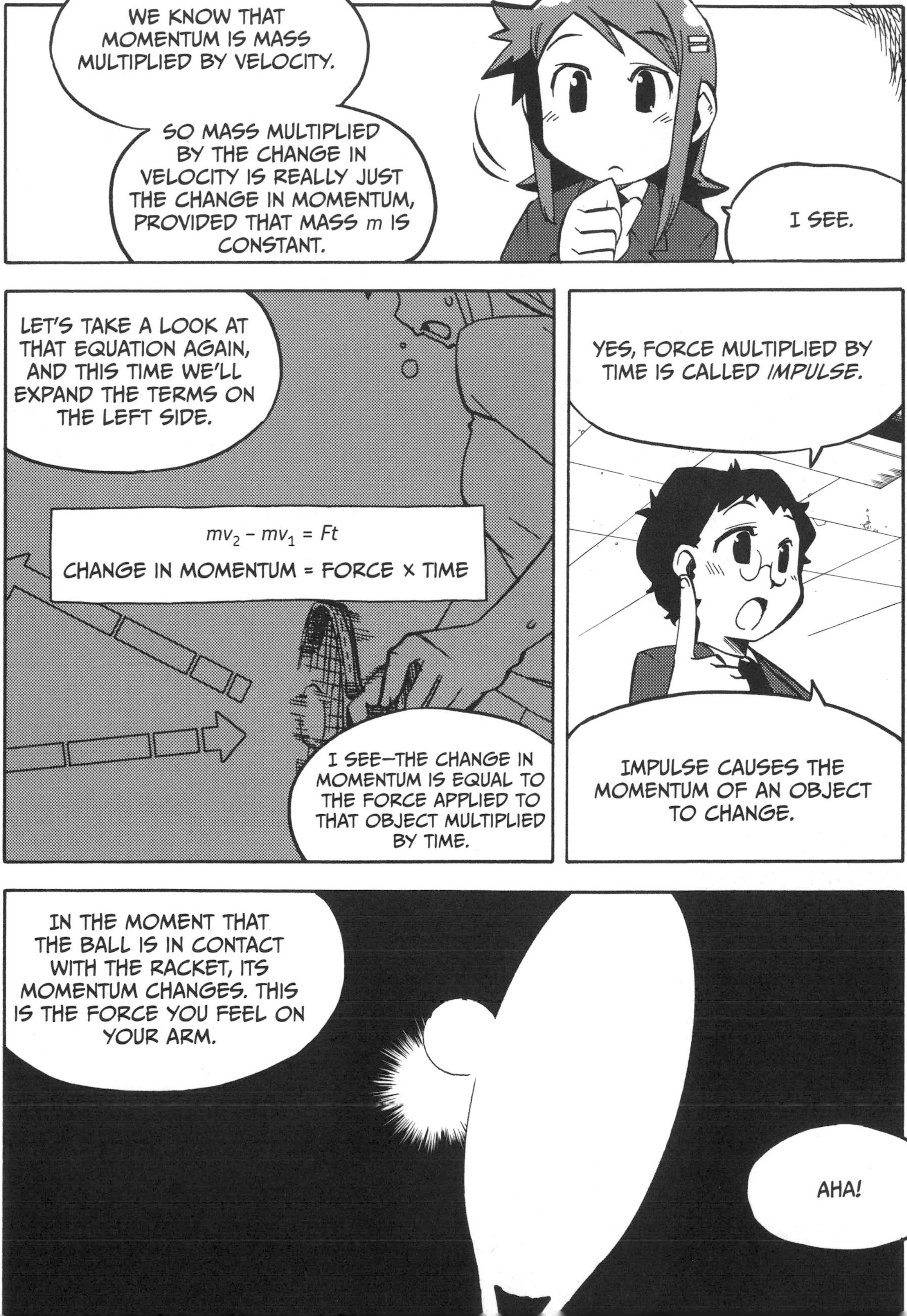
WE KNOW THAT MOMENTUM IS MASS MULTIPLIED BY VELOCITY.
SO MASS MULTIPLIED BY THE CHANGE IN VELOCITY IS REALLY JUST THE CHANGE IN MOMENTUM, PROVIDED THAT MASS *m* IS CONSTANT.
I SEE.
LET'S TAKE A LOOK AT THAT EQUATION AGAIN, AND THIS TIME WE'LL EXPAND THE TERMS ON THE LEFT SIDE.
$mv_2 - mv_1 = Ft$
CHANGE IN MOMENTUM = FORCE × TIME
I SEE—THE CHANGE IN MOMENTUM IS EQUAL TO THE FORCE APPLIED TO THAT OBJECT MULTIPLIED BY TIME.
YES, FORCE MULTIPLIED BY TIME IS CALLED *IMPULSE*.
IMPULSE CAUSES THE MOMENTUM OF AN OBJECT TO CHANGE.
IN THE MOMENT THAT THE BALL IS IN CONTACT WITH THE RACKET, ITS MOMENTUM CHANGES. THIS IS THE FORCE YOU FEEL ON YOUR ARM.
AHA!

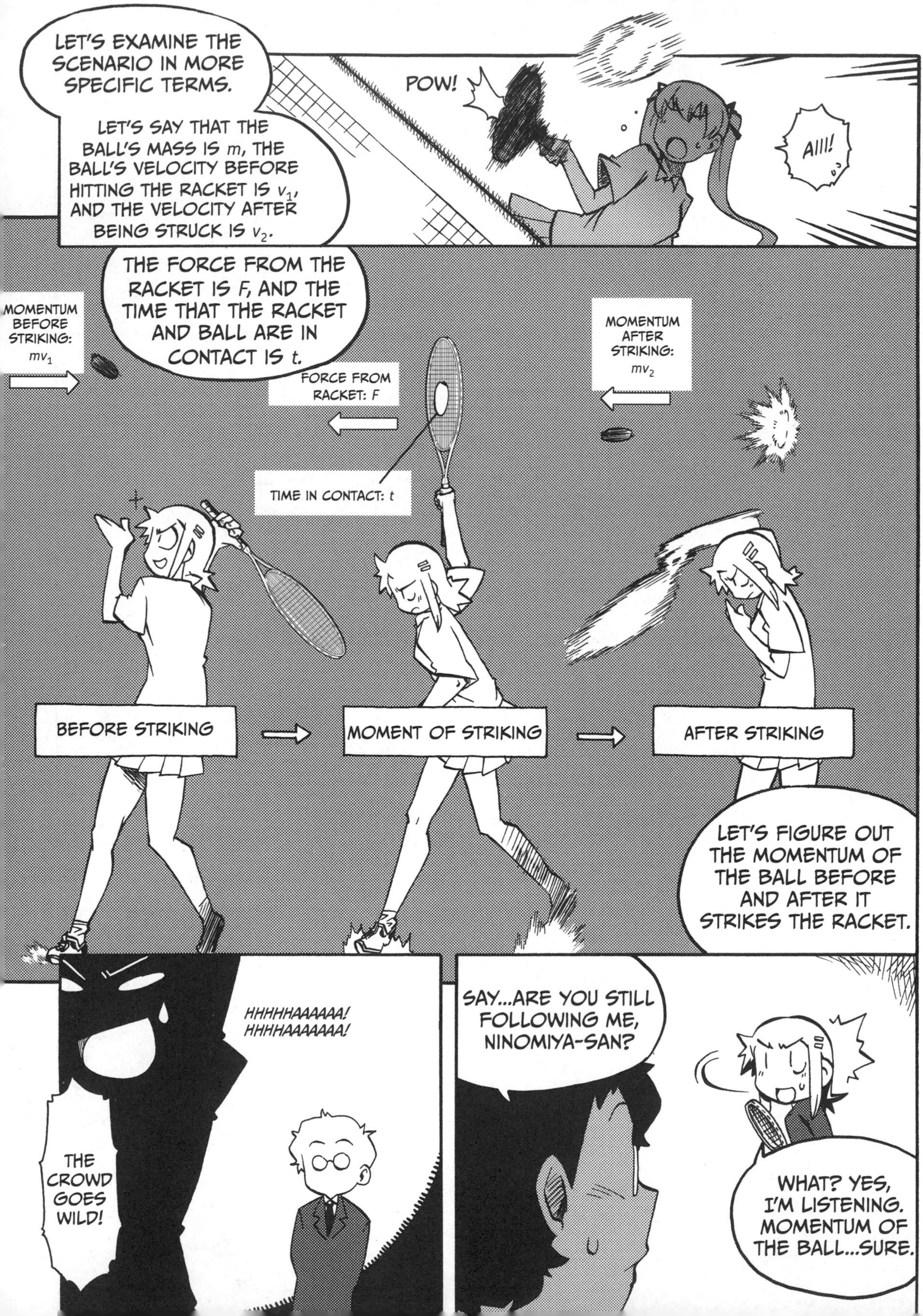
LET'S EXAMINE THE SCENARIO IN MORE SPECIFIC TERMS.
LET'S SAY THAT THE BALL'S MASS IS m, THE BALL'S VELOCITY BEFORE HITTING THE RACKET IS v_1, AND THE VELOCITY AFTER BEING STRUCK IS v_2.
POW!
Aiii!
THE FORCE FROM THE RACKET IS F, AND THE TIME THAT THE RACKET AND BALL ARE IN CONTACT IS t.
MOMENTUM BEFORE STRIKING: mv_1
FORCE FROM RACKET: F
TIME IN CONTACT: t
MOMENTUM AFTER STRIKING: mv_2
BEFORE STRIKING
MOMENT OF STRIKING
AFTER STRIKING
LET'S FIGURE OUT THE MOMENTUM OF THE BALL BEFORE AND AFTER IT STRIKES THE RACKET.
HHHHHAAAAAAA! HHHHAAAAAAA!
THE CROWD GOES WILD!
SAY...ARE YOU STILL FOLLOWING ME, NINOMIYA-SAN?
WHAT? YES, I'M LISTENING. MOMENTUM OF THE BALL...SURE.

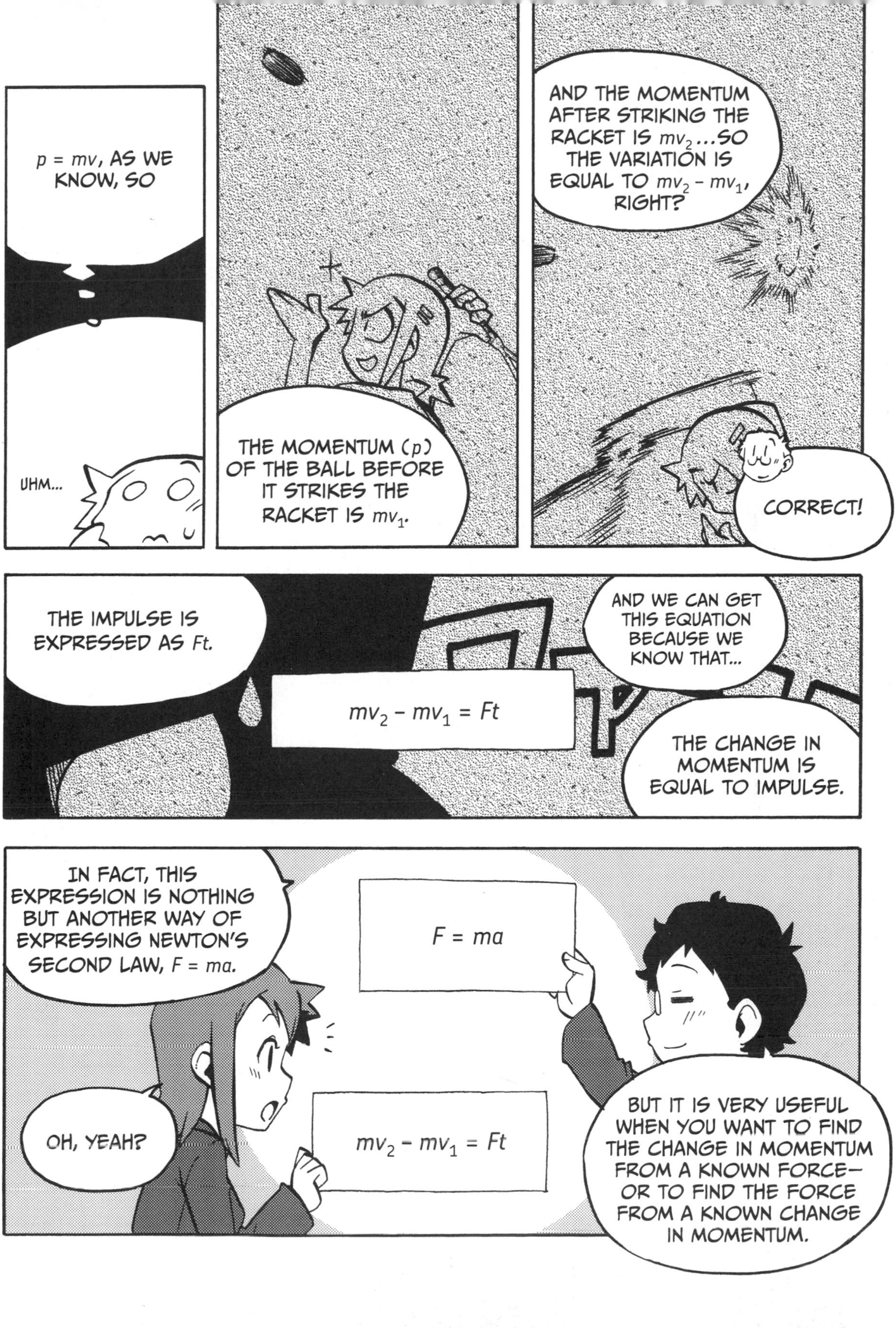

p = mv, AS WE KNOW, SO
UHM...
THE MOMENTUM (p) OF THE BALL BEFORE IT STRIKES THE RACKET IS mv_1.
AND THE MOMENTUM AFTER STRIKING THE RACKET IS mv_2...SO THE VARIATION IS EQUAL TO $mv_2 - mv_1$, RIGHT?
CORRECT!
THE IMPULSE IS EXPRESSED AS Ft.
$mv_2 - mv_1 = Ft$
AND WE CAN GET THIS EQUATION BECAUSE WE KNOW THAT...
THE CHANGE IN MOMENTUM IS EQUAL TO IMPULSE.
IN FACT, THIS EXPRESSION IS NOTHING BUT ANOTHER WAY OF EXPRESSING NEWTON'S SECOND LAW, F = ma.
F = ma
OH, YEAH?
$mv_2 - mv_1 = Ft$
BUT IT IS VERY USEFUL WHEN YOU WANT TO FIND THE CHANGE IN MOMENTUM FROM A KNOWN FORCE—OR TO FIND THE FORCE FROM A KNOWN CHANGE IN MOMENTUM.

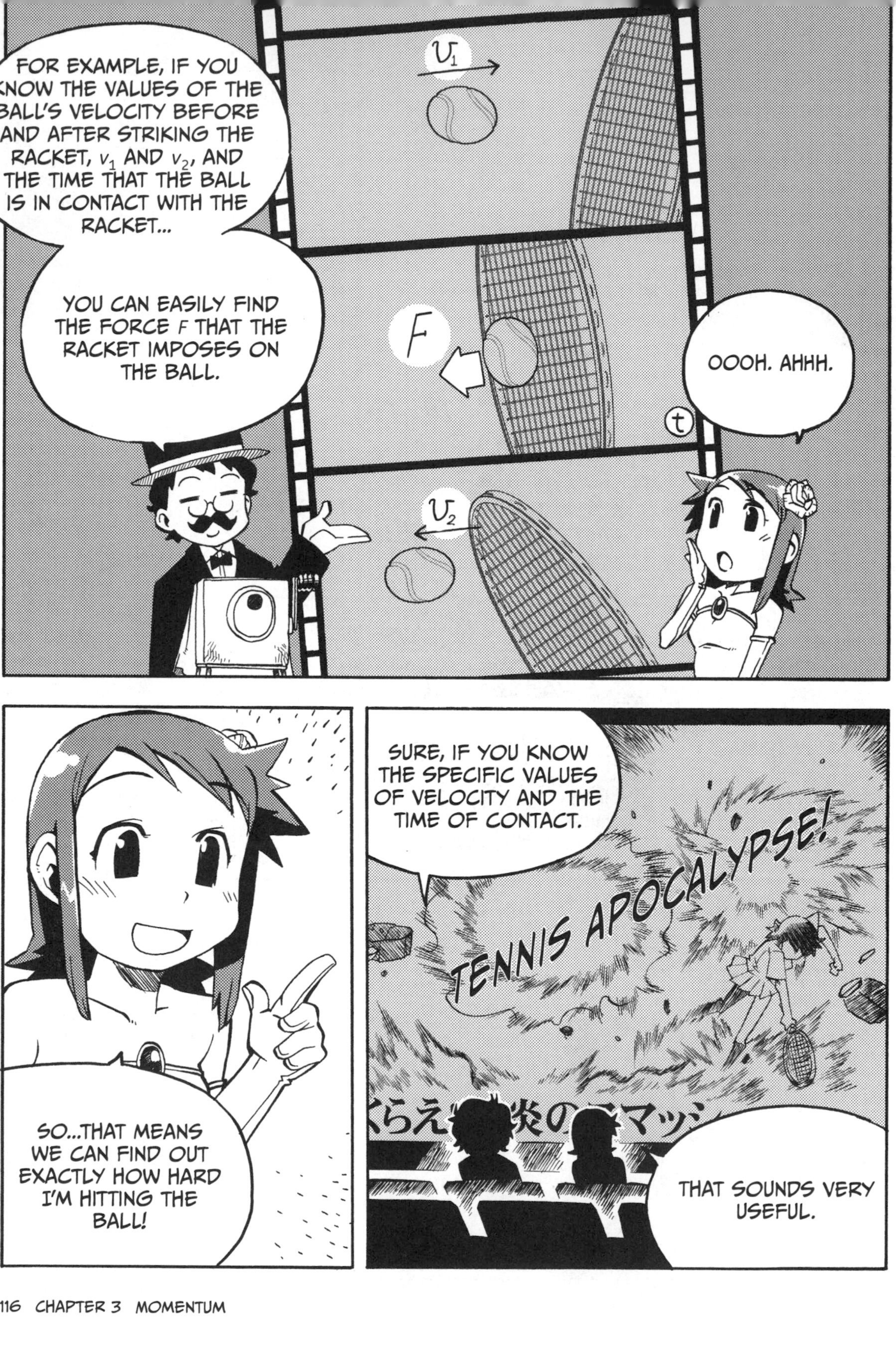
FOR EXAMPLE, IF YOU KNOW THE VALUES OF THE BALL'S VELOCITY BEFORE AND AFTER STRIKING THE RACKET, v_1 AND v_2, AND THE TIME THAT THE BALL IS IN CONTACT WITH THE RACKET...
YOU CAN EASILY FIND THE FORCE F THAT THE RACKET IMPOSES ON THE BALL.
v_1
F
v_2
OOOH. AHHH.
SO...THAT MEANS WE CAN FIND OUT EXACTLY HOW HARD I'M HITTING THE BALL!
SURE, IF YOU KNOW THE SPECIFIC VALUES OF VELOCITY AND THE TIME OF CONTACT.
TENNIS APOCALYPSE!
THAT SOUNDS VERY USEFUL.

LABORATORY

FINDING THE MOMENTUM OF A STROKE

 Let's actually analyze this scenario, Ninomiya-san, and find out the force you're applying to the ball. During your match with Sayaka, I filmed your motion with a high-speed camera. We'll analyze a time when you returned her smash.

 Here you go again. Yet another make-believe scenario.

 This time, I really did shoot the footage.

 What on earth . . . ?

 It's all in the name of science. Anyway, I analyzed the images and learned that the velocity of the ball when it hit the racket was about 100 km per hour, and you returned the ball at about 80 km per hour. And I measured the time that the ball was in contact with your racket—it was 0.01 second.

 So we should have all the numbers we need!

 Using these values, we can find the magnitude of the force your racket imposed on the ball. But it's actually not so simple. A graph of the force over time looks like this.

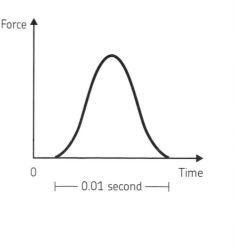

However, we'll assume an average magnitude of F in this example.

That makes the calculation much easier.

Force
F
Impulse
Time
0.01 second

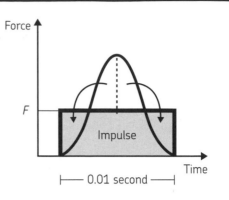

First, let's calculate the momentum of the ball before you hit it. The mass of a tennis ball is 0.06 kg. The velocity is negative 100 km per hour, as viewed from the direction of the return. As 1 km = 1000 m, and 1 hour = 3600 seconds, we'll convert our units for velocity into meters per second (m/s) as follows: 1 km/h = 1000 m / 3600 s. The calculation looks like this:

$$\frac{-100\ \text{km}}{\text{h}} \times \frac{1000\ \text{m}}{\text{km}} \times \frac{1\ \text{h}}{3600\ \text{s}} = -27.8\ \frac{\text{m}}{\text{s}}$$

$$p = mv$$

$$p = 0.06\ \text{kg} \times -27.8\ \text{m/s}$$

$$p = -1.7\ \text{kg} \times \text{m/s}$$

Now we know the ball's initial momentum. It's a little weird that the value is negative, but I guess it just indicates the direction from my point of view.

So now let's calculate the momentum of the ball after you've struck it. Given that the velocity of the ball afterwards is 80 km/h, and its orientation is positive, the result is as follows:

$$\frac{80\ \text{km}}{\text{h}} \times \frac{1000\ \text{m}}{\text{km}} \times \frac{1\ \text{h}}{3600\ \text{s}} = 22.2\ \frac{\text{m}}{\text{s}}$$

$$p = mv$$

$$p = 0.06\ \text{kg} \times 22.2\ \text{m/s}$$

$$p = 1.3\ \text{kg} \times \text{m/s}$$

Now we can find the change in these two values.

The change in momentum can be calculated like so:

$$1.3 \text{ kg} \times \text{m/s} - (-1.7 \text{ kg} \times \text{m/s}) = 3.0 \text{ kg} \times \text{m/s} = \Delta p$$

So that's the change in the ball's momentum. And since the force was working for 0.01 seconds, we can figure out the force, using this equation:

$$\Delta p = Ft \qquad \text{or} \qquad \frac{\Delta p}{t} = F$$

In our example, that means (3.0 kg × m/s) / 0.01 s = 300N. That's the force on my racket, I bet.

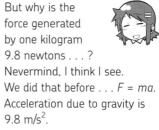

Yes, that's it. Since you probably don't know what a newton feels like, let's find the equivalent force generated by 1 kg weight, assuming that 1 kg is about equal to 9.8N:

$$300\text{N} \times \frac{1 \text{ kg}}{9.8\text{N}} = 30.6 \text{ kg}$$

But why is the force generated by one kilogram 9.8 newtons . . . ? Nevermind, I think I see. We did that before . . . $F = ma$. Acceleration due to gravity is 9.8 m/s².

Wow, that's a lot to lift!

Well, remember, the force from gravity is constant—this is just momentary. And you're using your muscles in a very different way, in a different direction.

THE CONSERVATION OF MOMENTUM

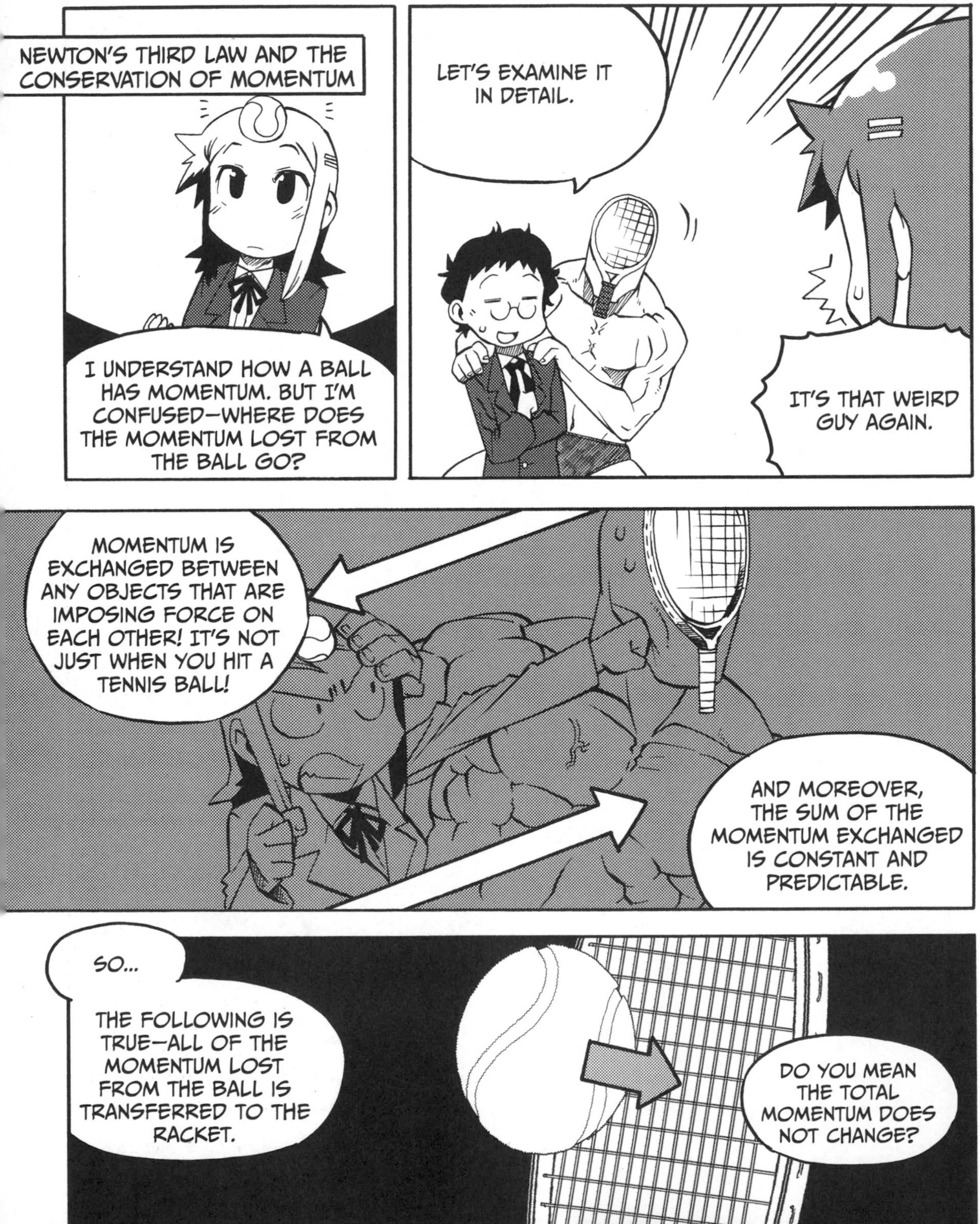

NEWTON'S THIRD LAW AND THE CONSERVATION OF MOMENTUM
I UNDERSTAND HOW A BALL HAS MOMENTUM. BUT I'M CONFUSED—WHERE DOES THE MOMENTUM LOST FROM THE BALL GO?
LET'S EXAMINE IT IN DETAIL.
IT'S THAT WEIRD GUY AGAIN.
MOMENTUM IS EXCHANGED BETWEEN ANY OBJECTS THAT ARE IMPOSING FORCE ON EACH OTHER! IT'S NOT JUST WHEN YOU HIT A TENNIS BALL!
AND MOREOVER, THE SUM OF THE MOMENTUM EXCHANGED IS CONSTANT AND PREDICTABLE.
SO...
THE FOLLOWING IS TRUE—ALL OF THE MOMENTUM LOST FROM THE BALL IS TRANSFERRED TO THE RACKET.
DO YOU MEAN THE TOTAL MOMENTUM DOES NOT CHANGE?

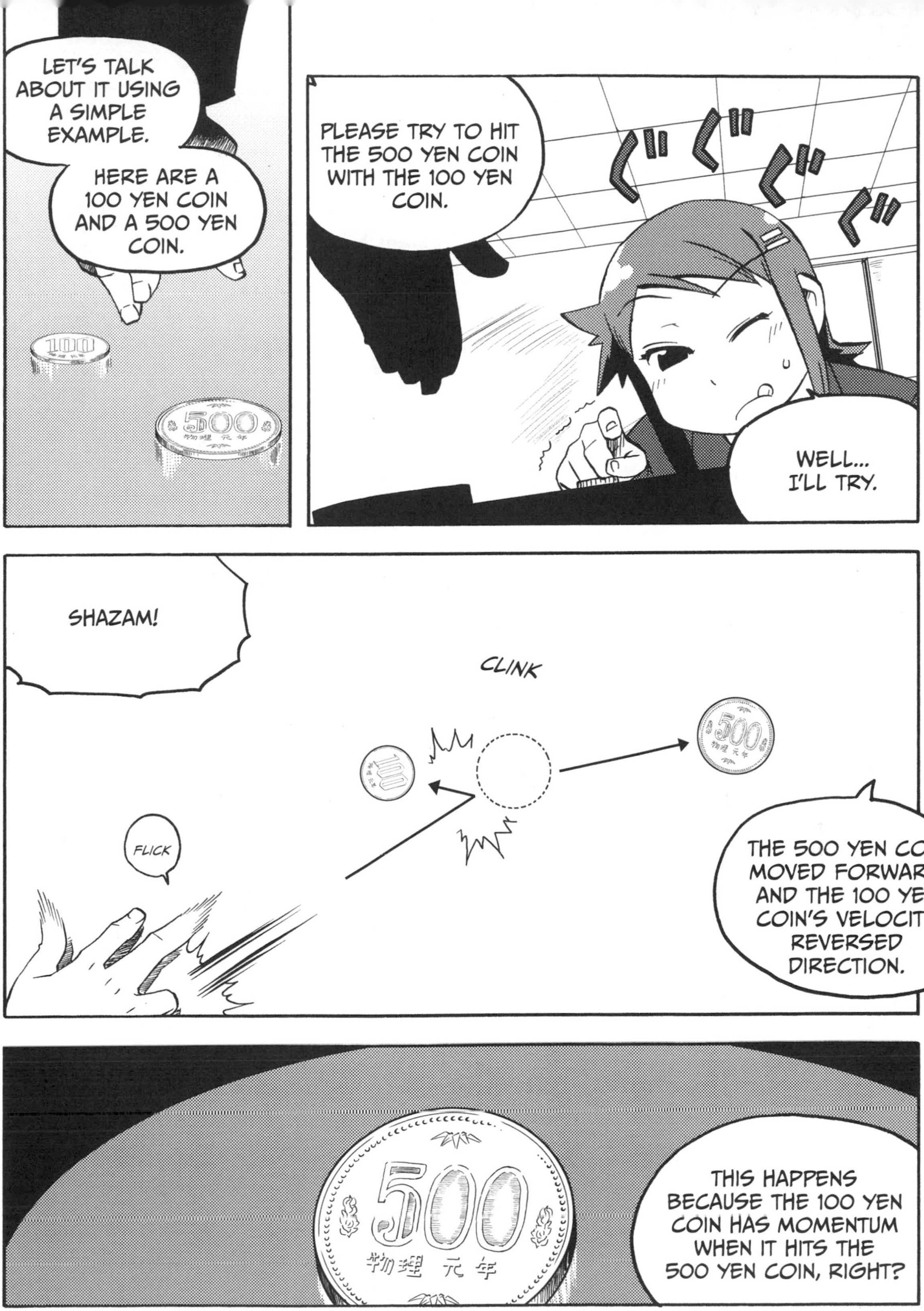
LET'S TALK ABOUT IT USING A SIMPLE EXAMPLE.
HERE ARE A 100 YEN COIN AND A 500 YEN COIN.
PLEASE TRY TO HIT THE 500 YEN COIN WITH THE 100 YEN COIN.
WELL... I'LL TRY.
SHAZAM!
CLINK
FLICK
THE 500 YEN COIN MOVED FORWARD, AND THE 100 YEN COIN'S VELOCITY REVERSED DIRECTION.
THIS HAPPENS BECAUSE THE 100 YEN COIN HAS MOMENTUM WHEN IT HITS THE 500 YEN COIN, RIGHT?

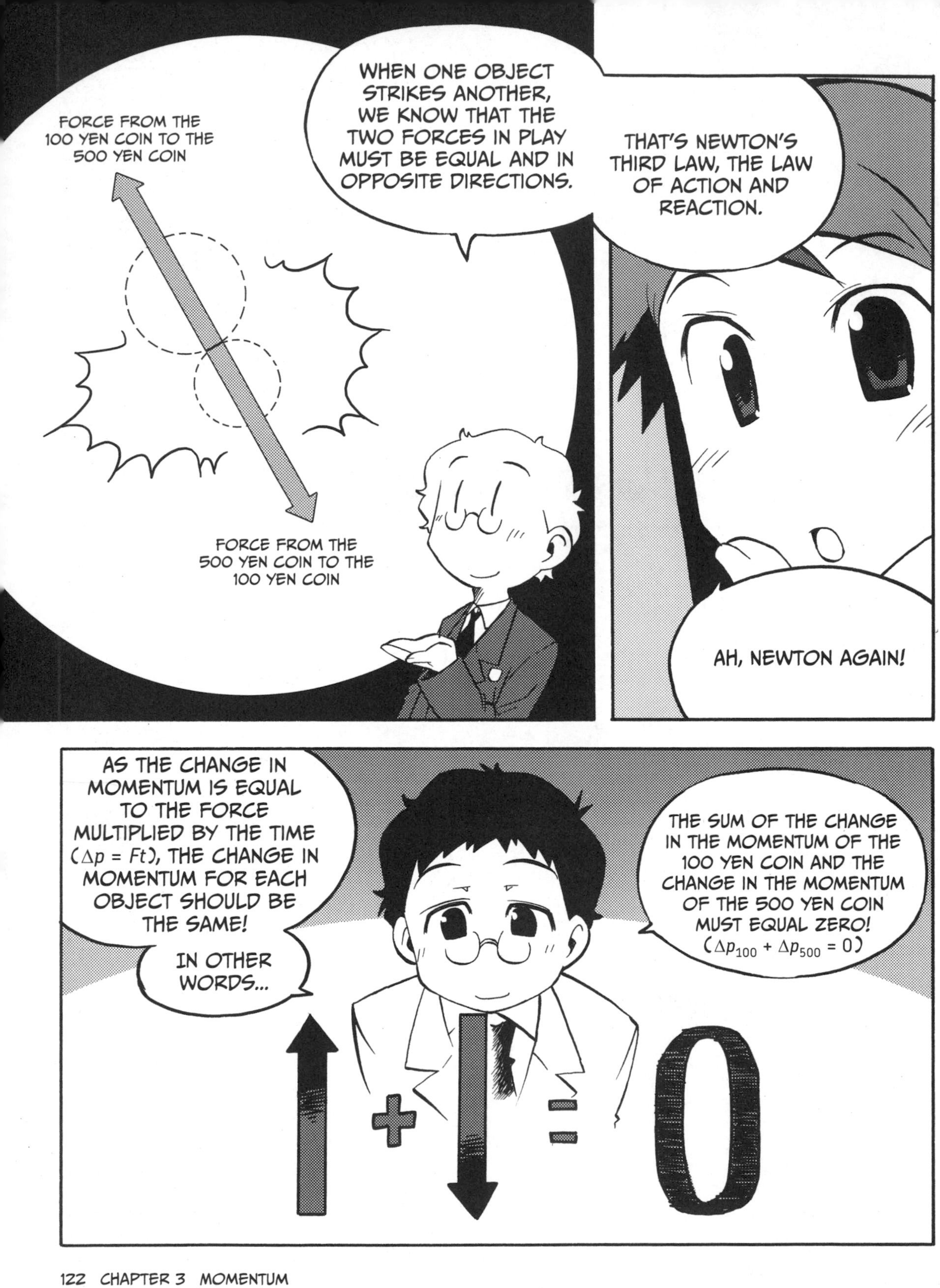
WHEN ONE OBJECT STRIKES ANOTHER, WE KNOW THAT THE TWO FORCES IN PLAY MUST BE EQUAL AND IN OPPOSITE DIRECTIONS.
FORCE FROM THE 100 YEN COIN TO THE 500 YEN COIN
FORCE FROM THE 500 YEN COIN TO THE 100 YEN COIN
THAT'S NEWTON'S THIRD LAW, THE LAW OF ACTION AND REACTION.
AH, NEWTON AGAIN!
AS THE CHANGE IN MOMENTUM IS EQUAL TO THE FORCE MULTIPLIED BY THE TIME ($\Delta p = Ft$), THE CHANGE IN MOMENTUM FOR EACH OBJECT SHOULD BE THE SAME!
IN OTHER WORDS...
THE SUM OF THE CHANGE IN THE MOMENTUM OF THE 100 YEN COIN AND THE CHANGE IN THE MOMENTUM OF THE 500 YEN COIN MUST EQUAL ZERO! ($\Delta p_{100} + \Delta p_{500} = 0$)
0

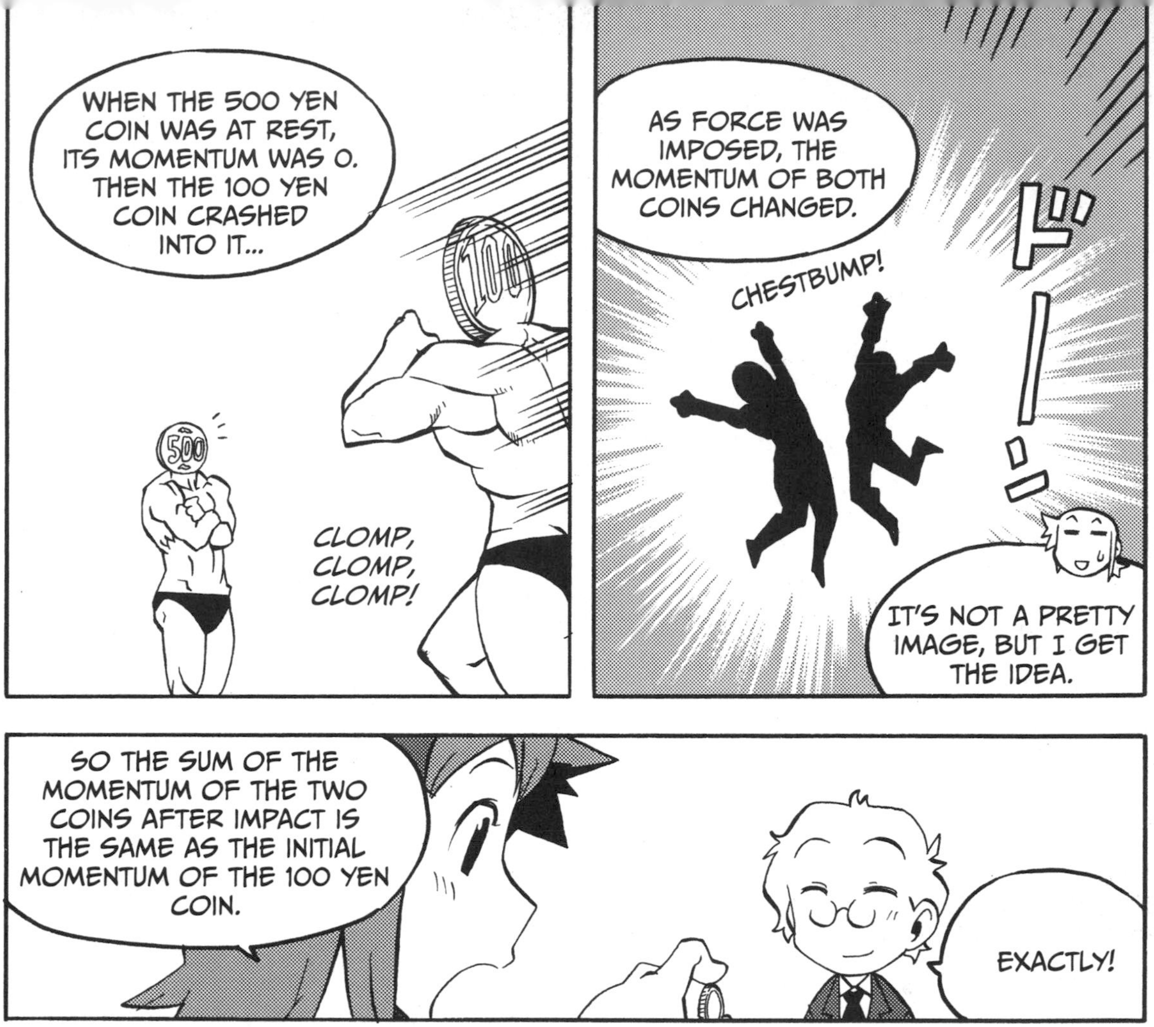
WHEN THE 500 YEN COIN WAS AT REST, ITS MOMENTUM WAS 0. THEN THE 100 YEN COIN CRASHED INTO IT...
CLOMP, CLOMP, CLOMP!
AS FORCE WAS IMPOSED, THE MOMENTUM OF BOTH COINS CHANGED.
CHESTBUMP!
ド一ン
IT'S NOT A PRETTY IMAGE, BUT I GET THE IDEA.
SO THE SUM OF THE MOMENTUM OF THE TWO COINS AFTER IMPACT IS THE SAME AS THE INITIAL MOMENTUM OF THE 100 YEN COIN.
EXACTLY!

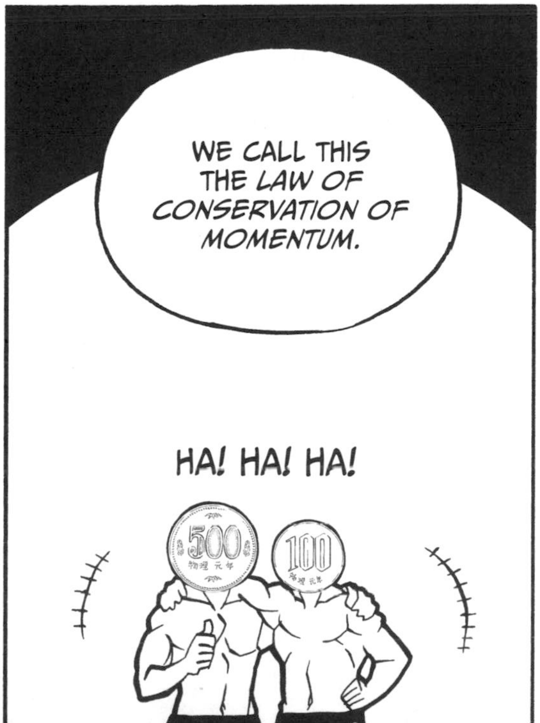
WE CALL THIS THE *LAW OF CONSERVATION OF MOMENTUM.*
HA! HA! HA!

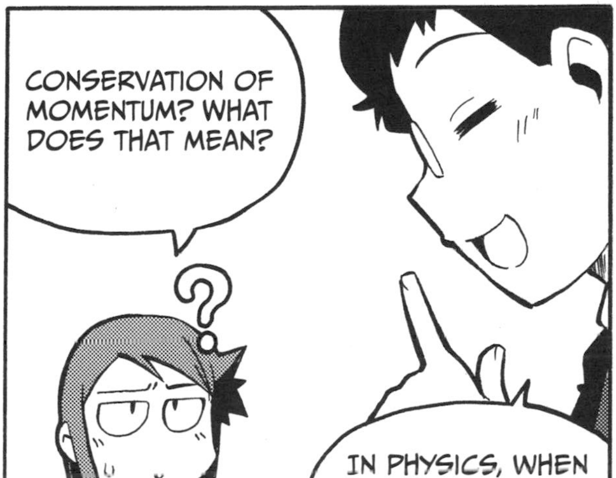
CONSERVATION OF MOMENTUM? WHAT DOES THAT MEAN?
?
IN PHYSICS, WHEN A QUANTITY DOES NOT CHANGE OVER TIME, IT IS CALLED *CONSERVATION.*

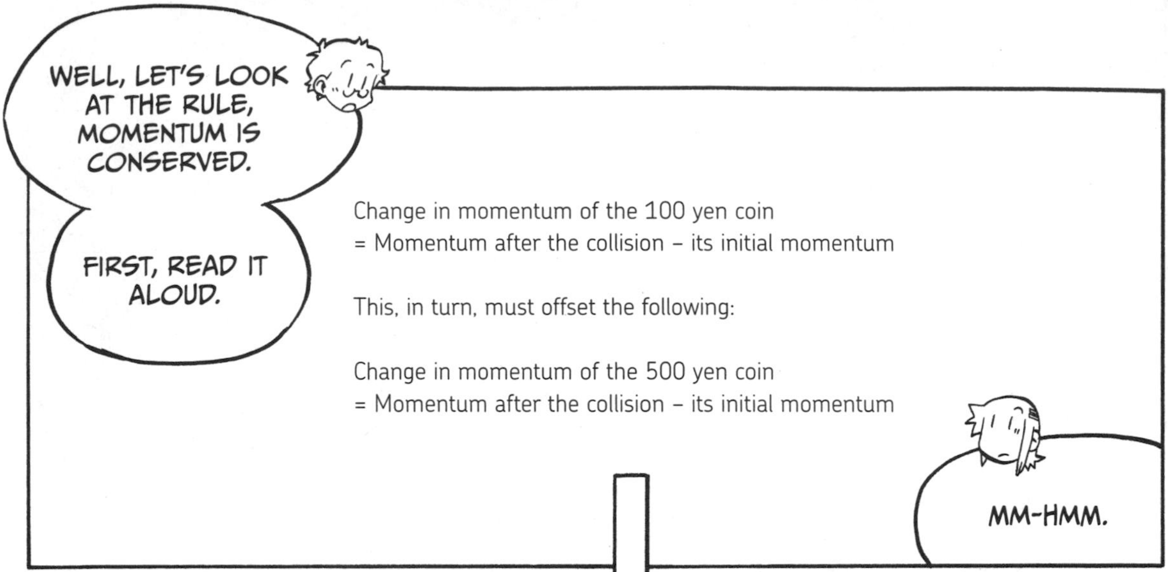
WELL, LET'S LOOK AT THE RULE, MOMENTUM IS CONSERVED.
FIRST, READ IT ALOUD.
Change in momentum of the 100 yen coin
= Momentum after the collision – its initial momentum
This, in turn, must offset the following:
Change in momentum of the 500 yen coin
= Momentum after the collision – its initial momentum
MM-HMM.

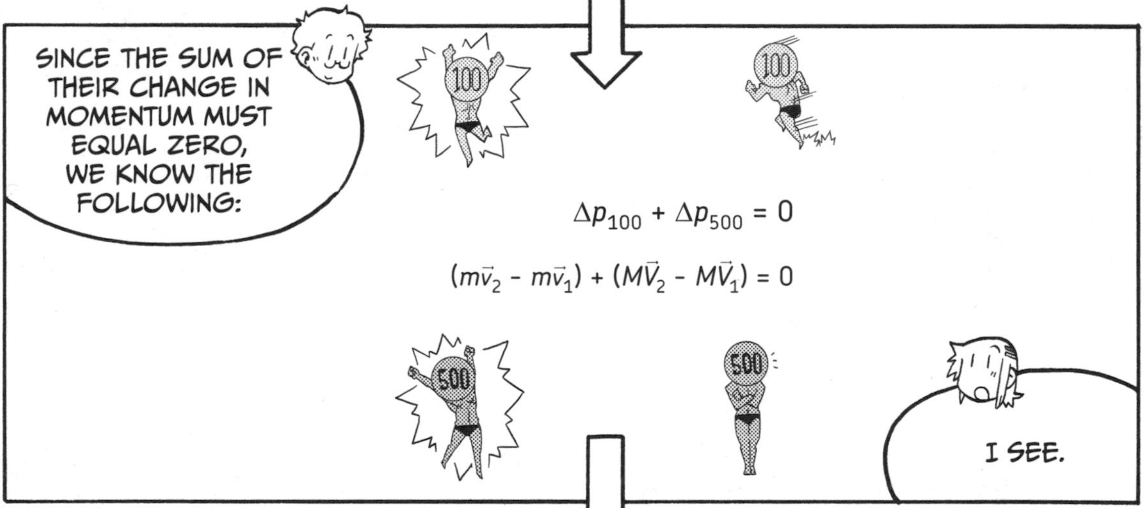
SINCE THE SUM OF THEIR CHANGE IN MOMENTUM MUST EQUAL ZERO, WE KNOW THE FOLLOWING:
100
100
$\Delta p_{100} + \Delta p_{500} = 0$
$(m\vec{v}_2 - m\vec{v}_1) + (M\vec{V}_2 - M\vec{V}_1) = 0$
500
500
I SEE.

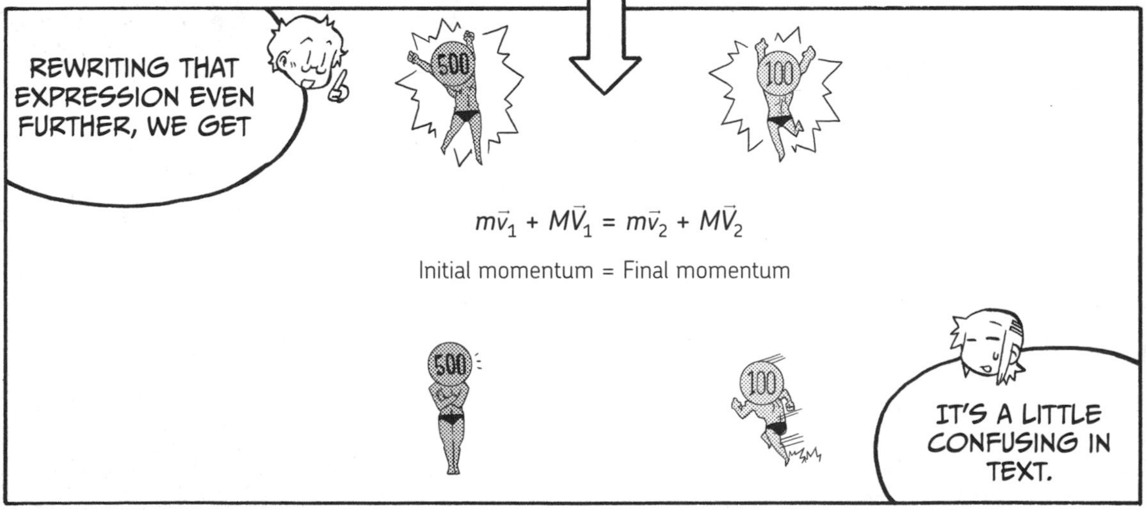
REWRITING THAT EXPRESSION EVEN FURTHER, WE GET
500
100
$m\vec{v}_1 + M\vec{V}_1 = m\vec{v}_2 + M\vec{V}_2$
Initial momentum = Final momentum
500
100
IT'S A LITTLE CONFUSING IN TEXT.

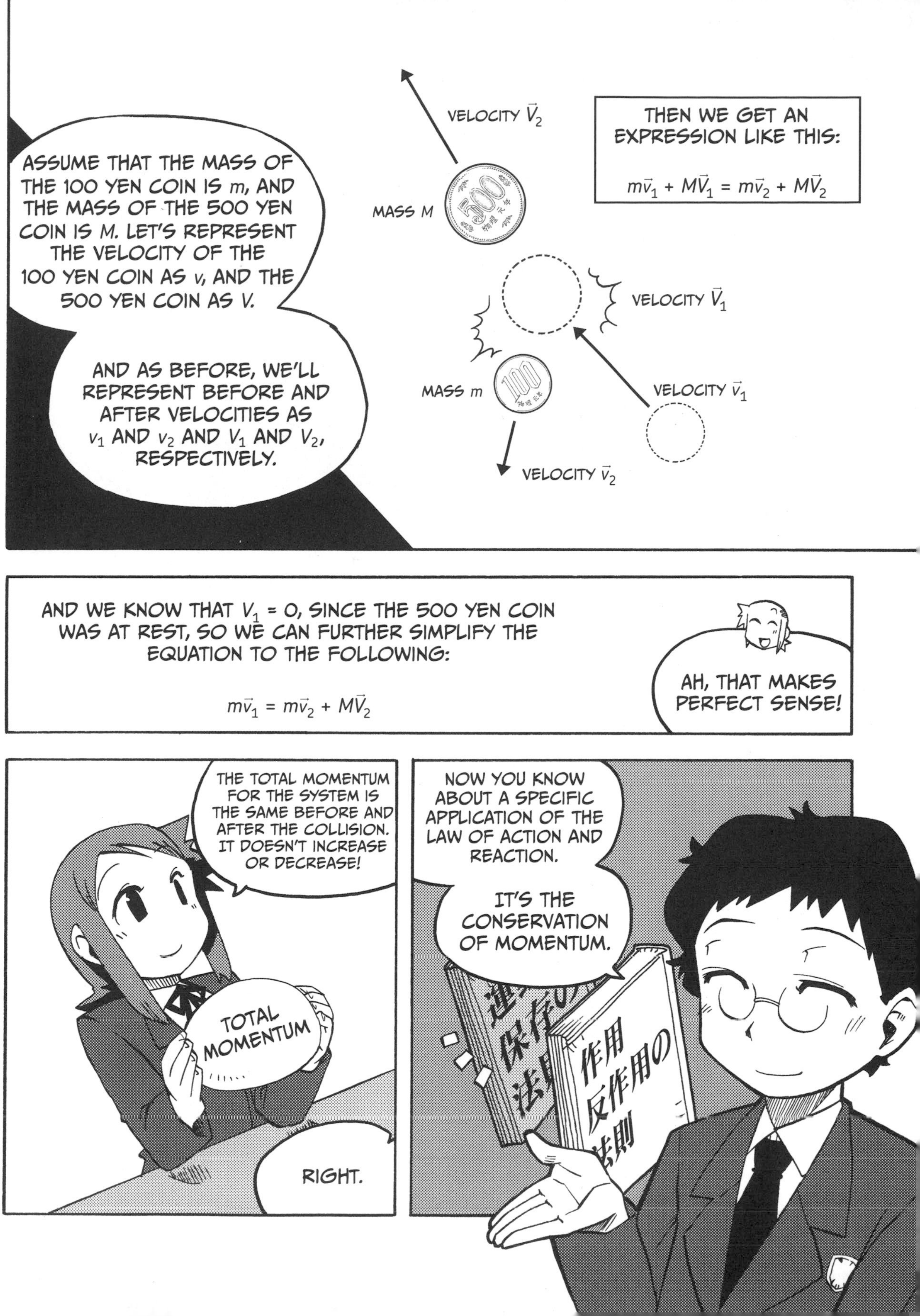

ASSUME THAT THE MASS OF THE 100 YEN COIN IS m, AND THE MASS OF THE 500 YEN COIN IS M. LET'S REPRESENT THE VELOCITY OF THE 100 YEN COIN AS v, AND THE 500 YEN COIN AS V.
AND AS BEFORE, WE'LL REPRESENT BEFORE AND AFTER VELOCITIES AS v_1 AND v_2 AND V_1 AND V_2, RESPECTIVELY.
VELOCITY $\vec{V}_2$
MASS M
VELOCITY $\vec{V}_1$
MASS m
VELOCITY $\vec{v}_1$
VELOCITY $\vec{v}_2$
THEN WE GET AN EXPRESSION LIKE THIS:
$m\vec{v}_1 + M\vec{V}_1 = m\vec{v}_2 + M\vec{V}_2$
AND WE KNOW THAT $V_1 = 0$, SINCE THE 500 YEN COIN WAS AT REST, SO WE CAN FURTHER SIMPLIFY THE EQUATION TO THE FOLLOWING:
$m\vec{v}_1 = m\vec{v}_2 + M\vec{V}_2$
AH, THAT MAKES PERFECT SENSE!
THE TOTAL MOMENTUM FOR THE SYSTEM IS THE SAME BEFORE AND AFTER THE COLLISION. IT DOESN'T INCREASE OR DECREASE!
TOTAL MOMENTUM
RIGHT.
NOW YOU KNOW ABOUT A SPECIFIC APPLICATION OF THE LAW OF ACTION AND REACTION.
IT'S THE CONSERVATION OF MOMENTUM.

LABORATORY

OUTER SPACE AND THE CONSERVATION OF MOMENTUM

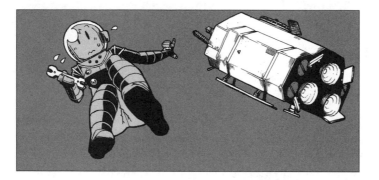

 Let's think about outer space for our next example of the conservation of momentum.

 What is this, space camp?

 Sigh. Let's just suppose you are an astronaut, Ninomiya-san. During vehicle repairs outside the space craft, your tether has become disconnected, leaving you floating away from your space shuttle. All you have in your hand is the wrench you've been using to repair your ship. How can you get back to your ship?

 Maybe I can swim back.

 Oh, ho ho ho, it's quite impossible to "swim" in a vacuum. Recall the first law of motion: An object at rest tends to stay at rest unless a force is imposed. No matter how hard you move your arms and legs, you won't have anything to push against. You'd just be rotating around your center of gravity, flailing your arms around.

 Oh no! Things are really looking bad!

Never give up hope! Your physics knowledge may save your life. You have that wrench, remember? Throw it in the direction opposite to the rocket. Thanks to the conservation of momentum, you will move.

Really? I'm gonna make it?

To confirm that this works, let's assume that you're at rest, in outer space. Then let's set the wrench's mass as *m* and assume you throw it away from you at velocity *v*. Your mass and subsequent velocity are represented by *M* and *V*.

Since we are starting with no momentum, the momentum of both objects afterward must equal zero, right?

Indeed! Given the law of conservation of momentum, the sum of the momentum of both bodies should equal zero. If we put that in an equation, it looks like this:

$$mv + MV = 0$$

To find *V*, or your velocity back to your ship, we rearrange the equation:

$$V = -\frac{m}{M} \times v$$

This value is negative because it indicates that your motion is in the opposite direction of the wrench.

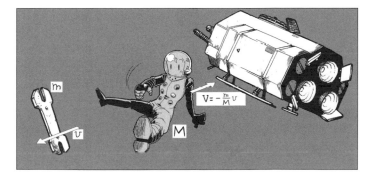

Can you see why you'd want to throw the wrench as hard as you could? The faster its *v*, the faster your *V*.

 Yes, that makes sense.

 Let's assign some numeric values and try to predict things. We'll say the wrench has a mass of 1 kg and give you a mass of 60 kg with that heavy space suit on. Assuming that the tool's velocity when thrown is 30 km/h, we get the following:

$$V = -\frac{1 \text{ kg}}{60 \text{ kg}} \times 30 \text{ km/h} = -0.5 \text{ km/h}$$

So that would be your velocity back to the ship.

 Let's say I have a whole toolbox. If I throw tools one after another, will I move faster?

 That's a great idea. Yes, you would go faster and faster that way. In fact, that's basically how a rocket moves. The exhaust that is belched out the rear of a rocket is equivalent to an object being thrown.

 Gee, I never thought of it that way.

 A rocket can continue to accelerate by belching exhaust continuously. As long as fuel continues to discharge, the rocket will accelerate. When the rocket stops discharging exhaust, the rocket's velocity becomes uniform.

REAL-WORLD EXPLORATIONS OF IMPULSE

REDUCING THE IMPACT

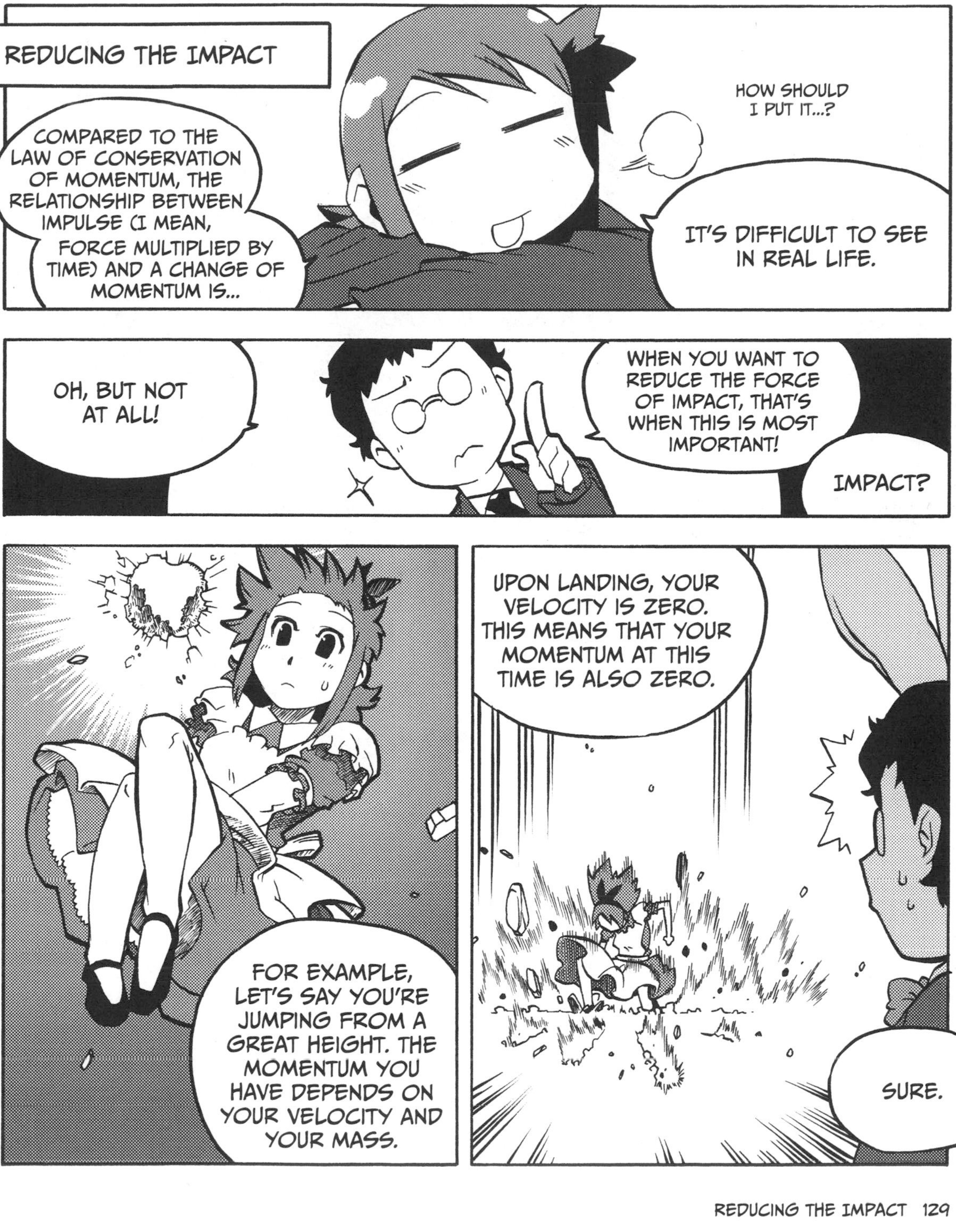
COMPARED TO THE LAW OF CONSERVATION OF MOMENTUM, THE RELATIONSHIP BETWEEN IMPULSE (I MEAN, FORCE MULTIPLIED BY TIME) AND A CHANGE OF MOMENTUM IS...
HOW SHOULD I PUT IT...?
IT'S DIFFICULT TO SEE IN REAL LIFE.
OH, BUT NOT AT ALL!
WHEN YOU WANT TO REDUCE THE FORCE OF IMPACT, THAT'S WHEN THIS IS MOST IMPORTANT!
IMPACT?
FOR EXAMPLE, LET'S SAY YOU'RE JUMPING FROM A GREAT HEIGHT. THE MOMENTUM YOU HAVE DEPENDS ON YOUR VELOCITY AND YOUR MASS.
UPON LANDING, YOUR VELOCITY IS ZERO. THIS MEANS THAT YOUR MOMENTUM AT THIS TIME IS ALSO ZERO.
SURE.

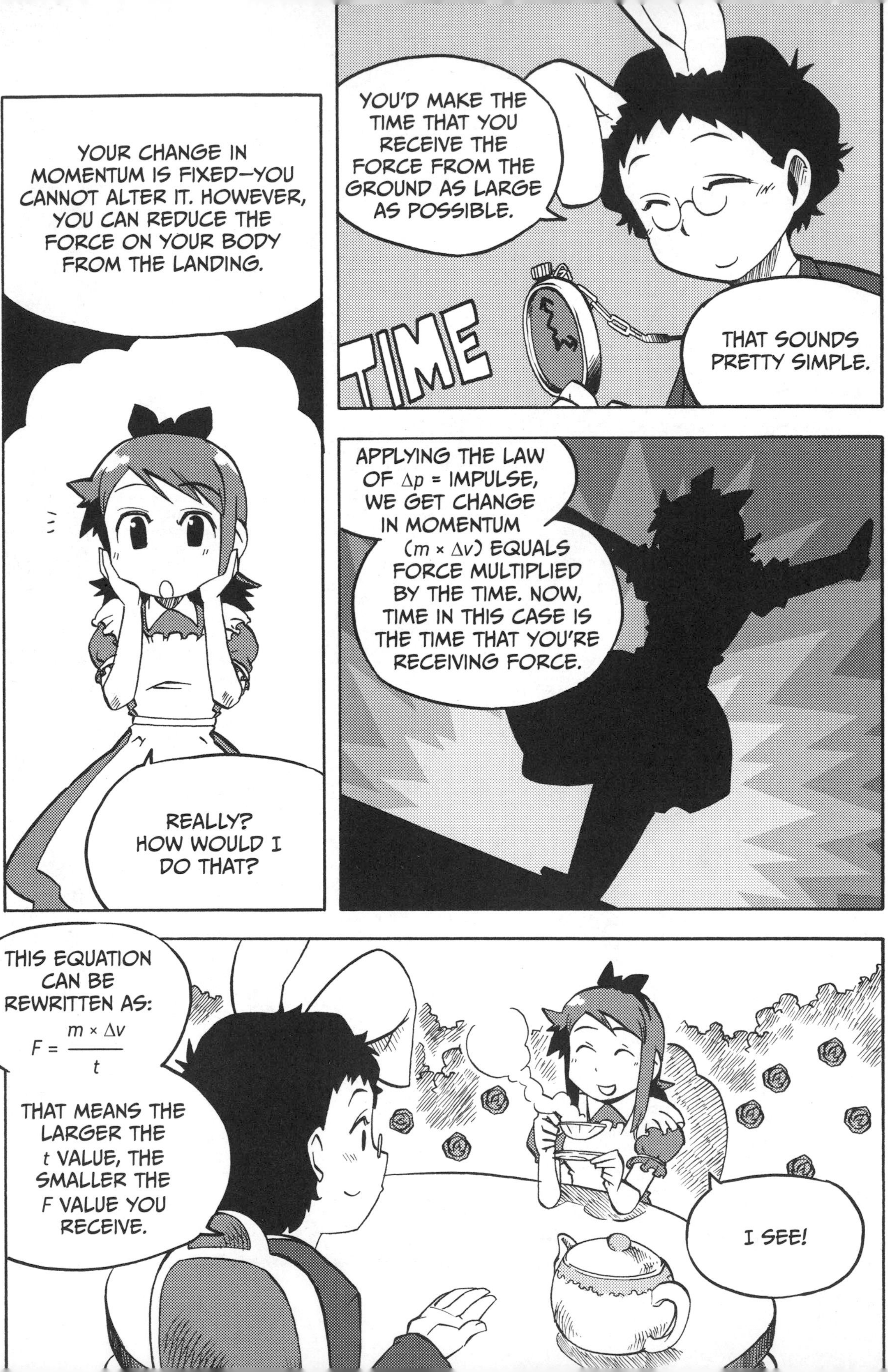

YOUR CHANGE IN MOMENTUM IS FIXED—YOU CANNOT ALTER IT. HOWEVER, YOU CAN REDUCE THE FORCE ON YOUR BODY FROM THE LANDING.
REALLY? HOW WOULD I DO THAT?
YOU'D MAKE THE TIME THAT YOU RECEIVE THE FORCE FROM THE GROUND AS LARGE AS POSSIBLE.
TIME
THAT SOUNDS PRETTY SIMPLE.
APPLYING THE LAW OF Δp = IMPULSE, WE GET CHANGE IN MOMENTUM ($m \times \Delta v$) EQUALS FORCE MULTIPLIED BY THE TIME. NOW, TIME IN THIS CASE IS THE TIME THAT YOU'RE RECEIVING FORCE.
THIS EQUATION CAN BE REWRITTEN AS: $F = \frac{m \times \Delta v}{t}$ THAT MEANS THE LARGER THE t VALUE, THE SMALLER THE F VALUE YOU RECEIVE.
I SEE!

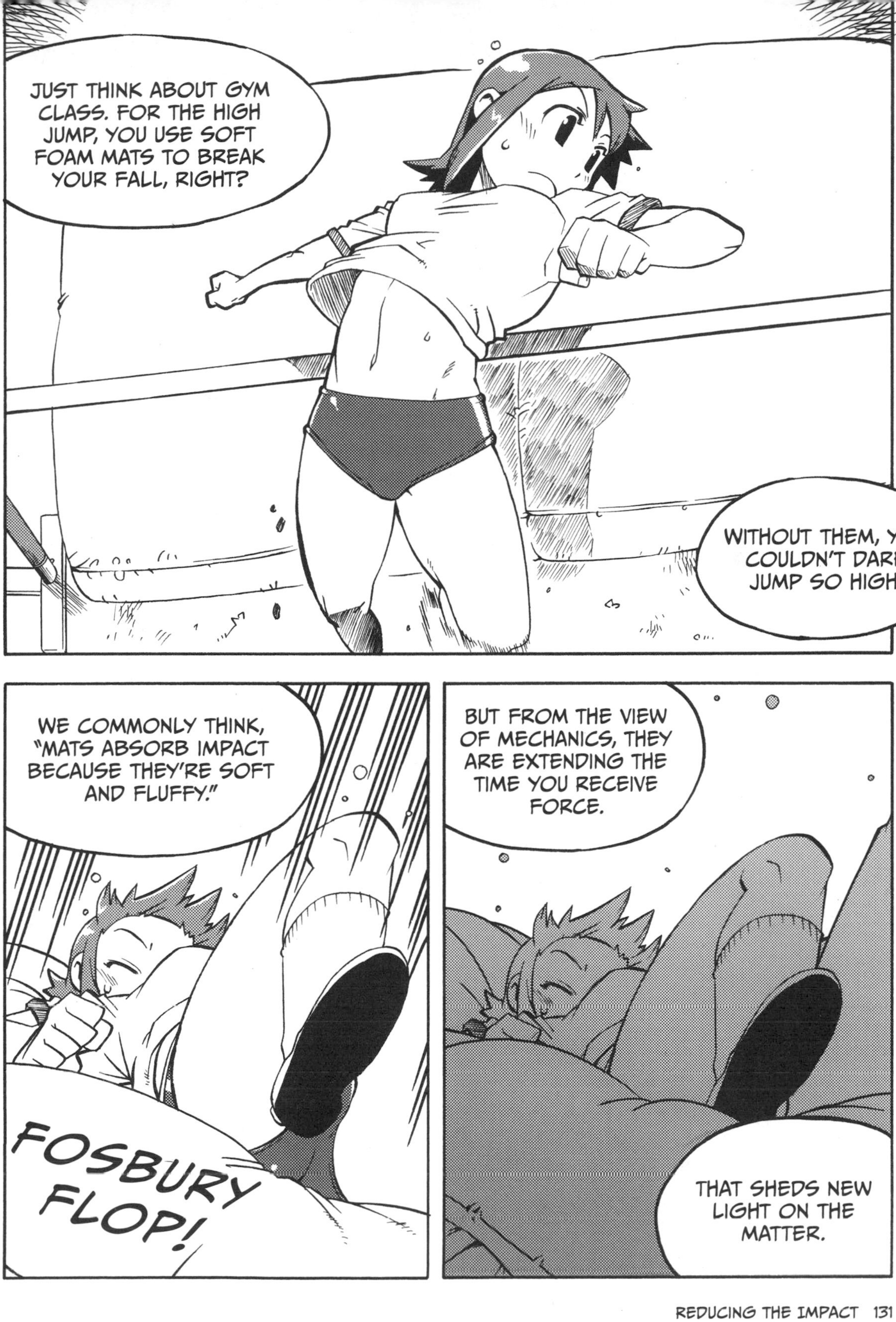
JUST THINK ABOUT GYM CLASS. FOR THE HIGH JUMP, YOU USE SOFT FOAM MATS TO BREAK YOUR FALL, RIGHT?
WITHOUT THEM, YOU COULDN'T DARE JUMP SO HIGH.
WE COMMONLY THINK, "MATS ABSORB IMPACT BECAUSE THEY'RE SOFT AND FLUFFY."
FOSBURY FLOP!
BUT FROM THE VIEW OF MECHANICS, THEY ARE EXTENDING THE TIME YOU RECEIVE FORCE.
THAT SHEDS NEW LIGHT ON THE MATTER.

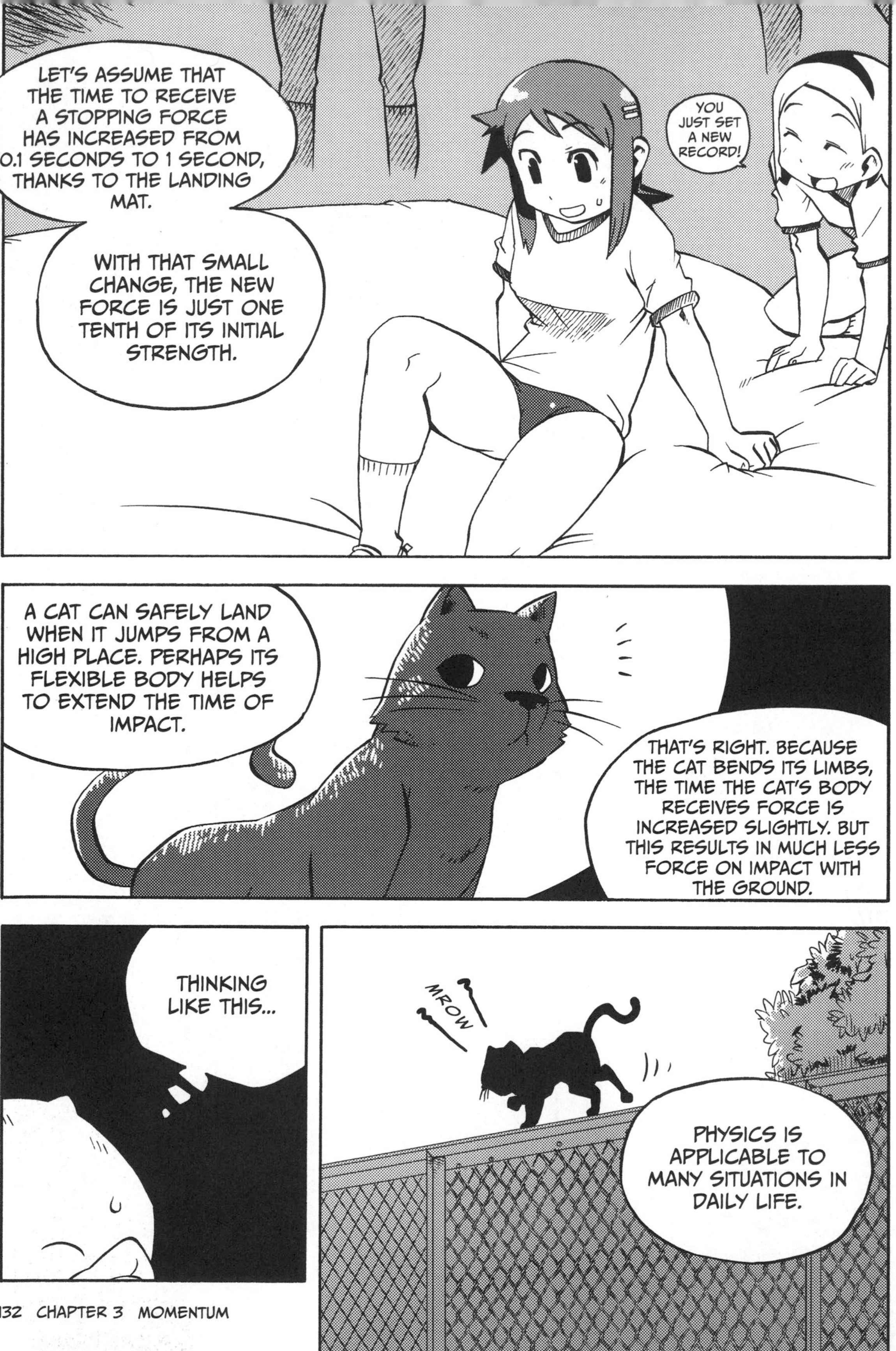

LET'S ASSUME THAT THE TIME TO RECEIVE A STOPPING FORCE HAS INCREASED FROM 0.1 SECONDS TO 1 SECOND, THANKS TO THE LANDING MAT.
WITH THAT SMALL CHANGE, THE NEW FORCE IS JUST ONE TENTH OF ITS INITIAL STRENGTH.
YOU JUST SET A NEW RECORD!
A CAT CAN SAFELY LAND WHEN IT JUMPS FROM A HIGH PLACE. PERHAPS ITS FLEXIBLE BODY HELPS TO EXTEND THE TIME OF IMPACT.
THAT'S RIGHT. BECAUSE THE CAT BENDS ITS LIMBS, THE TIME THE CAT'S BODY RECEIVES FORCE IS INCREASED SLIGHTLY. BUT THIS RESULTS IN MUCH LESS FORCE ON IMPACT WITH THE GROUND.
THINKING LIKE THIS...
MROW
PHYSICS IS APPLICABLE TO MANY SITUATIONS IN DAILY LIFE.

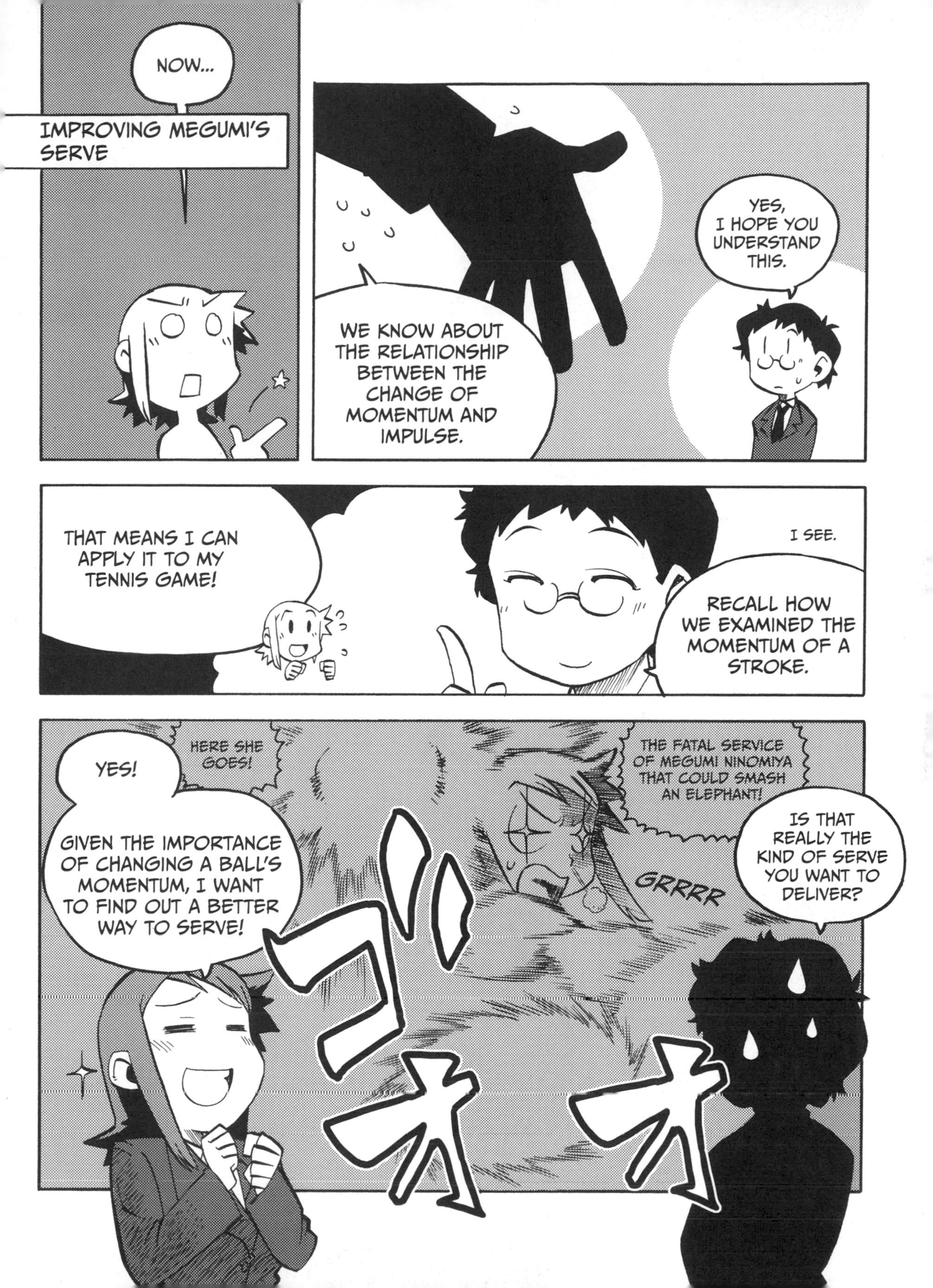
NOW...
IMPROVING MEGUMI'S SERVE
WE KNOW ABOUT THE RELATIONSHIP BETWEEN THE CHANGE OF MOMENTUM AND IMPULSE.
YES, I HOPE YOU UNDERSTAND THIS.
THAT MEANS I CAN APPLY IT TO MY TENNIS GAME!
I SEE.
RECALL HOW WE EXAMINED THE MOMENTUM OF A STROKE.
YES!
HERE SHE GOES!
GIVEN THE IMPORTANCE OF CHANGING A BALL'S MOMENTUM, I WANT TO FIND OUT A BETTER WAY TO SERVE!
THE FATAL SERVICE OF MEGUMI NINOMIYA THAT COULD SMASH AN ELEPHANT!
GRRRR
IS THAT REALLY THE KIND OF SERVE YOU WANT TO DELIVER?
ゴオオ

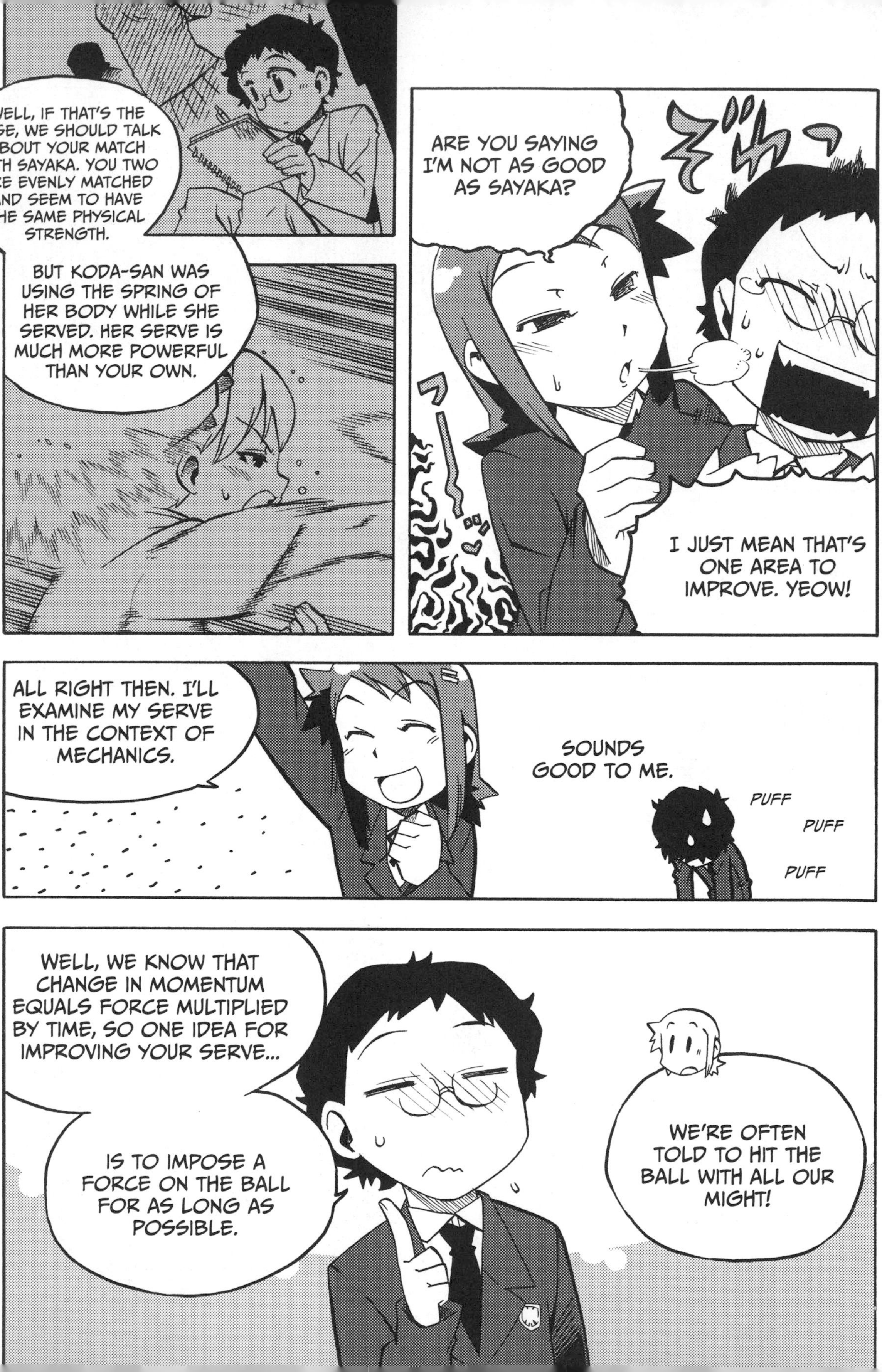
WELL, IF THAT'S THE CASE, WE SHOULD TALK ABOUT YOUR MATCH WITH SAYAKA. YOU TWO ARE EVENLY MATCHED AND SEEM TO HAVE THE SAME PHYSICAL STRENGTH.
BUT KODA-SAN WAS USING THE SPRING OF HER BODY WHILE SHE SERVED. HER SERVE IS MUCH MORE POWERFUL THAN YOUR OWN.
ARE YOU SAYING I'M NOT AS GOOD AS SAYAKA?
I JUST MEAN THAT'S ONE AREA TO IMPROVE. YEOW!
ALL RIGHT THEN. I'LL EXAMINE MY SERVE IN THE CONTEXT OF MECHANICS.
SOUNDS GOOD TO ME.
PUFF
PUFF
PUFF
WELL, WE KNOW THAT CHANGE IN MOMENTUM EQUALS FORCE MULTIPLIED BY TIME, SO ONE IDEA FOR IMPROVING YOUR SERVE...
IS TO IMPOSE A FORCE ON THE BALL FOR AS LONG AS POSSIBLE.
WE'RE OFTEN TOLD TO HIT THE BALL WITH ALL OUR MIGHT!

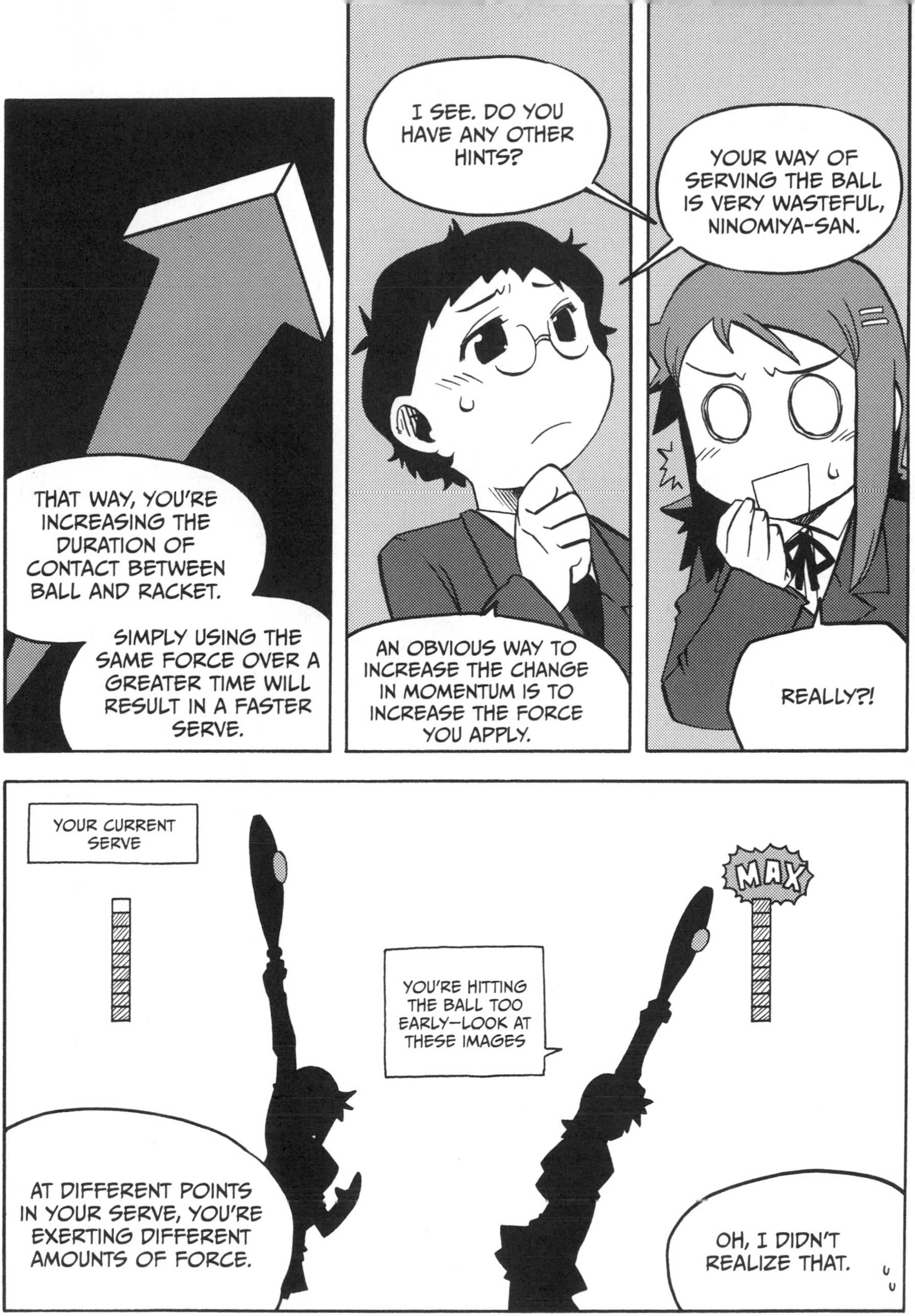
THAT WAY, YOU'RE INCREASING THE DURATION OF CONTACT BETWEEN BALL AND RACKET.
SIMPLY USING THE SAME FORCE OVER A GREATER TIME WILL RESULT IN A FASTER SERVE.
I SEE. DO YOU HAVE ANY OTHER HINTS?
AN OBVIOUS WAY TO INCREASE THE CHANGE IN MOMENTUM IS TO INCREASE THE FORCE YOU APPLY.
YOUR WAY OF SERVING THE BALL IS VERY WASTEFUL, NINOMIYA-SAN.
REALLY?!
YOUR CURRENT SERVE
MAX
YOU'RE HITTING THE BALL TOO EARLY—LOOK AT THESE IMAGES
AT DIFFERENT POINTS IN YOUR SERVE, YOU'RE EXERTING DIFFERENT AMOUNTS OF FORCE.
OH, I DIDN'T REALIZE THAT.

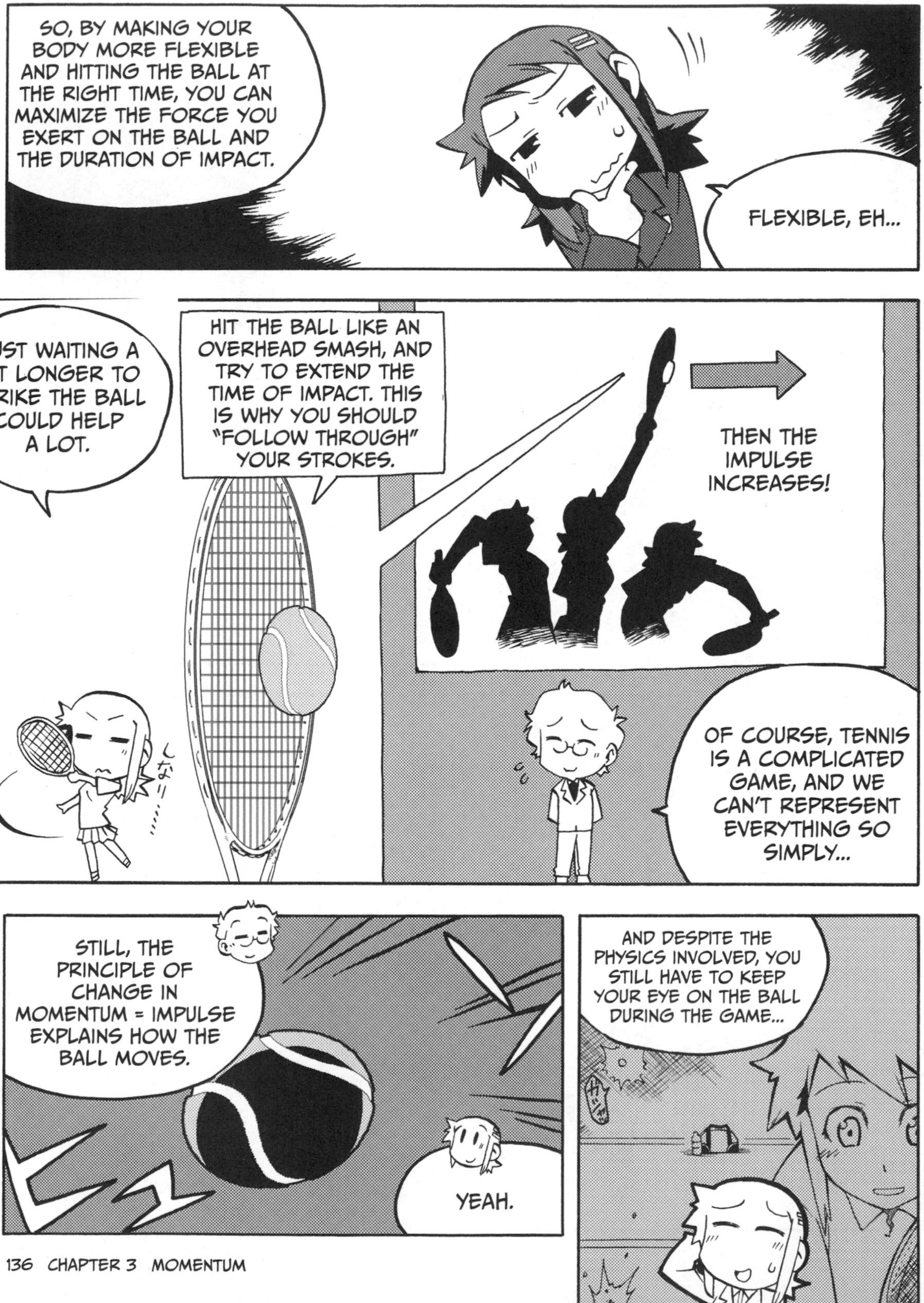
SO, BY MAKING YOUR BODY MORE FLEXIBLE AND HITTING THE BALL AT THE RIGHT TIME, YOU CAN MAXIMIZE THE FORCE YOU EXERT ON THE BALL AND THE DURATION OF IMPACT.
FLEXIBLE, EH...
JUST WAITING A BIT LONGER TO STRIKE THE BALL COULD HELP A LOT.
HIT THE BALL LIKE AN OVERHEAD SMASH, AND TRY TO EXTEND THE TIME OF IMPACT. THIS IS WHY YOU SHOULD "FOLLOW THROUGH" YOUR STROKES.
THEN THE IMPULSE INCREASES!
OF COURSE, TENNIS IS A COMPLICATED GAME, AND WE CAN'T REPRESENT EVERYTHING SO SIMPLY...
STILL, THE PRINCIPLE OF CHANGE IN MOMENTUM = IMPULSE EXPLAINS HOW THE BALL MOVES.
YEAH.
AND DESPITE THE PHYSICS INVOLVED, YOU STILL HAVE TO KEEP YOUR EYE ON THE BALL DURING THE GAME...

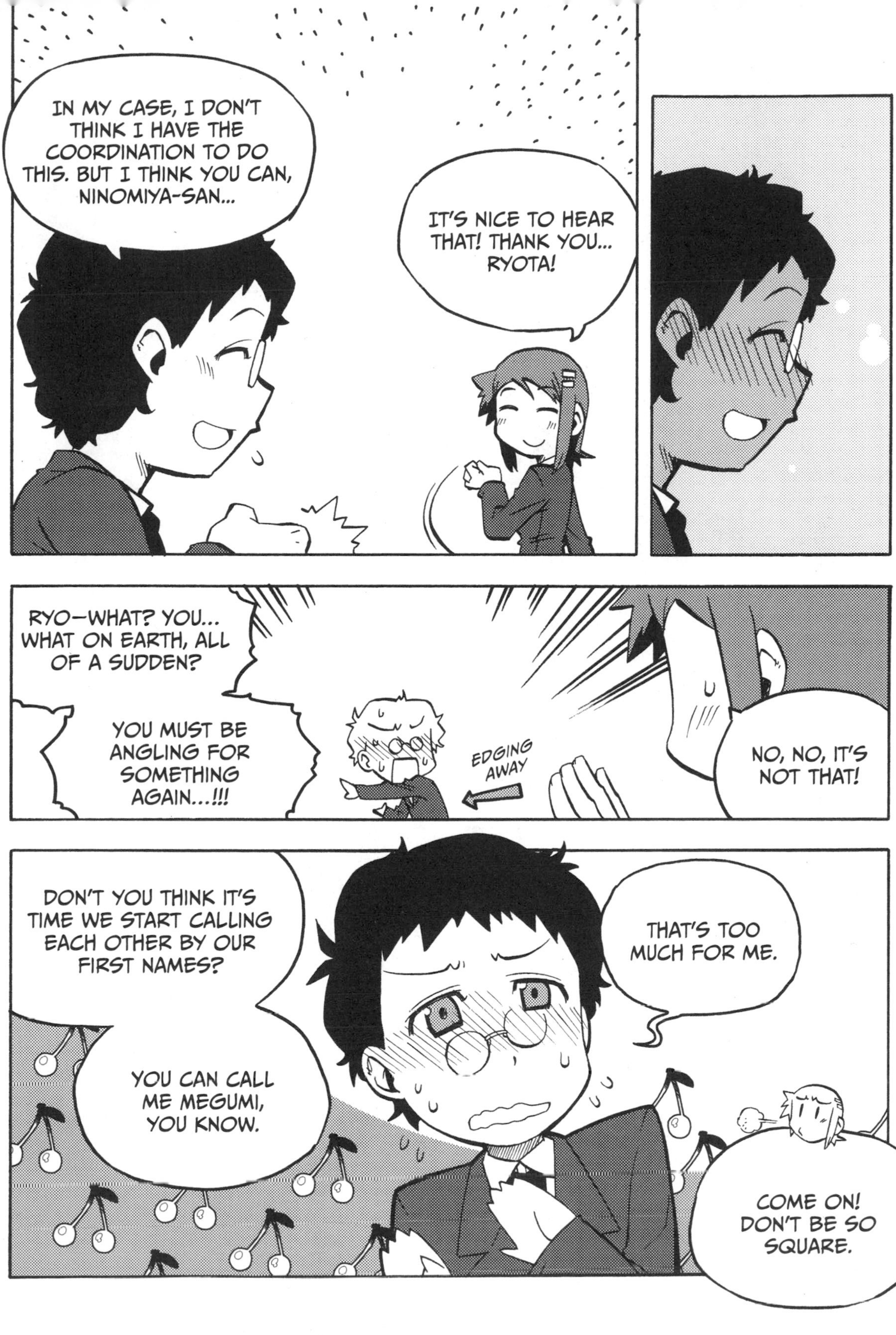
IN MY CASE, I DON'T THINK I HAVE THE COORDINATION TO DO THIS. BUT I THINK YOU CAN, NINOMIYA-SAN...
IT'S NICE TO HEAR THAT! THANK YOU... RYOTA!
RYO—WHAT? YOU... WHAT ON EARTH, ALL OF A SUDDEN?
YOU MUST BE ANGLING FOR SOMETHING AGAIN...!!!
EDGING AWAY
NO, NO, IT'S NOT THAT!
DON'T YOU THINK IT'S TIME WE START CALLING EACH OTHER BY OUR FIRST NAMES?
YOU CAN CALL ME MEGUMI, YOU KNOW.
THAT'S TOO MUCH FOR ME.
COME ON! DON'T BE SO SQUARE.

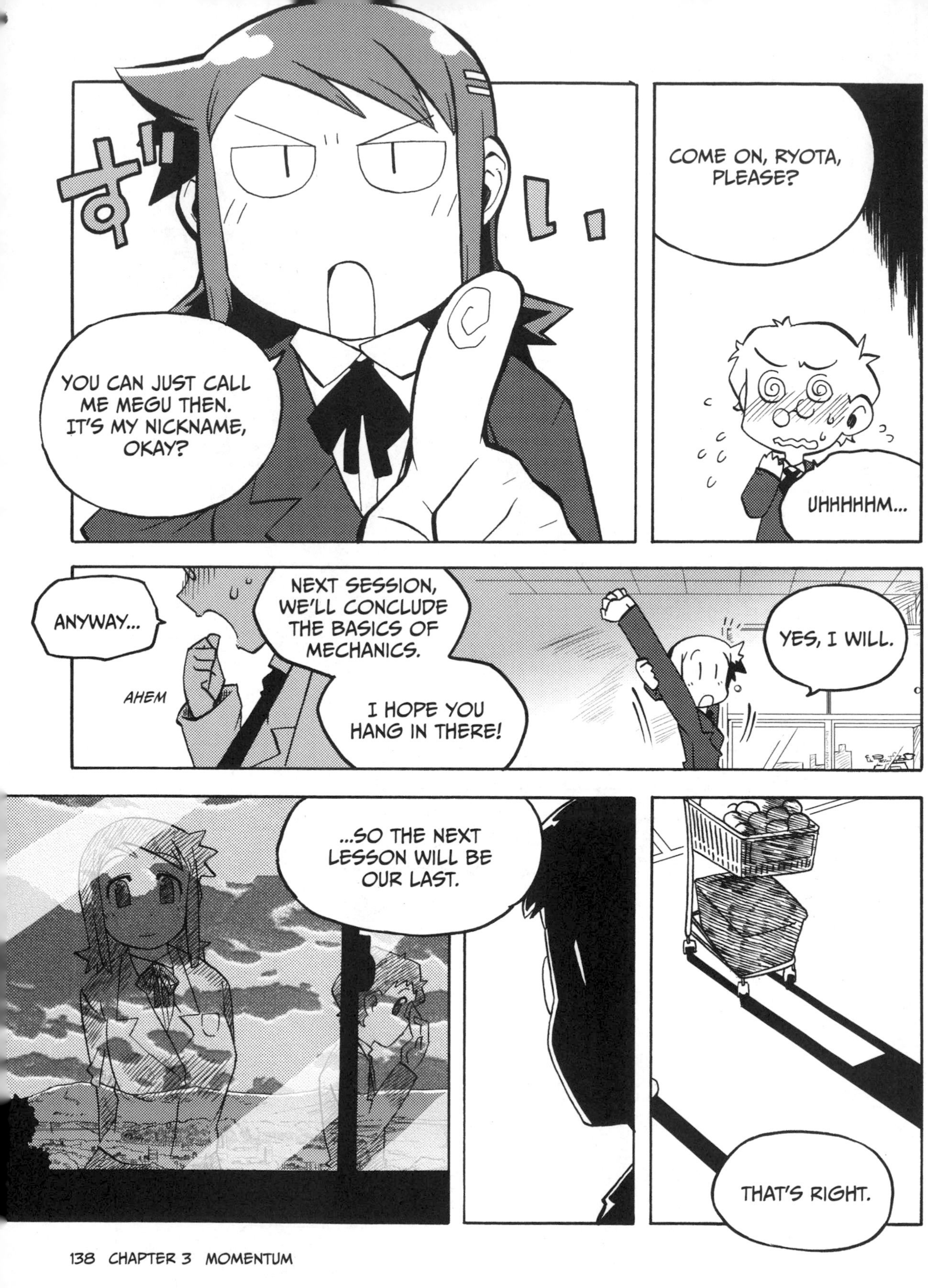
YOU CAN JUST CALL ME MEGU THEN. IT'S MY NICKNAME, OKAY?
COME ON, RYOTA, PLEASE?
UHHHHHM...
ANYWAY...
AHEM
NEXT SESSION, WE'LL CONCLUDE THE BASICS OF MECHANICS.
I HOPE YOU HANG IN THERE!
YES, I WILL.
...SO THE NEXT LESSON WILL BE OUR LAST.
THAT'S RIGHT.

MOMENTUM AND IMPULSE

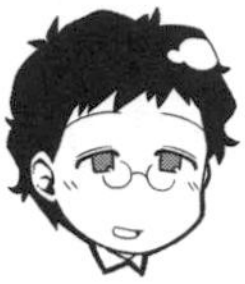

Momentum is a quantity representing the magnitude and orientation of the motion of an object. Assuming that an object with mass m and velocity $\vec{v}$ has momentum $\vec{p}$, the relationship among them can be represented as follows:

$$\vec{p} = m\vec{v}$$

Since velocity is a vector, momentum is also a vector. An object's momentum and velocity will have the same orientation.

As mentioned in Chapter 2, an object in motion does not have force inside of it—it has momentum. The momentum of an object varies as an external force is imposed on it, and a change in momentum is called *impulse*. So, let's derive the relationship between momentum and impulse, starting by examining Newton's second law.

Suppose a ball with mass m hits a racket. Assume $\vec{v}_1$ for the velocity of the ball before it hits the racket and $\vec{v}_2$ for the velocity of the ball after it hits the racket. Also assume $\vec{F}$ for the force imposed on the ball from the racket.

Given Newton's second law,

$$\vec{F} = m\vec{a}$$

the ball undergoes acceleration $\vec{a}$. Generally, force $\vec{F}$ is not constant, but for our purposes, let's assume that $\vec{F}$ is constant at its average value (see page 118). If $\vec{F}$ is assumed to be constant, then acceleration $\vec{a}$ is also constant. If we let t equal the time that the ball receives a force from the racket, acceleration $\vec{a}$ can be expressed as follows:

$$\vec{a} = \frac{(\vec{v}_2 - \vec{v}_1)}{t}$$

We can substitute this value for $\vec{a}$ into Newton's second law:

$$\vec{F} = m \times \frac{(\vec{v}_2 - \vec{v}_1)}{t}$$

If we multiply both sides by t, we get the following:

$$m\vec{v}_2 - m\vec{v}_1 = \vec{F}t$$

The expression $m\vec{v}_2 - m\vec{v}_1$ represents the object's change in momentum. When we call quantity $\vec{F}t$ the impulse, the following relationship is true:

change in momentum = impulse

Note that momentum $m\vec{v}_2 - m\vec{v}_1$ and impulse $\vec{F}t$ follow the rule of composition of vectors, as shown in the figure below.

$m\vec{v}_2$
$\vec{F}t$
$m\vec{v}_1$

$m\vec{v}_2$
$\vec{F}t$
$-m\vec{v}_1$

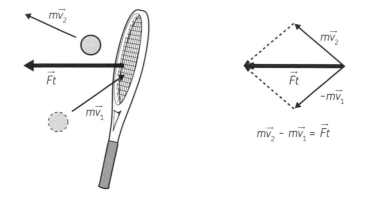

$$m\vec{v}_2 - m\vec{v}_1 = \vec{F}t$$

We can see from the way this equation is derived that the relational expression of change in momentum and impulse is an application of Newton's second law, in a situation where force is constant. When I said the equation for impulse is "nothing but another way of expressing Newton's second law" on page 115, this is what I meant.

IMPULSE AND MOMENTUM IN OUR LIVES

As we learned on page 129, the relationship between the change in momentum and impulse is useful when we want to determine how to reduce the impact of a collision.

In order to minimize the force imposed on an object while the object is in motion and up until the moment the motion stops, we must maximize the collision period because of the following relationship:

change in momentum of an object = imposed force × time duration of imposed force

Assume you are jumping from a high place, and your velocity immediately before landing is v. Once you land and are in a stationary state, the change in your momentum is mv. (How do we know this? Well, at rest, you must have no momentum at all, since you have no velocity: $m \times 0 = 0$.) This change in momentum is generated by the force from the ground, and your body must withstand this *impact force* that it receives. If we assume F for the impact force and t for the period of time the force is being withstood, the following expression is true:

$$mv = Ft$$

When mv is constant, F becomes smaller as t becomes larger. For example, the mats used for the high jump function as a tool for extending the time period t from the point where the body's impact on the mat starts to the point where momentum mv becomes zero. As the body sinks into the mat, the jumper continues to receive force F. As Ft is constant, the greater time period t, the smaller force F becomes.

We can find examples of the fact that a change in momentum equals impulse everywhere in our daily lives. When catching a ball, we tend to unconsciously withdraw our hand.

We are actually trying to reduce the force by extending the time duration from the point of the ball's contact with the hand to the point where the ball stops. Similarly, the gloves used in baseball and boxing extend the time period of the impact and reduce the force. Ukemi (the judo art of reacting to an attack by falling strategically), the crumple zones of modern cars, and air bags are all designed to reduce the impact of the force accompanying the change in momentum by extending the time of collision. Similarly, the safety ropes used in rock climbing are designed to stretch when a climber falls, so the collision time will be longer. This also prevents a sudden force from being imposed on the climber's waist. It would be very dangerous to use a rope that does not stretch instead of a special rope for climbing.

DERIVING THE LAW OF CONSERVATION OF MOMENTUM

Let's derive the law of conservation of momentum by applying our knowledge that the change in momentum equals impulse to two colliding objects.

After collision
$m\vec{v}_2$
$M\vec{V}_2$
$m\vec{v}_1$
$M\vec{V}_1$
Object A
Object B
Before collision

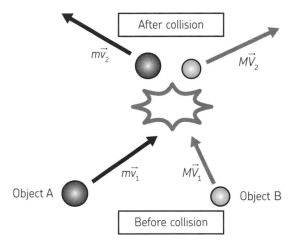

As in the preceding figure, assume that objects A and B collide without any external force being imposed and without the dissipation of any momentum in their impact.

First, let's focus on object A (the object on the left in the preceding figure). Assume m for the mass of object A and $\vec{v}_1$ and $\vec{v}_2$ for its velocity before and after collision. Also assume $\vec{F}$ for the force received by object A from object B. The relational expression showing that the change in momentum equals impulse can then be written as follows:

$$m\vec{v}_2 - m\vec{v}_1 = \vec{F}t$$

Here, t represents the time of the collision of objects A and B, and force approaches a constant value. Create an equation for object B (the object on the right in the preceding figure) using the knowledge that the change in momentum equals impulse. Assume M for the mass of object B, $\vec{V}_1$ and $\vec{V}_2$ for the velocity before and after collision, and $\vec{f}$ for the force received by object B from object A:

$$M\vec{V}_2 - M\vec{V}_1 = \vec{f}t$$

Note that the collision time is equal for both objects. It must be, since A cannot touch B without B touching A. But wait, why is the force the same for both as well? It's simply the law of action and reaction—$\vec{f} = \vec{F}$!

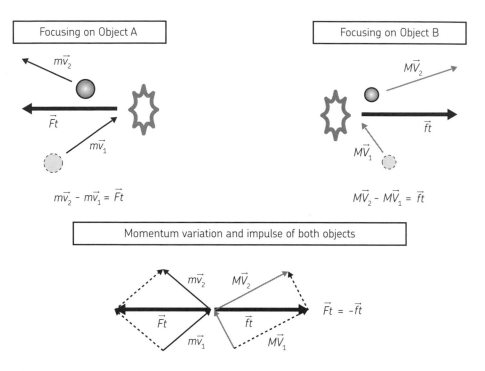

Focusing on Object A

$m\vec{v}_2$
$\vec{F}t$
$m\vec{v}_1$

$$m\vec{v}_2 - m\vec{v}_1 = \vec{F}t$$

Focusing on Object B

$M\vec{V}_2$
$\vec{f}t$
$M\vec{V}_1$

$$M\vec{V}_2 - M\vec{V}_1 = \vec{f}t$$

Momentum variation and impulse of both objects

$m\vec{v}_2$
$M\vec{V}_2$
$\vec{F}t$
$\vec{f}t$
$m\vec{v}_1$
$M\vec{V}_1$

$$\vec{F}t = -\vec{f}t$$

Substitute the previous two expressions representing the relationship between the change in momentum and impulse into the preceding expression to get the following:

$$M\vec{V}_2 - M\vec{V}_1 = -(m\vec{v}_2 - m\vec{v}_1)$$

Consolidate this expression:

$$m\vec{v}_2 + M\vec{V}_2 = m\vec{v}_1 + M\vec{V}_1$$

The momentum of the objects before impact must be equal to their momentum afterward. This is the law of conservation of momentum shown on page 125.*

* We can omit the vector signs in the case of a collision between objects moving on the same straight line.

$m\vec{v}_2$
$M\vec{V}_2$
$\vec{Ft}$
$\vec{ft}$
$m\vec{v}_1$
$M\vec{V}_1$
$M\vec{V}_2$
$M\vec{V}_1$
$m\vec{v}_2$
$m\vec{v}_1$

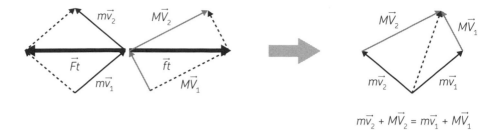

$$m\vec{v}_2 + M\vec{V}_2 = m\vec{v}_1 + M\vec{V}_1$$

These equations can be represented as vectors, as shown in the preceding figure on the left. The vectors can be rearranged to combine the change in momentum and impulse for objects 1 and 2, respectively, with the law of action and reaction, to get the vector on the right.

ELASTIC AND INELASTIC COLLISION

It's important to note that problems involving collision cannot always be solved using the law of conservation of momentum alone. In the real world, we must consider the dissipation of kinetic energy and other factors. We'll learn more about kinetic energy in the next chapter.

However, we can easily apply the law of conservation of momentum in two ideal situations—a perfectly *elastic* collision, or a perfectly *inelastic* one. The first example here was a perfectly elastic one—two objects that move separately after their collision, losing no energy in the process. Think of an elastic collision as something like two super-balls hitting each other—in the real world, the collision of atoms is said to be elastic. Now let's take a look at an example to better understand what an inelastic collision is.

An *inelastic collision* is one where the colliding objects combine to form a singular object in motion after their collision. An example of this would be a tackle in football, where after striking each other, the two players travel together as one.

Object A
v
m
Object B
V_1
M
Combined mass of A + B
V_2
$m + M$

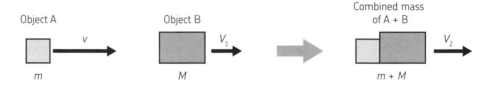

In this example, assume that object A with mass m and velocity v is combined with object B with mass M and velocity V_1. At this time, we get the following equation:

$$p = (m + M)V_2$$

The two objects achieve velocity V_2 after they are combined. Applying the law of conservation of momentum, we get the following equation:

$$mv + MV_1 = (m + M)V_2$$

Therefore, the velocity after the two objects are combined is as follows:

$$\frac{mv + MV_1}{m + M} = V_2$$

UNITS FOR MOMENTUM

Let's think about the unit we use to measure momentum. Recall that force is represented in newtons (N), but momentum doesn't use a special unit of measure. But from the equation momentum = mass × velocity, you can tell that:

units for momentum = units for mass × units for velocity
= (kg) × (m/s) = (kg × m/s)

You can also use the fact that a change in momentum equals impulse to determine the units for momentum. The units for momentum are the same as the units for impulse. Therefore, the following expression is also true:

units for momentum = units for impulse = units for force × units for time
= (N) × (s) = (N × s)

This seems different from the units we just calculated, (kg × m/s). However, given (N) = (kg × m/s^2), you get:

(kg × m/s^2) × (s) = (kg × m/s)

Both units are identical. We've learned that the units for momentum are (kg × m/s), or (N × s).

LAW OF CONSERVATION OF MOMENTUM FOR VECTORS

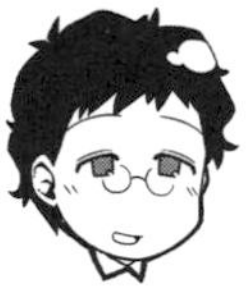

Since momentum is a vector, to follow the law of conservation of momentum, we must consider the orientation of momentum as well. In other words, when momentum is conserved, we must conserve both its orientation and its magnitude. Therefore, if the orientation of momentum changes (as in the example of a collision of coins on page 121), you need to calculate this change by dividing momentum into separate horizontal and vertical components, as vectors.

Assume a perfectly elastic collision in which object A collides with stationary object B, as shown in the following figure.

Assume m for the mass of object A, v_1 and v_2 for its velocity before and after the collision, M for the mass of object B, and V for its velocity after the collision. Place the x-axis on the vector representing the velocity of object A before the collision, assume θ and φ for the angles made by object A and object B after the collision, respectively, and assume $v_1 = |v_1|$, $v_2 = |v_2|$, $V_2 = |V_2|$.

We'll then split the velocities into their constituent parts, like so, in the form of $v = (v_x, v_y)$:

$$v = (v_1, 0),\ v_2 = (v_2 \cos \theta,\ v_2 \sin \theta),\ V_1 = (V \cos \varphi,\ -V \sin \varphi)$$

Now that we've done that, consider that the law of momentum must hold true in both the x and y directions. Note that the object initially has *no* momentum in the y direction. So that means that the following must be true:

For the x direction: $mv_1 = mv_2 \cos \theta + MV \cos \varphi$

For the y direction: $0 = mv_2 \sin \theta - MV \sin \varphi$

When a 500 yen coin collides with a 100 yen coin, the 100 yen coin often bounces backward. In this case, $\theta > 90°$, so $\cos \theta < 0$. The following figure shows an example where $\theta < 90°$.

m
$m\vec{v}_1$
M
(x-axis)
COLLISION!
Object A
Object B
(at rest)
m
Object A
$m\vec{v}_2$
θ
(x-axis)
φ
$M\vec{V}_2$
M
Object B

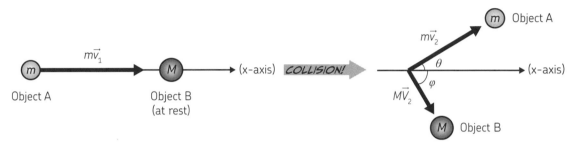

Let's see how we split the objects' momentum into horizontal and vertical parts.

Object A
m
$m\vec{v}_2$
$m\vec{v}_2 \sin \theta$
θ
$m\vec{v}_2 \cos \theta$
$M\vec{V}_2 \cos \varphi$
φ
$M\vec{V}_2$
$M\vec{V}_2 \sin \varphi$
M
Object B

$m\vec{v}_2$
$M\vec{V}_2$
θ
φ
$m\vec{v}_1$

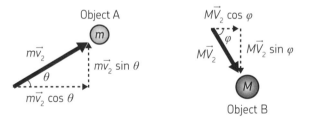

If we place these head-to-tail, we can visually see what we already know: Momentum has been conserved in the system.

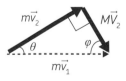

In other words, in the y direction, the momentum of objects 1 and 2 must offset each other. And the sum of their momentum in the x directions must equal $m\vec{v}_1$.

We need to know more than the law of conservation of momentum to predict the velocity and the angle at which the objects move after the collision. We'll look at this in more detail in the next chapter.

LAW OF ACTION AND REACTION VS. LAW OF CONSERVATION OF MOMENTUM

WARNING: CALCULUS AHEAD!

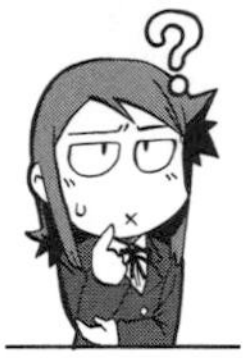
?

Using differential and integral calculus, we can easily derive the law of conservation of momentum. Assume v and m for the velocity and mass of object 1 and V and M for those of object 2. Suppose no external force is working on these objects. Assuming $\vec{F}_{m \to M}$ for the force imposed on object 2 by object 1 and $\vec{F}_{M \to m}$ for the force imposed on object 1 by object 2, we can apply Newton's second law as follows:

$$m\frac{d\vec{v}}{dt} = \vec{F}_{M \to m} \quad \text{and} \quad M\frac{d\vec{V}}{dt} = \vec{F}_{m \to M}$$

Substitute these two equations into the following equation for the law of action and reaction:

$$\vec{F}_{M \to m} = -\vec{F}_{m \to M}$$

The following will result:

$$m\frac{d\vec{v}}{dt} = -M\frac{d\vec{V}}{dt}$$

As mass is a constant, the above expression can be transformed into the following:

$$\frac{d(m\vec{v})}{dt} = -\frac{d(M\vec{V})}{dt}$$

Consolidate these two equations:

$$\frac{d}{dt}(m\vec{v} + M\vec{V}) = 0$$

This equation indicates that the sum of the momentum of objects 1 and 2 ($m\vec{v} + M\vec{V}$) will not change over time. From this equation, you can derive the law of conservation of momentum:

$$m\vec{v} + M\vec{V} = \text{constant}$$

A derivative of zero means that the momentum does not change! The law of conservation of momentum is derived from both the law of action and reaction and Newton's second law. So you can also say that the law of conservation of momentum stems from the law of action and reaction.

You can use the same method to derive the law of conservation of momentum for three or more objects.

PROPULSION OF A ROCKET

In the Laboratory section on page 126, we learned that an astronaut in space will move in the opposite direction of an item he has thrown. This phenomenon occurs according to the same principles that drive rocket propulsion. A rocket increases its velocity by belching exhaust at a high rate out of its engine, and it moves in the opposite direction of its exhaust. Let's look into this phenomenon in depth.

First, assume that a stationary rocket in outer space discharges a small object with mass m at a velocity of v. Then assume M for the sum of mass of the small object and the rocket and V_1 for the velocity of the rocket after the exhaust discharge. Given the law of conservation of momentum (and knowing that these velocities are in the exact opposite direction), you get the following equation:

$$0 = (M - m)V_1 - mv$$

❶ $\quad V_1 = \dfrac{mv}{M - m}$

We've solved for the rocket's subsequent motion, V_1. Now, suppose this rocket discharges another object of mass m at *relative velocity* (velocity as viewed from the rocket) $-v$ and in the same direction as the previous discharge. At this time, assuming V_2 for the rocket's velocity and noting that the total mass of the rocket before and after discharging the second object is $M - m$ and $M - 2m$, respectively, you get the following equation:

$$(M - m)V_1 = (M - 2m)V_2 + m(V_1 - v)$$

Note that the small object moves at a velocity of $V_1 - v$ when the rocket is advancing at velocity V_1. From the expression above, you can find the value of V_2 as follows:

❷ $\quad V_2 = V_1 + \dfrac{mv}{M - 2m}$

Substituting the value of V_2 from that equation into this equation, we find the following:

$$V_2 = \dfrac{mv}{M - m} + \dfrac{mv}{M - 2m}$$

❸ $\quad V_2 = mv\left(\dfrac{1}{M - m} + \dfrac{1}{M - 2m}\right)$

We've found the velocity of the rocket after discharging two small objects.

A real rocket will continue to discharge small objects—so let's derive a general expression for the rocket's velocity after discharging n small objects. Let's assume that the rocket continues to discharge small objects with mass m at relative velocity v.

V_{n-1}
$M - (n - 1)m$
$V_{n-1} - v$
m
V_n
$M - nm$

As viewed from the rocket with velocity V_{n-1}, the small object is discharged at velocity $-v$ to the rear of the rocket.

Assuming V_n for the velocity of the rocket when it discharges n small objects, the law of conservation of momentum is expressed as follows:

$$[M - (n - 1)m]\, V_{n-1} = (M - nm)V_n + m(V_{n-1} - v)$$

Thus, V_n is expressed as follows:

$$V_n = V_{n-1} + \frac{m}{M - nm} v$$

By using this expression repeatedly, you can find the following:

❹ $$V_n = \left(\frac{1}{M - m} + \ldots + \frac{1}{M - nm}\right) mv = \sum_{k=1}^{n} \frac{m}{M - km} v$$

A real rocket continuously discharges exhaust from its rear engines, so we will transform expression ❹ for such a case. Assume that the rocket emits a jet of small mass Δm at one-minute intervals Δt at relative velocity $-v$. Assuming t for the time from the stationary state to the nth jet exhaust, $t = n\Delta t$ is true. Assume $V(t)$ is a function describing the rocket's velocity with respect to time, and transform expression ❹ into $m \to \Delta m$, $V_n = \to V(t)$ to find the following:

WARNING: CALCULUS AHEAD!

❺ $$V(t) = \sum_{k=1}^{n} \frac{\Delta m}{M - (\Delta m / \Delta t)(k\Delta t)} v$$

At a point where the jet interval Δt is divided into infinitely small sections—that is, when $\Delta t \to 0$, you can find the sum using integral calculus.* To work with integral calculus, note the following transformations: n becomes ∞ and $\Delta m / \Delta t$ becomes dm / dt (mass lost in

* The expression $\Delta t \to 0$ can be read aloud as "the change in time approaches zero."

unit time, or mass that is discharged in the form of fuel exhaust). Transform the equation as follows: $\Delta m \rightarrow (dm / dt)\, dt$. The following equation will result:

$$\textbf{\textcircled{6}} \quad V(t) = v \int_0^t \frac{1}{M - (dm / dt)\, t} \left(\frac{dm}{dt}\right) dt$$

$$= v \int_0^t \frac{1}{M\, (dm / dt)^{-1} - t}\, dt$$

If exhaust discharge in unit time is uniform, the following is true:

$$dm / dt = \alpha \qquad \text{(a constant value)}$$

This means that alpha (α) is a measure of how much mass the engine is discharging per unit time:

$$\textbf{\textcircled{7}} \quad V(t) = v \int_0^t \frac{1}{(M / \alpha) - t}\, dt = v \left[-\log_e (M / \alpha - t)\right]_0^t$$

$$= v \log_e \left(\frac{M}{M - \alpha t}\right)$$

Expression $\textbf{\textcircled{7}}$ represents the velocity of a rocket with initial velocity $V(0) = 0$. Note that αt is the total mass of the exhaust emitted by the rocket in time interval t. Therefore, assuming that the initial total mass of fuel carried in the rocket is m_0, the rocket consumes all the fuel in time t $(t = m_0 / \alpha)$ and then shifts to uniform motion from accelerated motion (as shown in the following figure).

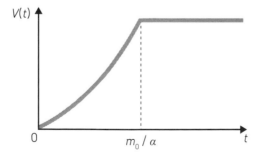

4

ENERGY

WORK AND ENERGY

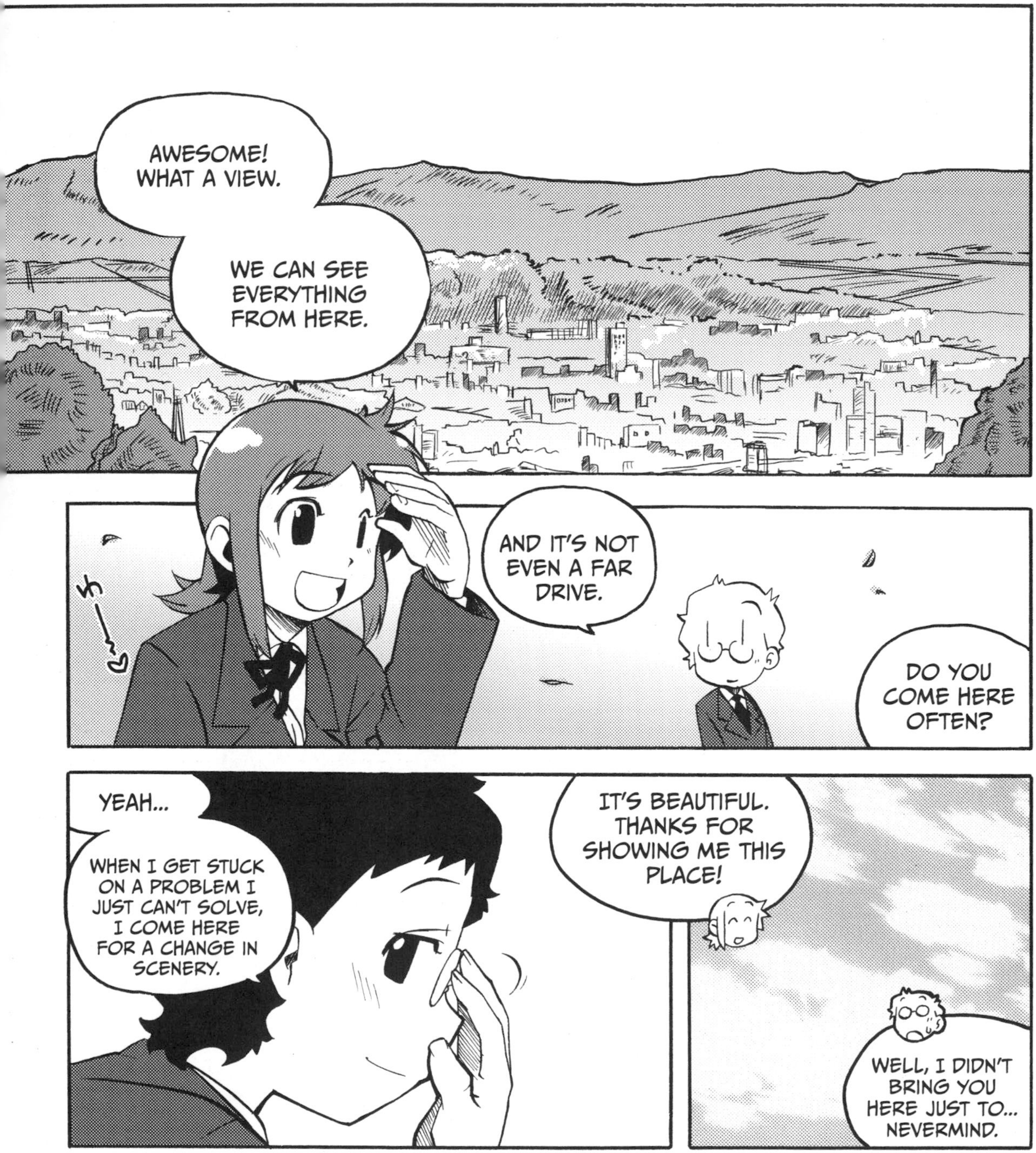

AWESOME! WHAT A VIEW.
WE CAN SEE EVERYTHING FROM HERE.
AND IT'S NOT EVEN A FAR DRIVE.
DO YOU COME HERE OFTEN?
YEAH...
WHEN I GET STUCK ON A PROBLEM I JUST CAN'T SOLVE, I COME HERE FOR A CHANGE IN SCENERY.
IT'S BEAUTIFUL. THANKS FOR SHOWING ME THIS PLACE!
WELL, I DIDN'T BRING YOU HERE JUST TO... NEVERMIND.

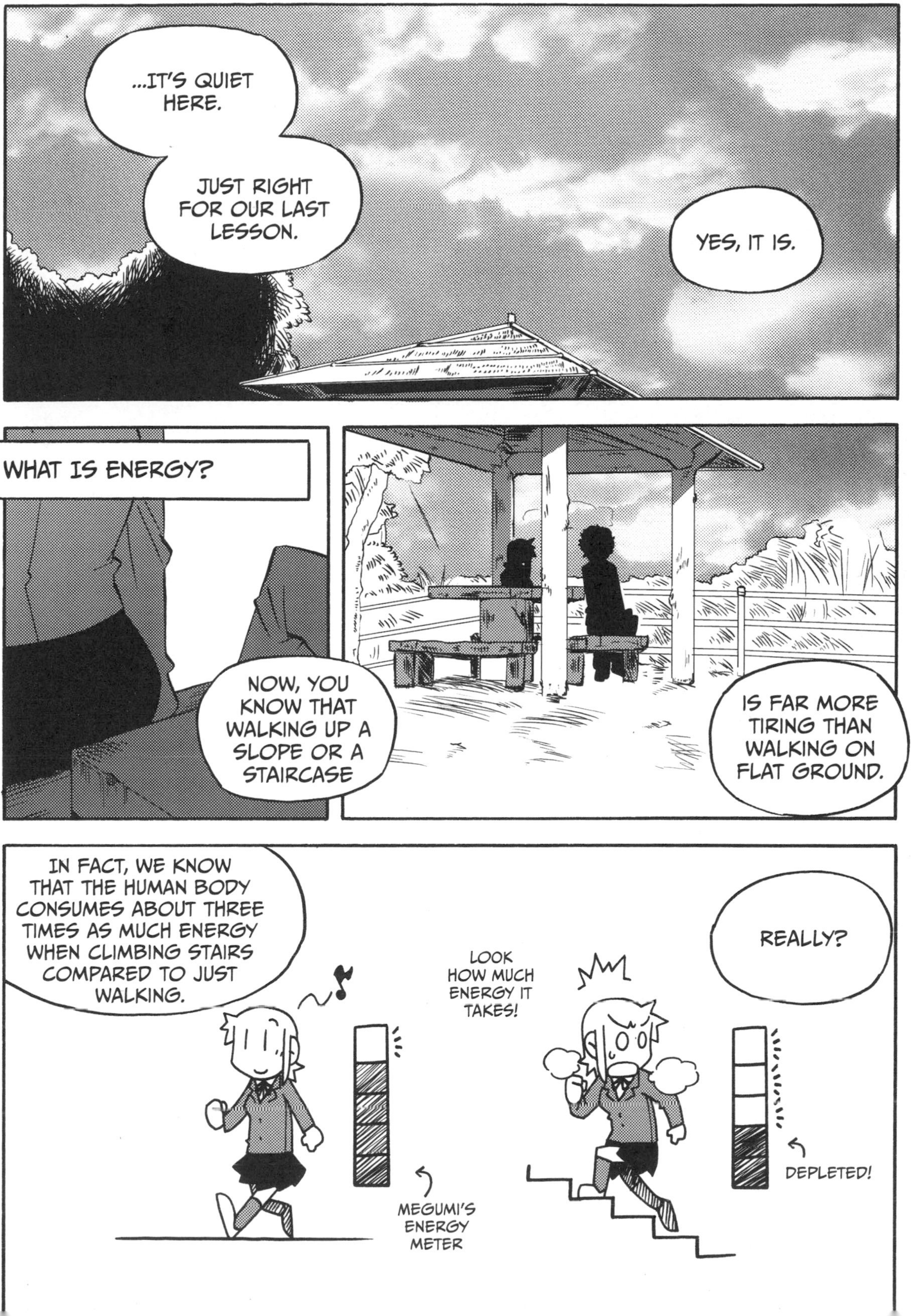

...IT'S QUIET HERE.
JUST RIGHT FOR OUR LAST LESSON.
YES, IT IS.
WHAT IS ENERGY?
NOW, YOU KNOW THAT WALKING UP A SLOPE OR A STAIRCASE
IS FAR MORE TIRING THAN WALKING ON FLAT GROUND.
IN FACT, WE KNOW THAT THE HUMAN BODY CONSUMES ABOUT THREE TIMES AS MUCH ENERGY WHEN CLIMBING STAIRS COMPARED TO JUST WALKING.
MEGUMI'S ENERGY METER
LOOK HOW MUCH ENERGY IT TAKES!
REALLY?
DEPLETED!

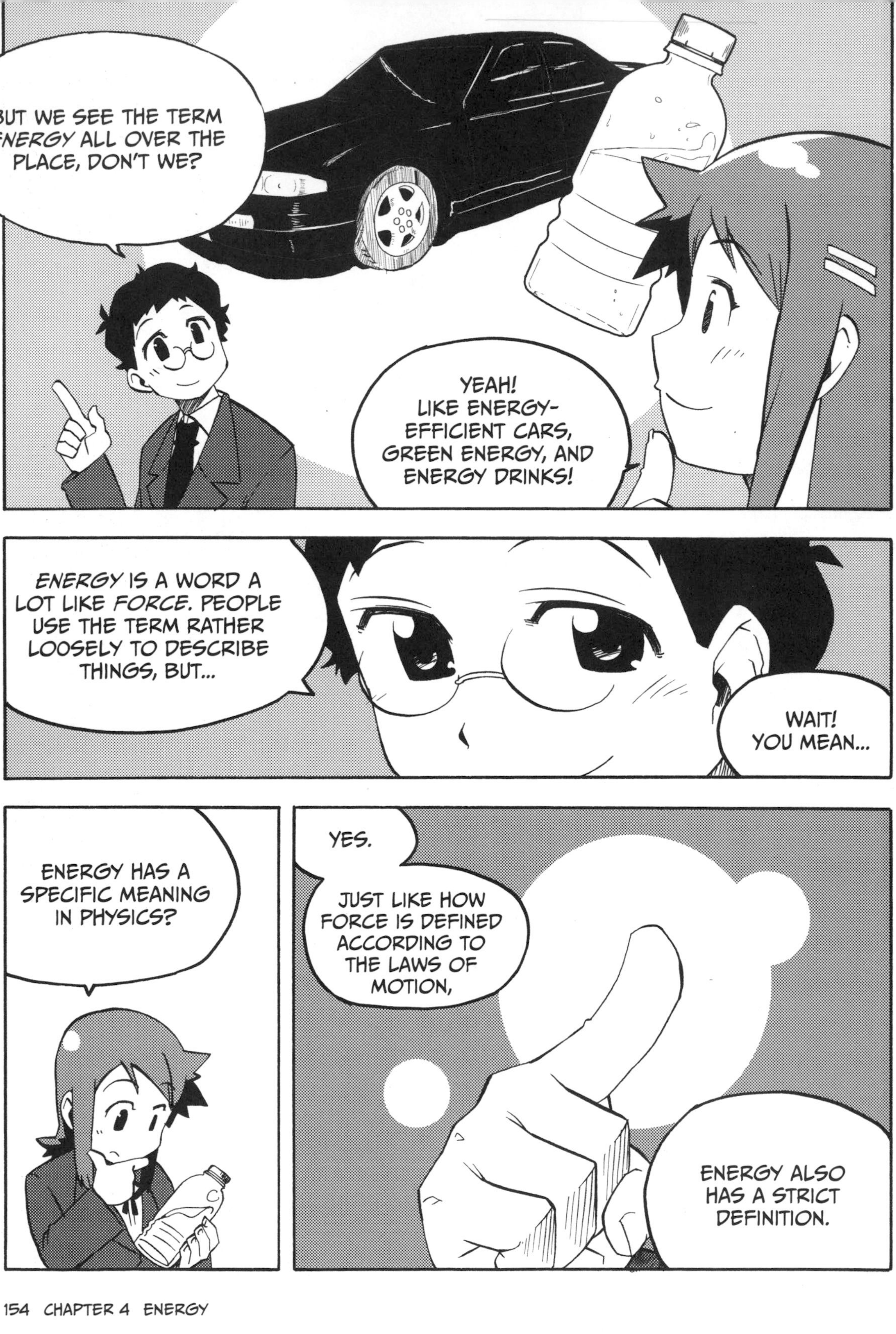
BUT WE SEE THE TERM ENERGY ALL OVER THE PLACE, DON'T WE?
YEAH! LIKE ENERGY-EFFICIENT CARS, GREEN ENERGY, AND ENERGY DRINKS!
ENERGY IS A WORD A LOT LIKE FORCE. PEOPLE USE THE TERM RATHER LOOSELY TO DESCRIBE THINGS, BUT...
WAIT! YOU MEAN...
ENERGY HAS A SPECIFIC MEANING IN PHYSICS?
YES.
JUST LIKE HOW FORCE IS DEFINED ACCORDING TO THE LAWS OF MOTION,
ENERGY ALSO HAS A STRICT DEFINITION.

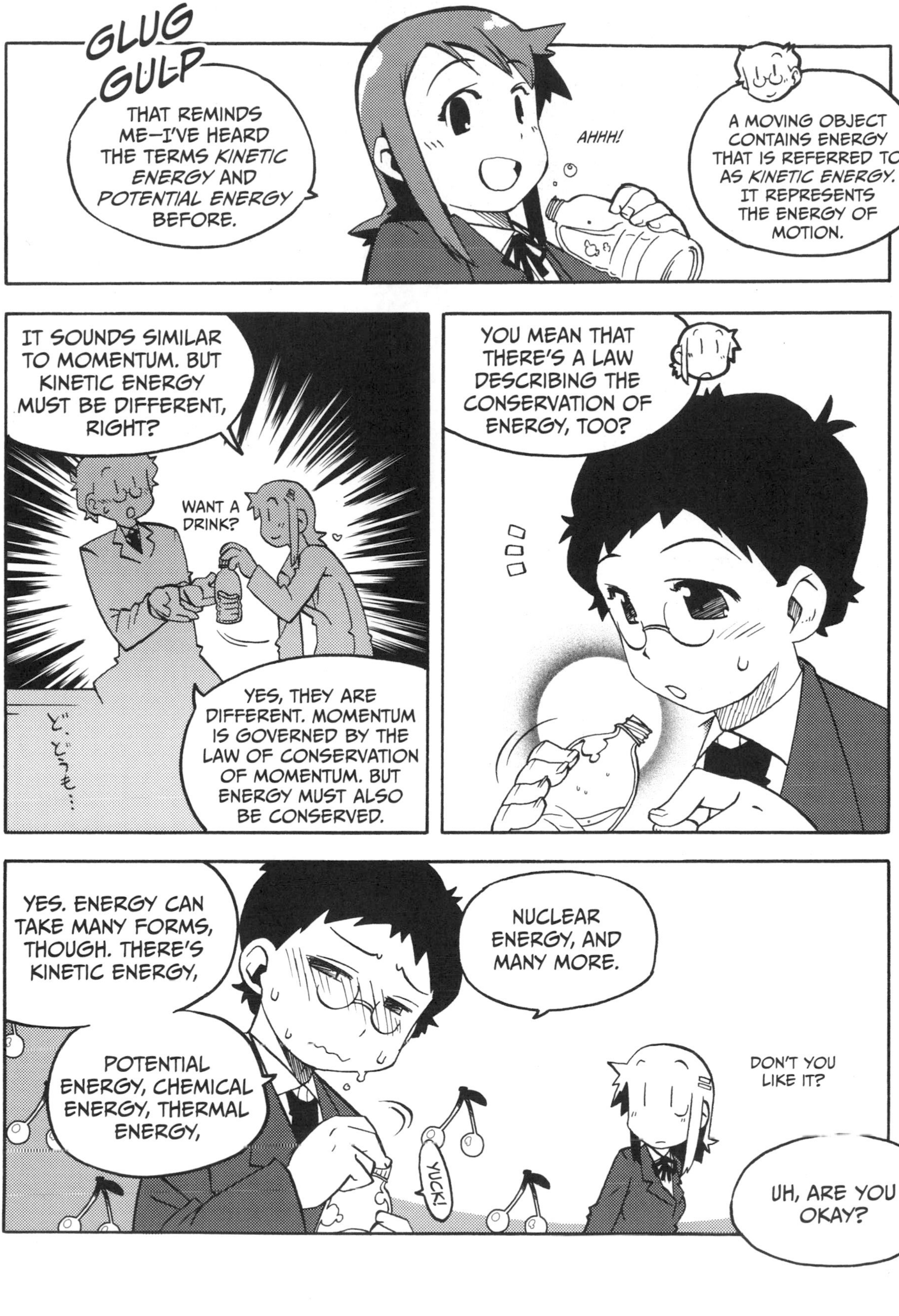
GLUG GULP
THAT REMINDS ME—I'VE HEARD THE TERMS *KINETIC ENERGY* AND *POTENTIAL ENERGY* BEFORE.
AHHH!
A MOVING OBJECT CONTAINS ENERGY THAT IS REFERRED TO AS *KINETIC ENERGY*. IT REPRESENTS THE ENERGY OF MOTION.
IT SOUNDS SIMILAR TO MOMENTUM. BUT KINETIC ENERGY MUST BE DIFFERENT, RIGHT?
WANT A DRINK?
ど、どうも…
YES, THEY ARE DIFFERENT. MOMENTUM IS GOVERNED BY THE LAW OF CONSERVATION OF MOMENTUM. BUT ENERGY MUST ALSO BE CONSERVED.
YOU MEAN THAT THERE'S A LAW DESCRIBING THE CONSERVATION OF ENERGY, TOO?
YES. ENERGY CAN TAKE MANY FORMS, THOUGH. THERE'S KINETIC ENERGY,
POTENTIAL ENERGY, CHEMICAL ENERGY, THERMAL ENERGY,
NUCLEAR ENERGY, AND MANY MORE.
YUCK!
DON'T YOU LIKE IT?
UH, ARE YOU OKAY?

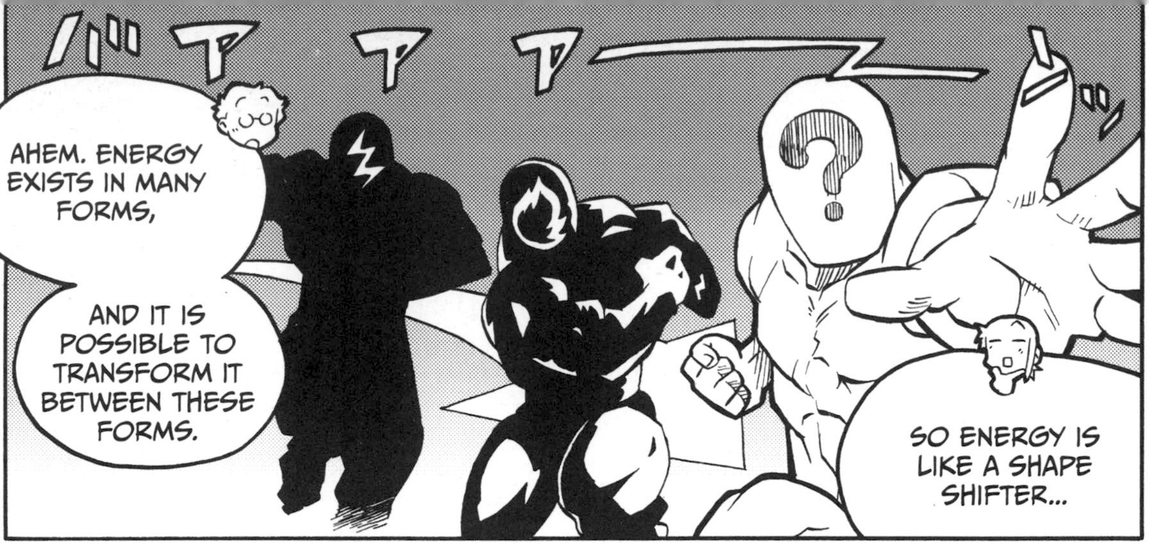
AHEM. ENERGY EXISTS IN MANY FORMS,
AND IT IS POSSIBLE TO TRANSFORM IT BETWEEN THESE FORMS.
SO ENERGY IS LIKE A SHAPE SHIFTER...

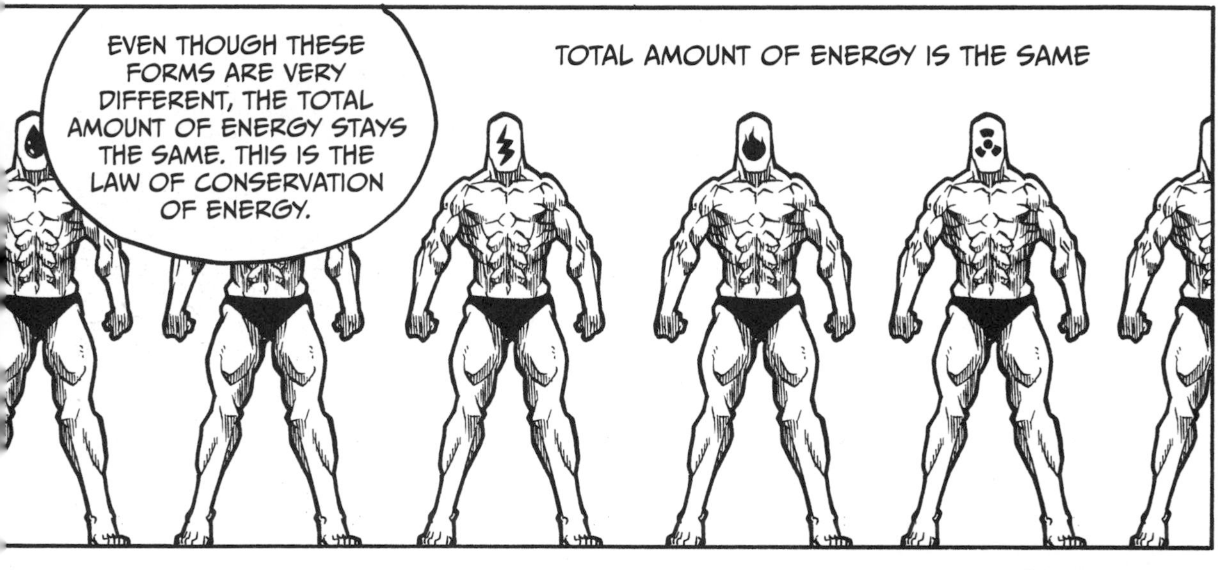
EVEN THOUGH THESE FORMS ARE VERY DIFFERENT, THE TOTAL AMOUNT OF ENERGY STAYS THE SAME. THIS IS THE LAW OF CONSERVATION OF ENERGY.
TOTAL AMOUNT OF ENERGY IS THE SAME

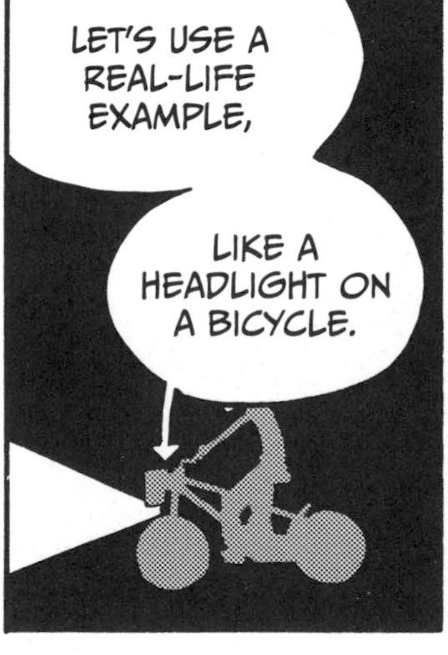
LET'S USE A REAL-LIFE EXAMPLE,
LIKE A HEADLIGHT ON A BICYCLE.

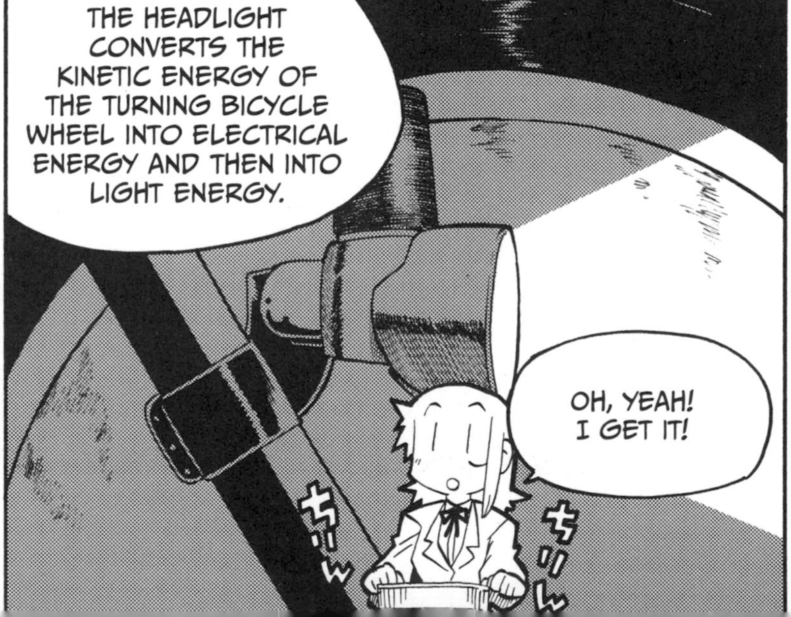
THE HEADLIGHT CONVERTS THE KINETIC ENERGY OF THE TURNING BICYCLE WHEEL INTO ELECTRICAL ENERGY AND THEN INTO LIGHT ENERGY.
OH, YEAH! I GET IT!

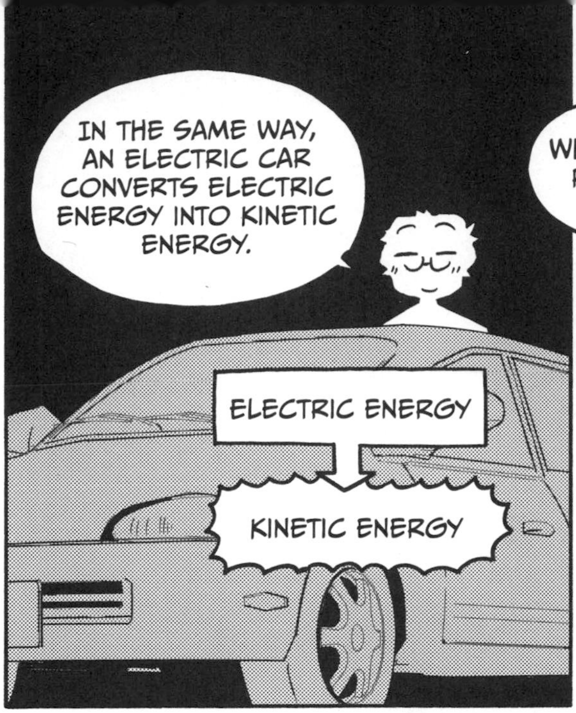
IN THE SAME WAY, AN ELECTRIC CAR CONVERTS ELECTRIC ENERGY INTO KINETIC ENERGY.
ELECTRIC ENERGY
KINETIC ENERGY
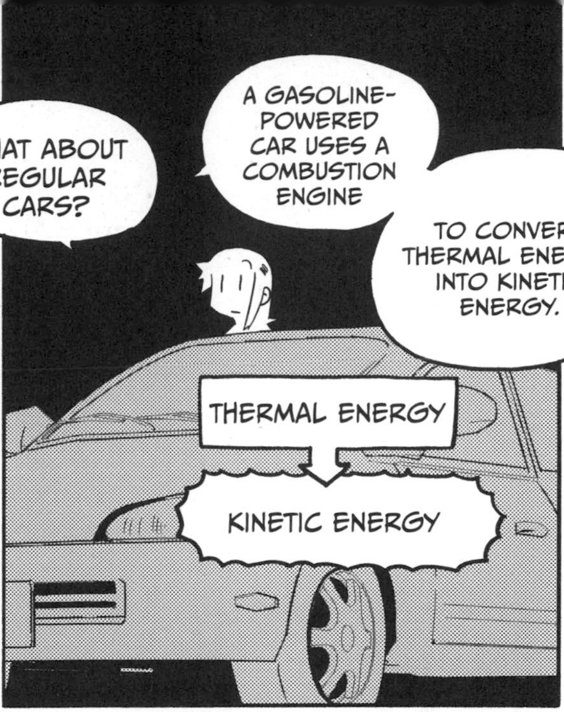
WHAT ABOUT REGULAR CARS?
A GASOLINE-POWERED CAR USES A COMBUSTION ENGINE
TO CONVERT THERMAL ENERGY INTO KINETIC ENERGY.
THERMAL ENERGY
KINETIC ENERGY
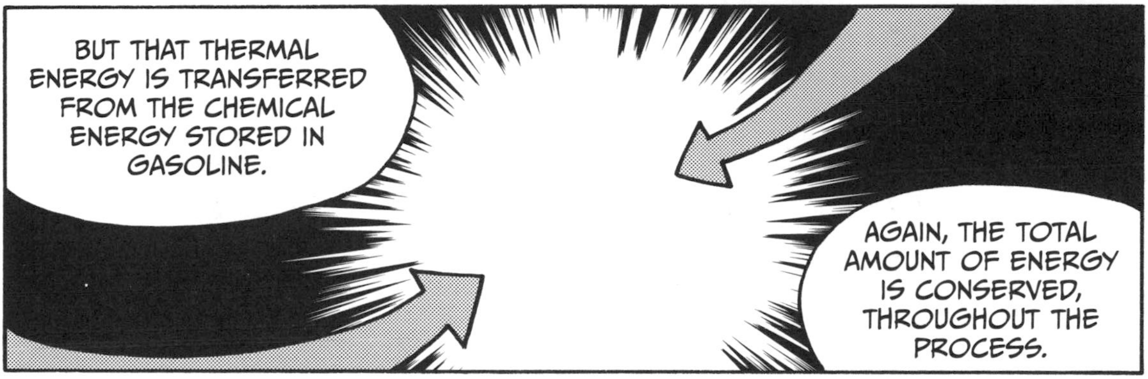
BUT THAT THERMAL ENERGY IS TRANSFERRED FROM THE CHEMICAL ENERGY STORED IN GASOLINE.
AGAIN, THE TOTAL AMOUNT OF ENERGY IS CONSERVED, THROUGHOUT THE PROCESS.
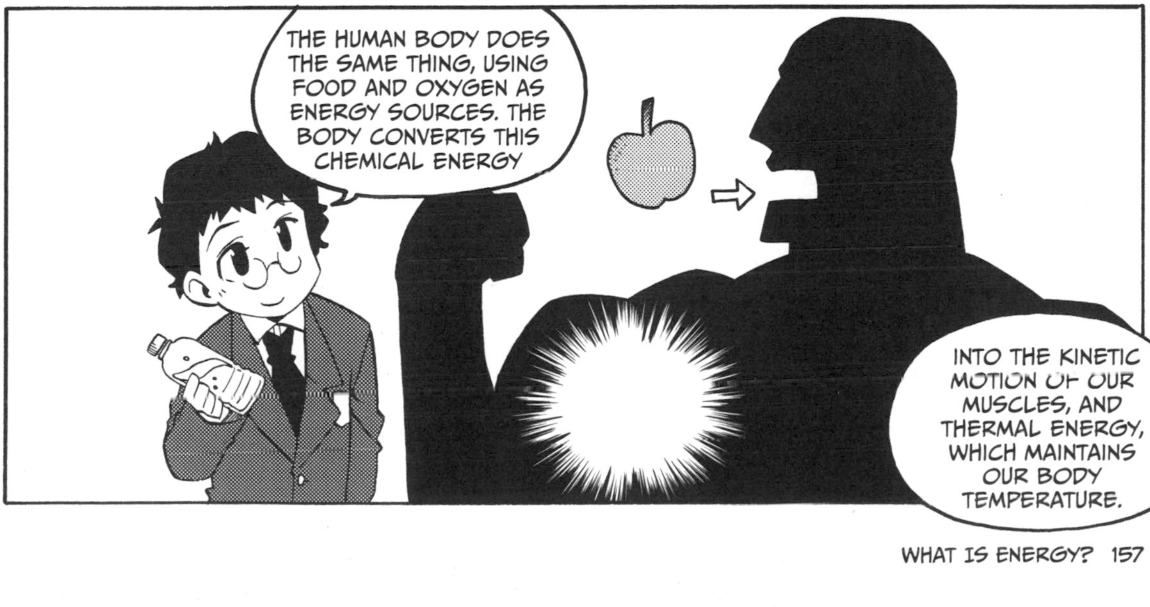
THE HUMAN BODY DOES THE SAME THING, USING FOOD AND OXYGEN AS ENERGY SOURCES. THE BODY CONVERTS THIS CHEMICAL ENERGY
INTO THE KINETIC MOTION OF OUR MUSCLES, AND THERMAL ENERGY, WHICH MAINTAINS OUR BODY TEMPERATURE.

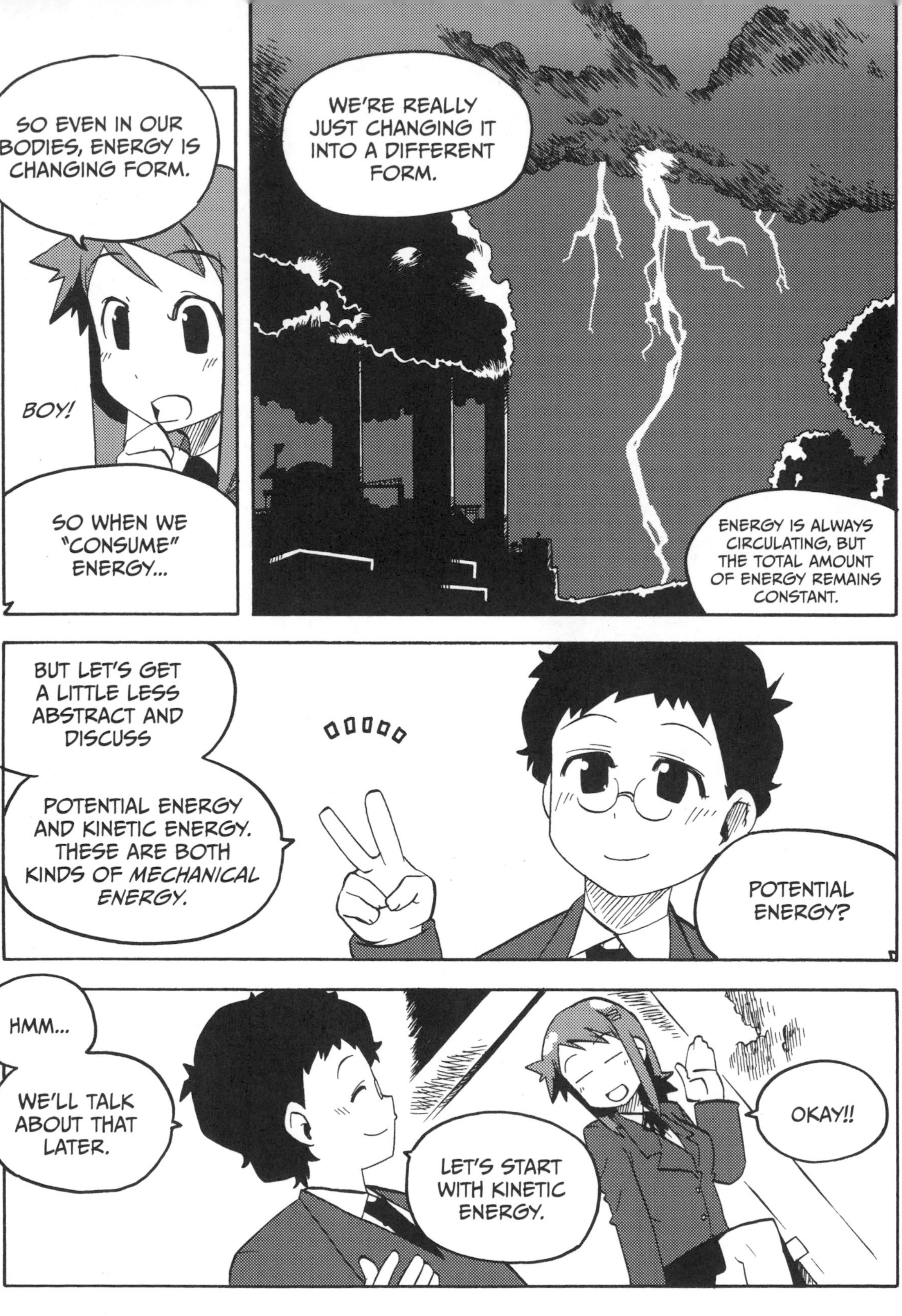
SO EVEN IN OUR BODIES, ENERGY IS CHANGING FORM.
BOY!
SO WHEN WE "CONSUME" ENERGY...
WE'RE REALLY JUST CHANGING IT INTO A DIFFERENT FORM.
ENERGY IS ALWAYS CIRCULATING, BUT THE TOTAL AMOUNT OF ENERGY REMAINS CONSTANT.
BUT LET'S GET A LITTLE LESS ABSTRACT AND DISCUSS
POTENTIAL ENERGY AND KINETIC ENERGY. THESE ARE BOTH KINDS OF *MECHANICAL* ENERGY.
POTENTIAL ENERGY?
HMM...
WE'LL TALK ABOUT THAT LATER.
LET'S START WITH KINETIC ENERGY.
OKAY!!

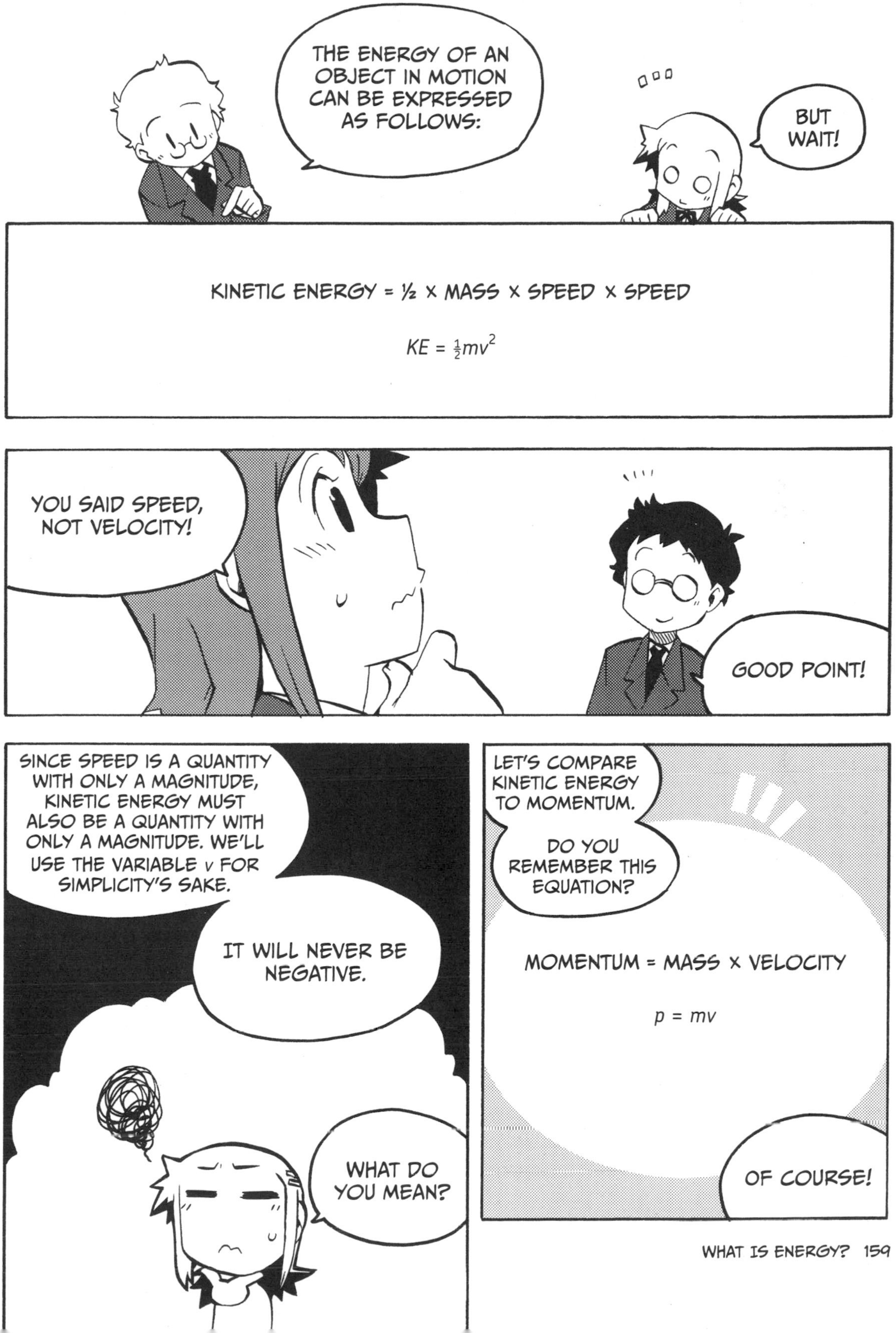
THE ENERGY OF AN OBJECT IN MOTION CAN BE EXPRESSED AS FOLLOWS:
BUT WAIT!
KINETIC ENERGY = ½ × MASS × SPEED × SPEED
$KE = \frac{1}{2}mv^2$
YOU SAID SPEED, NOT VELOCITY!
GOOD POINT!
SINCE SPEED IS A QUANTITY WITH ONLY A MAGNITUDE, KINETIC ENERGY MUST ALSO BE A QUANTITY WITH ONLY A MAGNITUDE. WE'LL USE THE VARIABLE v FOR SIMPLICITY'S SAKE.
IT WILL NEVER BE NEGATIVE.
WHAT DO YOU MEAN?
LET'S COMPARE KINETIC ENERGY TO MOMENTUM.
DO YOU REMEMBER THIS EQUATION?
MOMENTUM = MASS × VELOCITY
$p = mv$
OF COURSE!

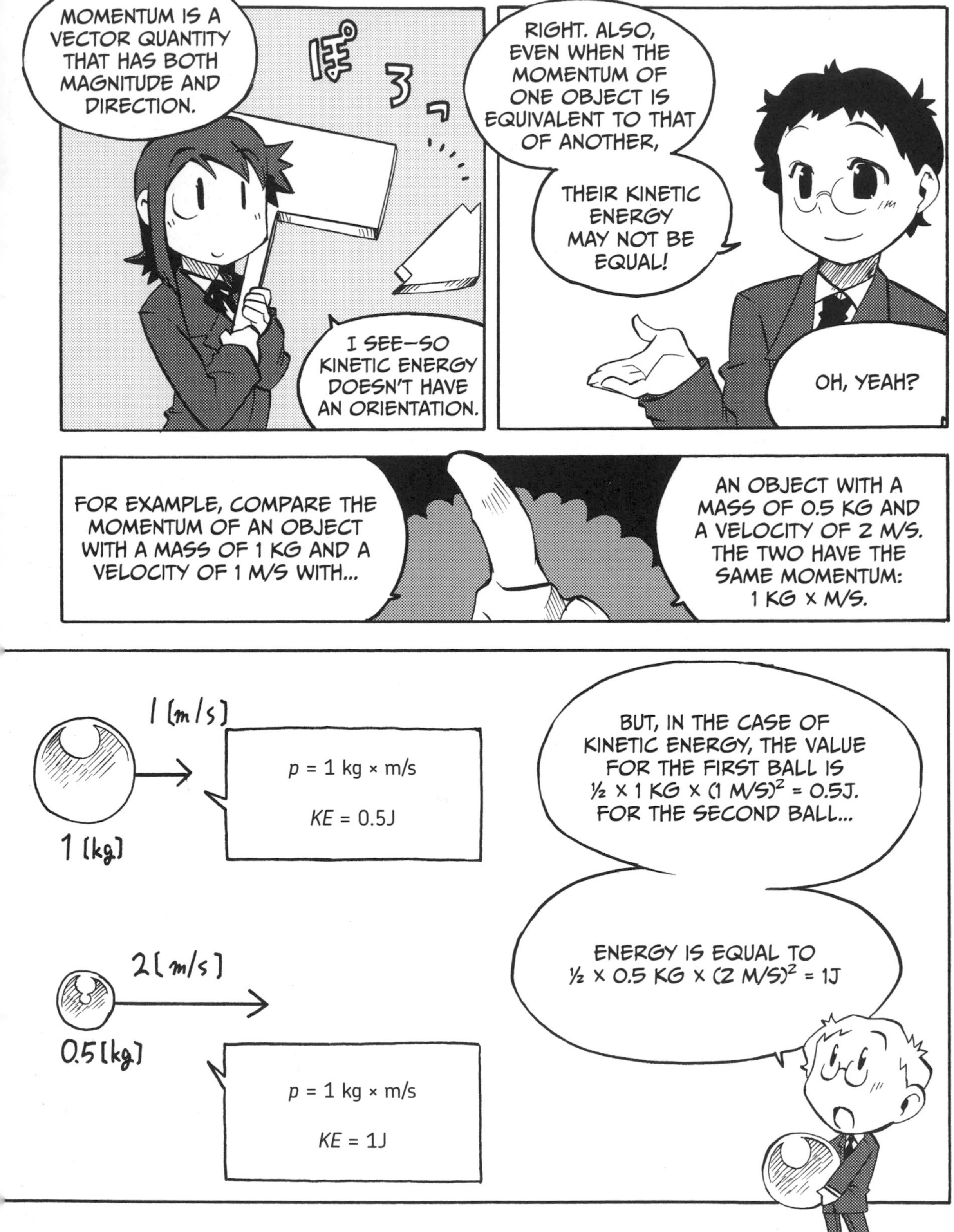
MOMENTUM IS A VECTOR QUANTITY THAT HAS BOTH MAGNITUDE AND DIRECTION.
ぽろっ
I SEE—SO KINETIC ENERGY DOESN'T HAVE AN ORIENTATION.
RIGHT. ALSO, EVEN WHEN THE MOMENTUM OF ONE OBJECT IS EQUIVALENT TO THAT OF ANOTHER,
THEIR KINETIC ENERGY MAY NOT BE EQUAL!
OH, YEAH?
FOR EXAMPLE, COMPARE THE MOMENTUM OF AN OBJECT WITH A MASS OF 1 KG AND A VELOCITY OF 1 M/S WITH...
AN OBJECT WITH A MASS OF 0.5 KG AND A VELOCITY OF 2 M/S. THE TWO HAVE THE SAME MOMENTUM: 1 KG × M/S.
1 [m/s]
1 [kg]
p = 1 kg × m/s
KE = 0.5J
2 [m/s]
0.5 [kg]
p = 1 kg × m/s
KE = 1J
BUT, IN THE CASE OF KINETIC ENERGY, THE VALUE FOR THE FIRST BALL IS ½ × 1 KG × (1 M/S)^2 = 0.5J. FOR THE SECOND BALL...
ENERGY IS EQUAL TO ½ × 0.5 KG × (2 M/S)^2 = 1J

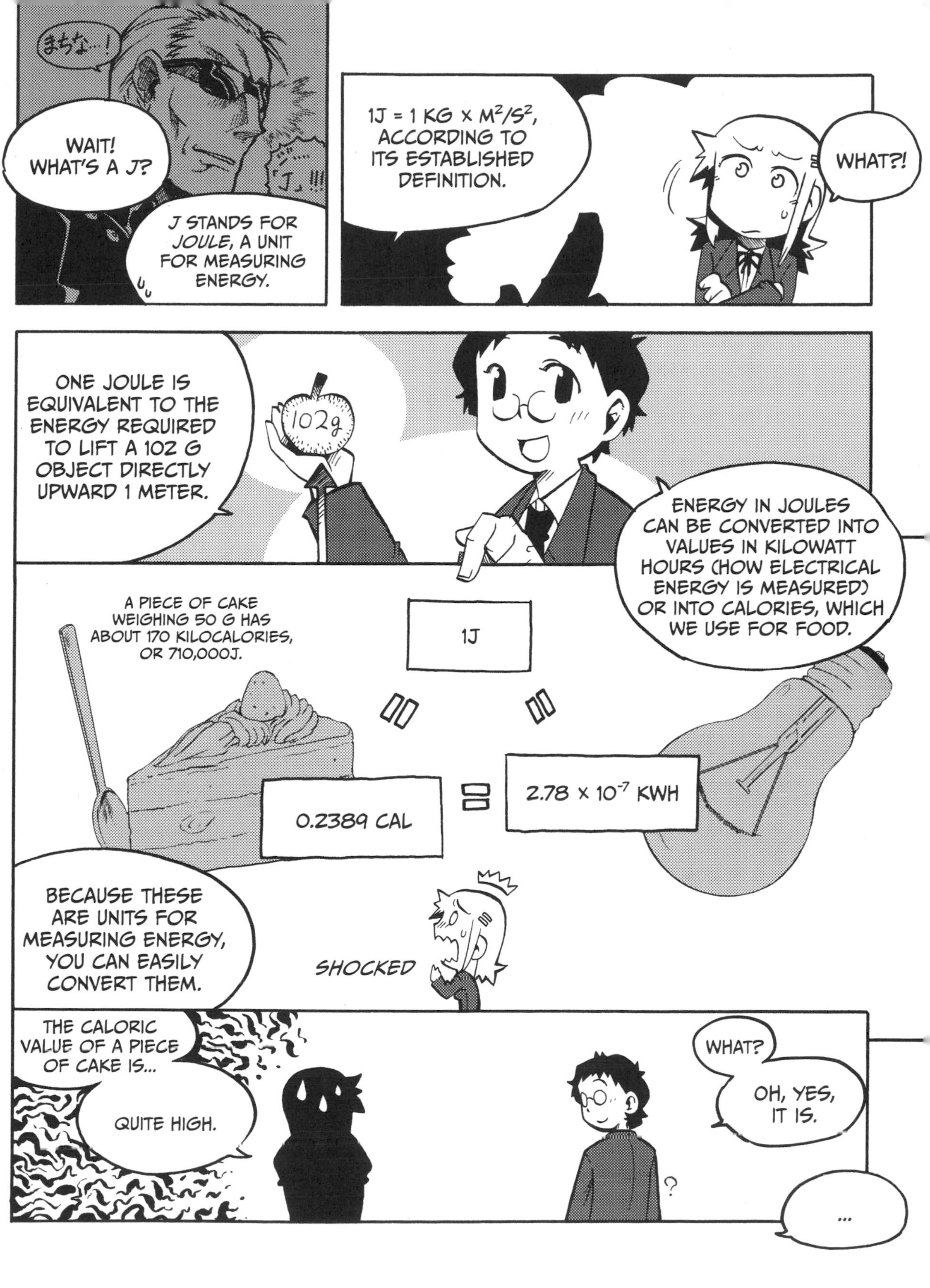
まちな…!
WAIT! WHAT'S A J?
「J」!!!
J STANDS FOR *JOULE*, A UNIT FOR MEASURING ENERGY.
$1J = 1\ KG \times M^2/S^2$, ACCORDING TO ITS ESTABLISHED DEFINITION.
WHAT?!
ONE JOULE IS EQUIVALENT TO THE ENERGY REQUIRED TO LIFT A 102 G OBJECT DIRECTLY UPWARD 1 METER.
102g
ENERGY IN JOULES CAN BE CONVERTED INTO VALUES IN KILOWATT HOURS (HOW ELECTRICAL ENERGY IS MEASURED) OR INTO CALORIES, WHICH WE USE FOR FOOD.
A PIECE OF CAKE WEIGHING 50 G HAS ABOUT 170 KILOCALORIES, OR 710,000J.
1J
=
=
0.2389 CAL
=
2.78×10^{-7} KWH
BECAUSE THESE ARE UNITS FOR MEASURING ENERGY, YOU CAN EASILY CONVERT THEM.
SHOCKED
THE CALORIC VALUE OF A PIECE OF CAKE IS...
QUITE HIGH.
WHAT?
OH, YES, IT IS.
?
...

LABORATORY

WHAT'S THE DIFFERENCE BETWEEN MOMENTUM AND KINETIC ENERGY?

 The difference between momentum and kinetic energy is easy to see when we consider two or more objects together.

 Oh, yeah?

 Let's recall the scenario where you were stranded outside your spaceship (page 126), and you used the law of conservation of momentum to return to the ship. Your momentum changed as a result of the momentum of the wrench, which you threw in the opposite direction. And, as I'm sure you recall, we use the equation $p = mv$ to express the relationship between momentum, mass, and velocity.

 Sure, I remember.

 Before you threw the wrench, the momentum for both objects was zero (as $v = 0$). After throwing the wrench, given the law of conservation of momentum, we know the following:

the sum of the momentum of the wrench and astronaut
$$= mv + MV = 0$$

Thus, we know that $mv = -MV$. In other words, the momentum of the wrench (mv) and your momentum (MV) are equivalent in magnitude and opposite in direction. They must equal zero when added together.

 Since momentum is a vector, it has an orientation! So two momentums with equivalent magnitude and opposite directions will cancel each other out.

Now, let's think about the kinetic energy of the wrench and that of the astronaut. Before throwing the wrench, both are stationary, and the momentum is zero for both objects. After throwing the wrench, the sum of the energy of the two objects in motion is *not* zero:

$$KE_{\text{wrench}} + KE_{\text{astronaut}} = \tfrac{1}{2}mv^2 + \tfrac{1}{2}MV^2 > 0$$

But you said energy is always conserved!

This kinetic energy was generated when you threw the tool. Consider the law of conservation of energy—the amount of energy lost in your body should be the same as the amount of kinetic energy gained in these two objects.

Well, okay.

While it's difficult to accurately measure the energy expended by the human body, we can say that it's possible to determine a decrease of energy in the body by finding the energy transferred by that body.

In other words, I know that my body has lost at least as much energy as I have gained in the objects I've thrown, right?

Yes, that's it. Now you need to remember, we must keep in mind the differences between energy and momentum.

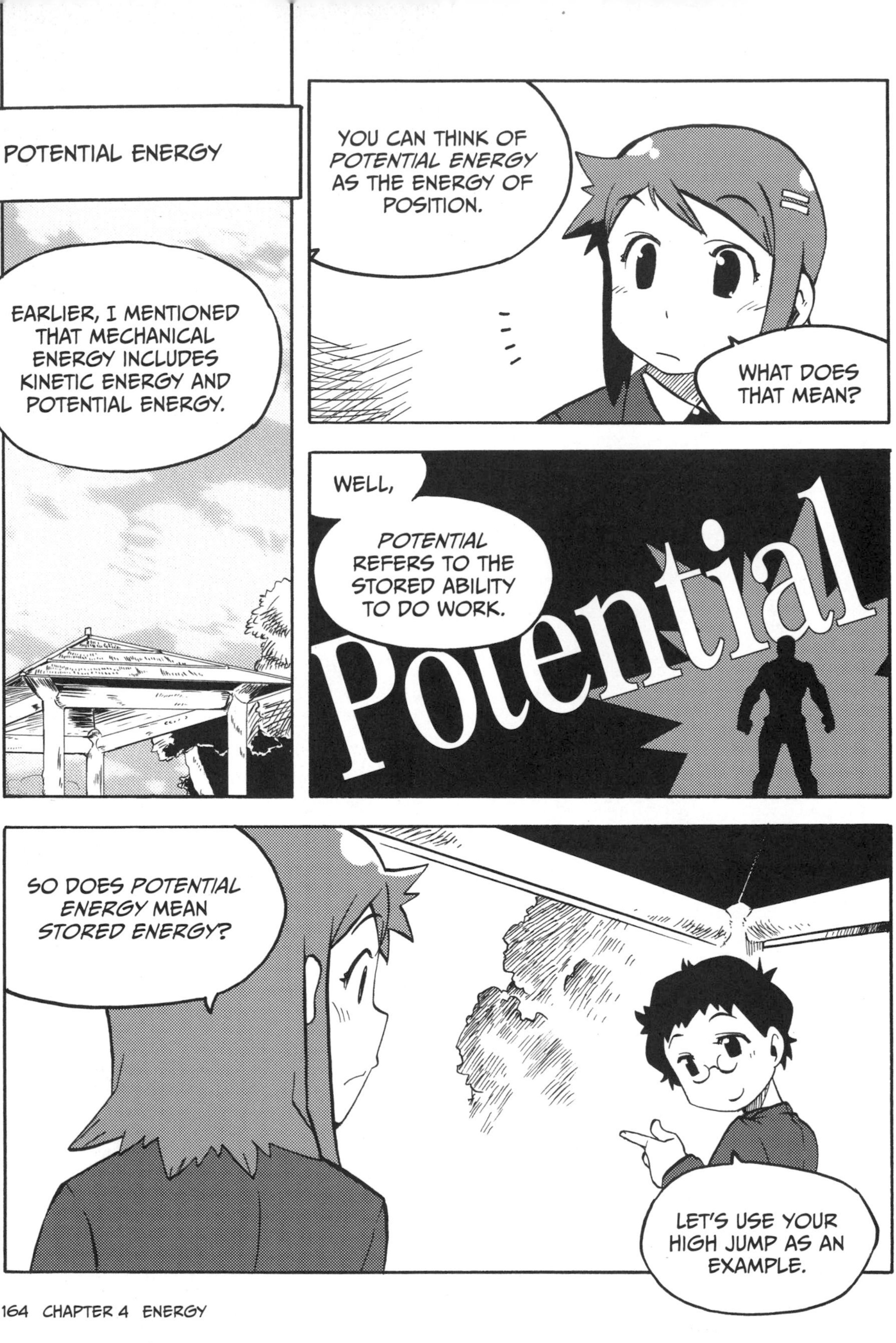
POTENTIAL ENERGY
EARLIER, I MENTIONED THAT MECHANICAL ENERGY INCLUDES KINETIC ENERGY AND POTENTIAL ENERGY.
YOU CAN THINK OF *POTENTIAL ENERGY* AS THE ENERGY OF POSITION.
WHAT DOES THAT MEAN?
WELL,
POTENTIAL REFERS TO THE STORED ABILITY TO DO WORK.
Potential
SO DOES *POTENTIAL ENERGY* MEAN *STORED ENERGY*?
LET'S USE YOUR HIGH JUMP AS AN EXAMPLE.

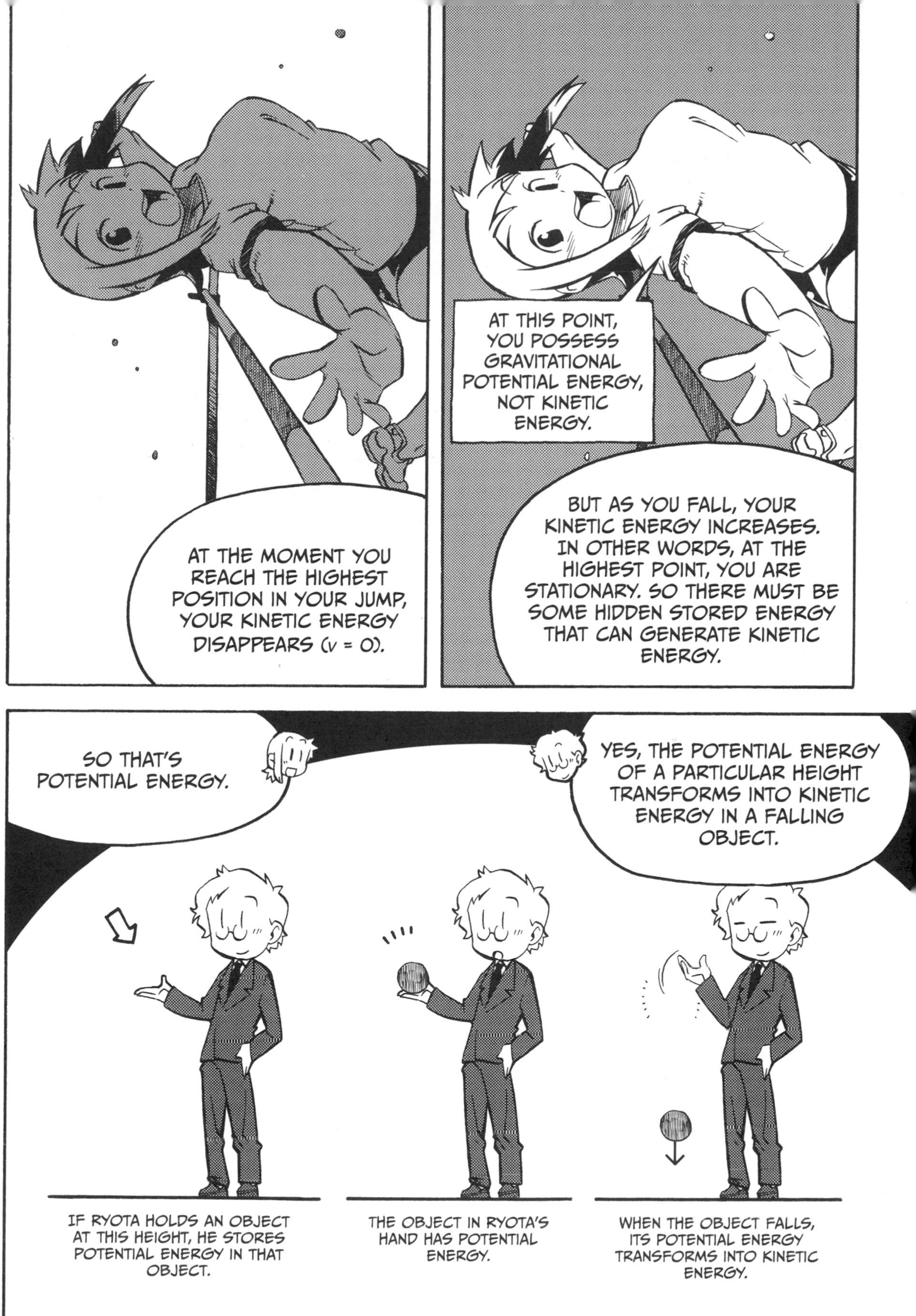
AT THE MOMENT YOU REACH THE HIGHEST POSITION IN YOUR JUMP, YOUR KINETIC ENERGY DISAPPEARS (v = 0).
AT THIS POINT, YOU POSSESS GRAVITATIONAL POTENTIAL ENERGY, NOT KINETIC ENERGY.
BUT AS YOU FALL, YOUR KINETIC ENERGY INCREASES. IN OTHER WORDS, AT THE HIGHEST POINT, YOU ARE STATIONARY. SO THERE MUST BE SOME HIDDEN STORED ENERGY THAT CAN GENERATE KINETIC ENERGY.
SO THAT'S POTENTIAL ENERGY.
YES, THE POTENTIAL ENERGY OF A PARTICULAR HEIGHT TRANSFORMS INTO KINETIC ENERGY IN A FALLING OBJECT.
IF RYOTA HOLDS AN OBJECT AT THIS HEIGHT, HE STORES POTENTIAL ENERGY IN THAT OBJECT.
THE OBJECT IN RYOTA'S HAND HAS POTENTIAL ENERGY.
WHEN THE OBJECT FALLS, ITS POTENTIAL ENERGY TRANSFORMS INTO KINETIC ENERGY.

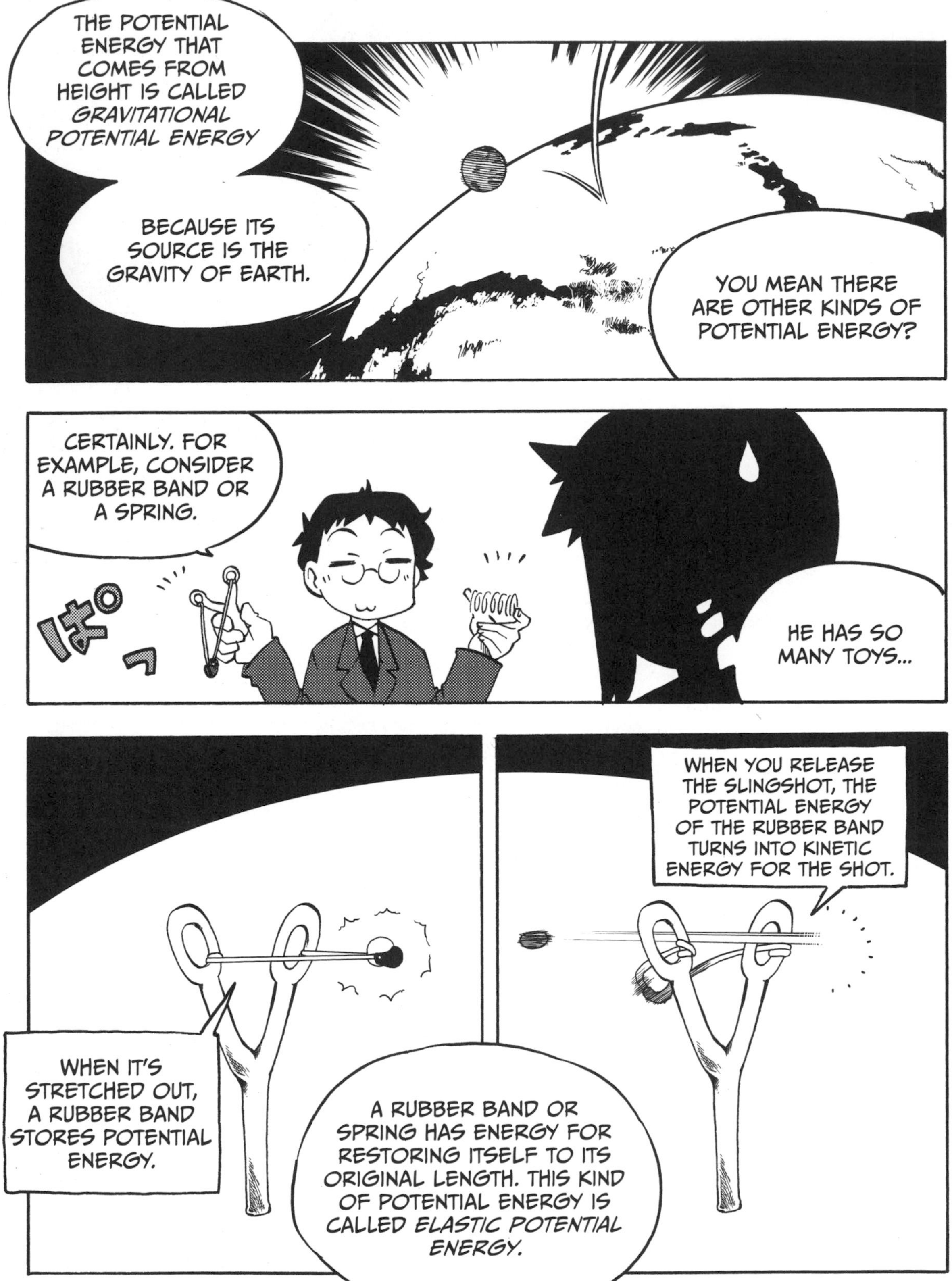
THE POTENTIAL ENERGY THAT COMES FROM HEIGHT IS CALLED *GRAVITATIONAL POTENTIAL ENERGY*
BECAUSE ITS SOURCE IS THE GRAVITY OF EARTH.
YOU MEAN THERE ARE OTHER KINDS OF POTENTIAL ENERGY?
CERTAINLY. FOR EXAMPLE, CONSIDER A RUBBER BAND OR A SPRING.
ぽっ
HE HAS SO MANY TOYS...
WHEN IT'S STRETCHED OUT, A RUBBER BAND STORES POTENTIAL ENERGY.
WHEN YOU RELEASE THE SLINGSHOT, THE POTENTIAL ENERGY OF THE RUBBER BAND TURNS INTO KINETIC ENERGY FOR THE SHOT.
A RUBBER BAND OR SPRING HAS ENERGY FOR RESTORING ITSELF TO ITS ORIGINAL LENGTH. THIS KIND OF POTENTIAL ENERGY IS CALLED *ELASTIC POTENTIAL ENERGY*.

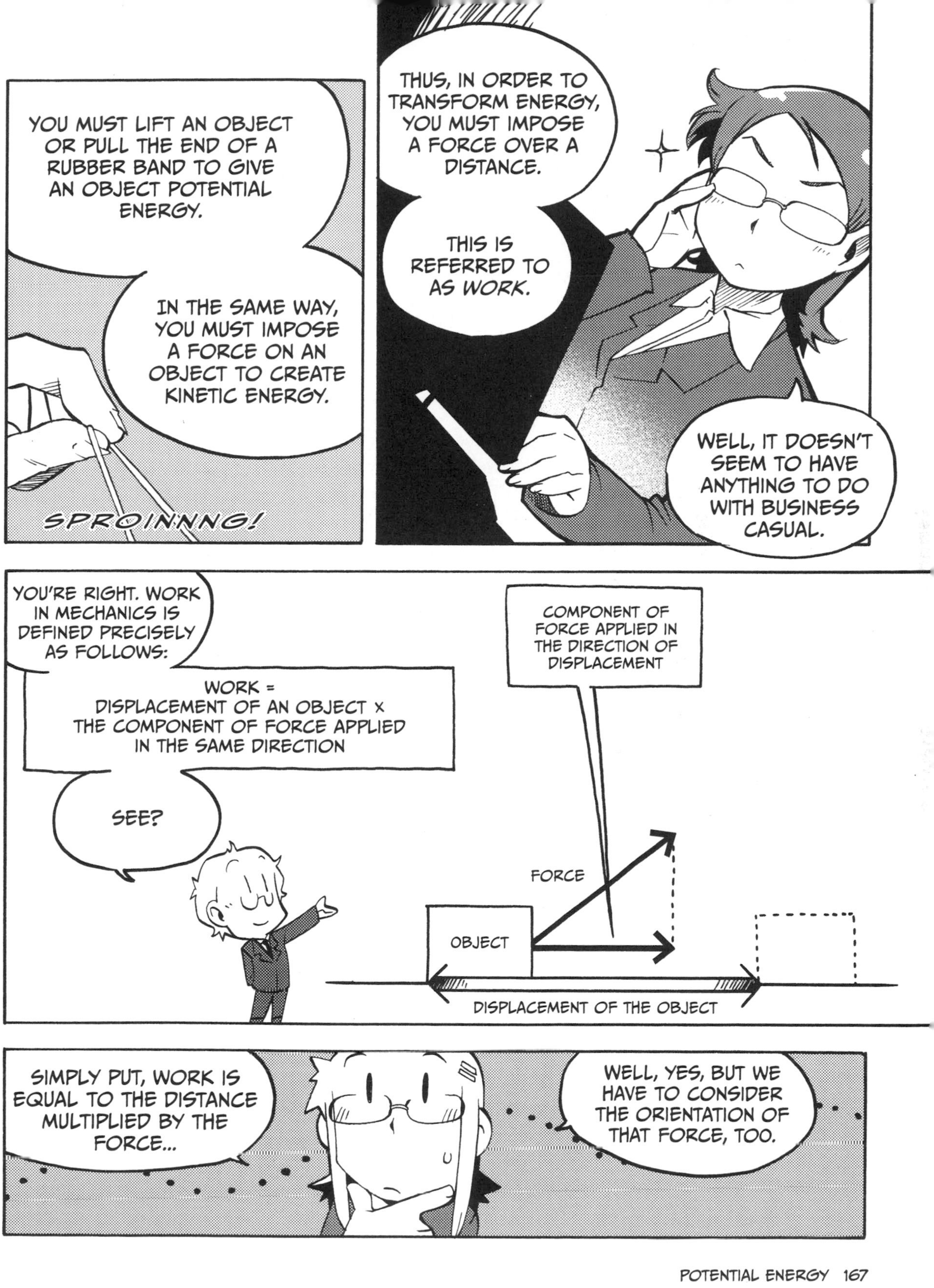
YOU MUST LIFT AN OBJECT OR PULL THE END OF A RUBBER BAND TO GIVE AN OBJECT POTENTIAL ENERGY.
IN THE SAME WAY, YOU MUST IMPOSE A FORCE ON AN OBJECT TO CREATE KINETIC ENERGY.
SPROINNNG!
THUS, IN ORDER TO TRANSFORM ENERGY, YOU MUST IMPOSE A FORCE OVER A DISTANCE.
THIS IS REFERRED TO AS *WORK*.
WELL, IT DOESN'T SEEM TO HAVE ANYTHING TO DO WITH BUSINESS CASUAL.
YOU'RE RIGHT. WORK IN MECHANICS IS DEFINED PRECISELY AS FOLLOWS:
WORK = DISPLACEMENT OF AN OBJECT × THE COMPONENT OF FORCE APPLIED IN THE SAME DIRECTION
SEE?
COMPONENT OF FORCE APPLIED IN THE DIRECTION OF DISPLACEMENT
FORCE
OBJECT
DISPLACEMENT OF THE OBJECT
SIMPLY PUT, WORK IS EQUAL TO THE DISTANCE MULTIPLIED BY THE FORCE...
WELL, YES, BUT WE HAVE TO CONSIDER THE ORIENTATION OF THAT FORCE, TOO.

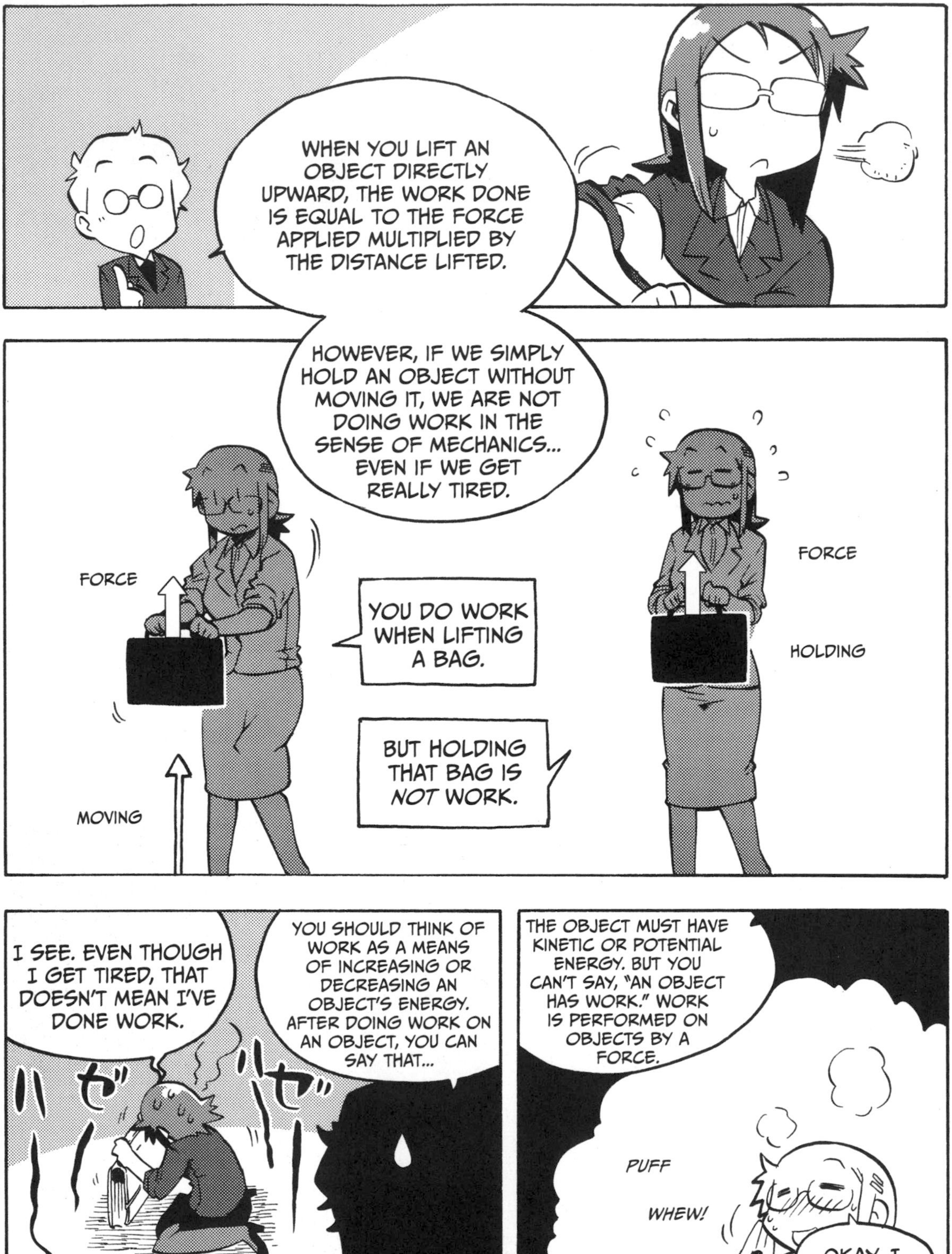
WHEN YOU LIFT AN OBJECT DIRECTLY UPWARD, THE WORK DONE IS EQUAL TO THE FORCE APPLIED MULTIPLIED BY THE DISTANCE LIFTED.
HOWEVER, IF WE SIMPLY HOLD AN OBJECT WITHOUT MOVING IT, WE ARE NOT DOING WORK IN THE SENSE OF MECHANICS... EVEN IF WE GET REALLY TIRED.
FORCE
MOVING
YOU DO WORK WHEN LIFTING A BAG.
BUT HOLDING THAT BAG IS *NOT* WORK.
FORCE
HOLDING
I SEE. EVEN THOUGH I GET TIRED, THAT DOESN'T MEAN I'VE DONE WORK.
RYOTA'S BAG IS HEAVY!
YOU SHOULD THINK OF WORK AS A MEANS OF INCREASING OR DECREASING AN OBJECT'S ENERGY. AFTER DOING WORK ON AN OBJECT, YOU CAN SAY THAT...
THE OBJECT MUST HAVE KINETIC OR POTENTIAL ENERGY. BUT YOU CAN'T SAY, "AN OBJECT HAS WORK." WORK IS PERFORMED ON OBJECTS BY A FORCE.
PUFF
WHEW!
OKAY, I GET IT!

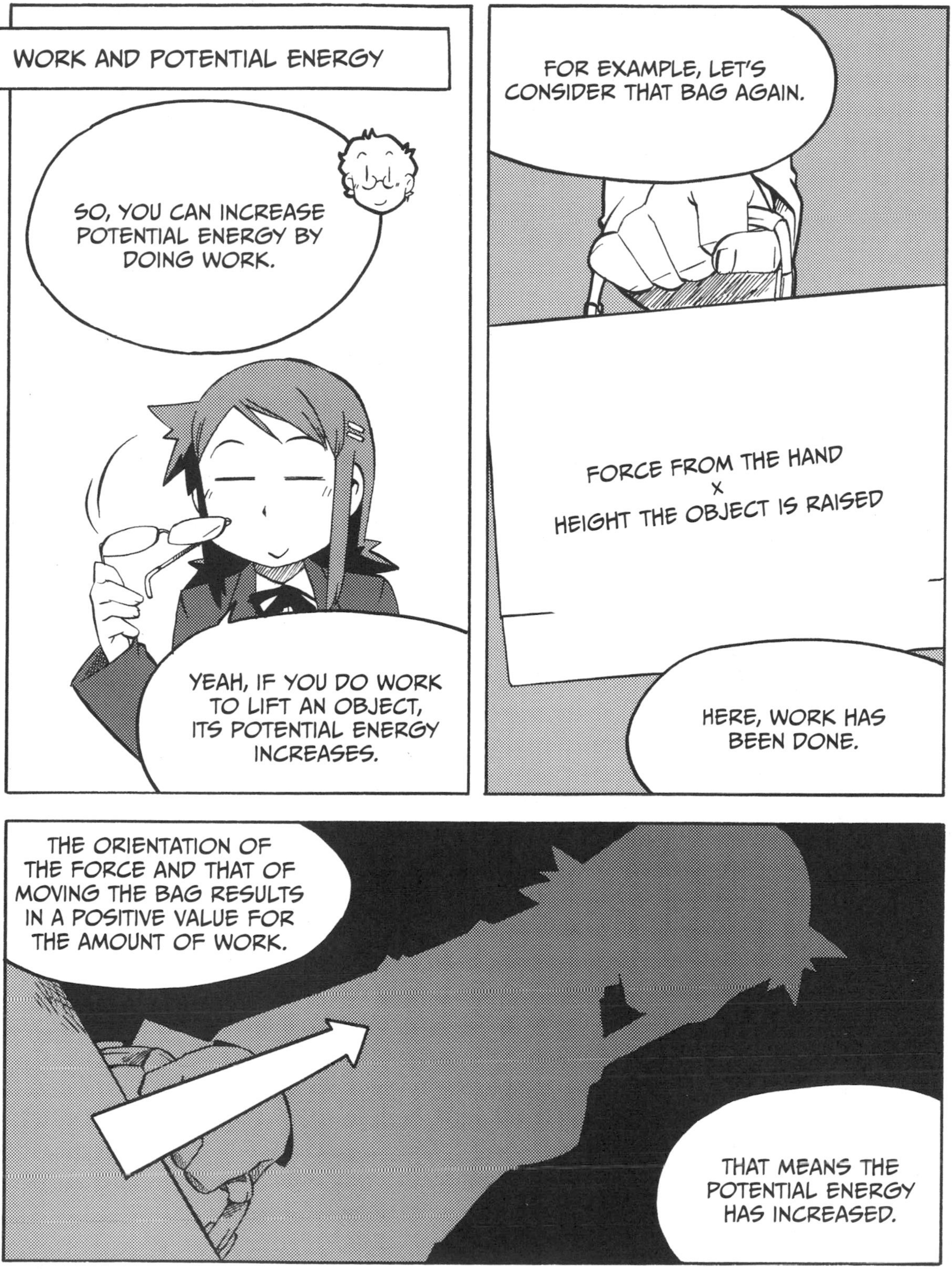
WORK AND POTENTIAL ENERGY
SO, YOU CAN INCREASE POTENTIAL ENERGY BY DOING WORK.
YEAH, IF YOU DO WORK TO LIFT AN OBJECT, ITS POTENTIAL ENERGY INCREASES.
FOR EXAMPLE, LET'S CONSIDER THAT BAG AGAIN.
FORCE FROM THE HAND × HEIGHT THE OBJECT IS RAISED
HERE, WORK HAS BEEN DONE.
THE ORIENTATION OF THE FORCE AND THAT OF MOVING THE BAG RESULTS IN A POSITIVE VALUE FOR THE AMOUNT OF WORK.
THAT MEANS THE POTENTIAL ENERGY HAS INCREASED.

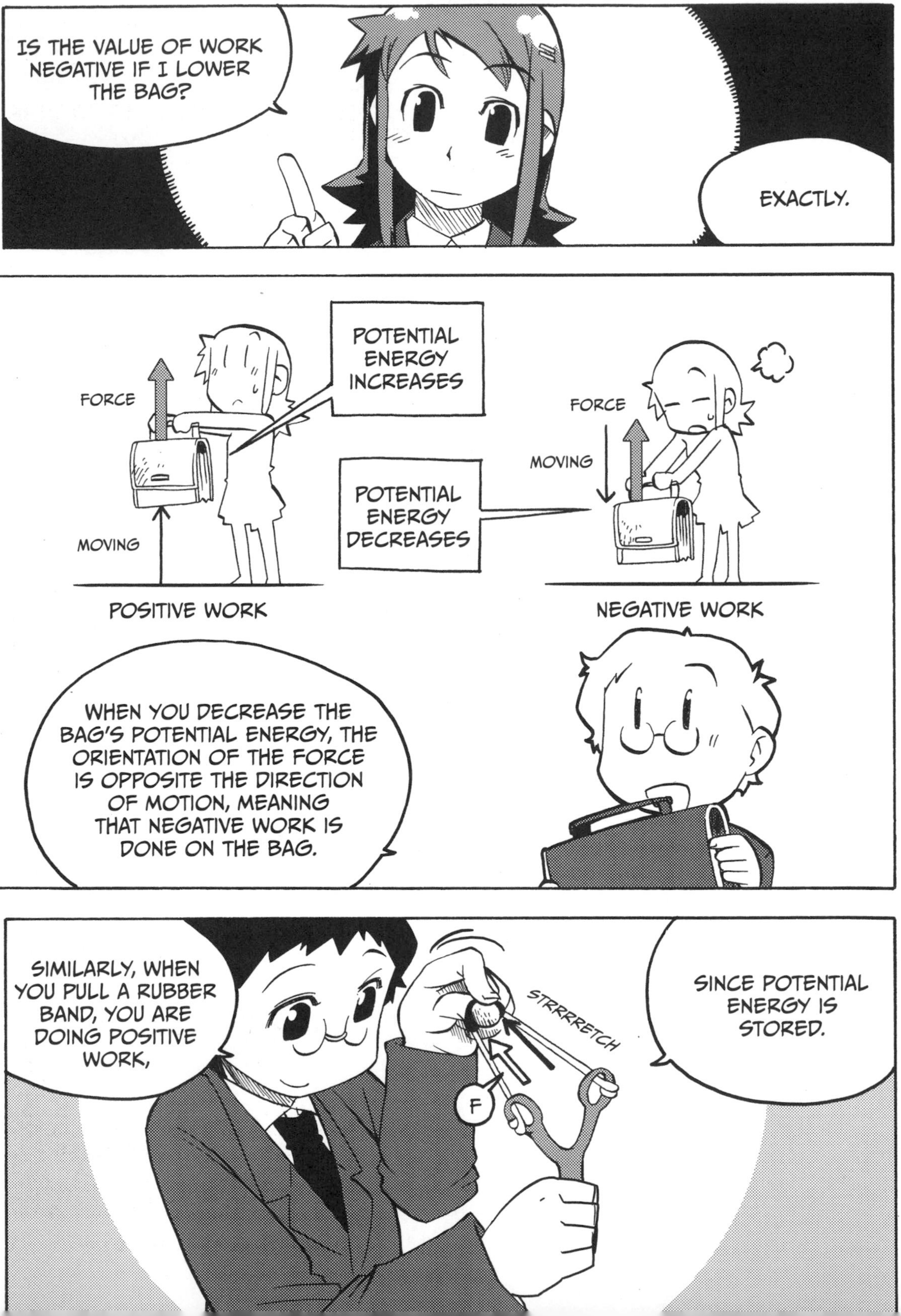
IS THE VALUE OF WORK NEGATIVE IF I LOWER THE BAG?
EXACTLY.
FORCE
MOVING
POTENTIAL ENERGY INCREASES
POSITIVE WORK
FORCE
MOVING
POTENTIAL ENERGY DECREASES
NEGATIVE WORK
WHEN YOU DECREASE THE BAG'S POTENTIAL ENERGY, THE ORIENTATION OF THE FORCE IS OPPOSITE THE DIRECTION OF MOTION, MEANING THAT NEGATIVE WORK IS DONE ON THE BAG.
SIMILARLY, WHEN YOU PULL A RUBBER BAND, YOU ARE DOING POSITIVE WORK,
STRRRRETCH
F
SINCE POTENTIAL ENERGY IS STORED.

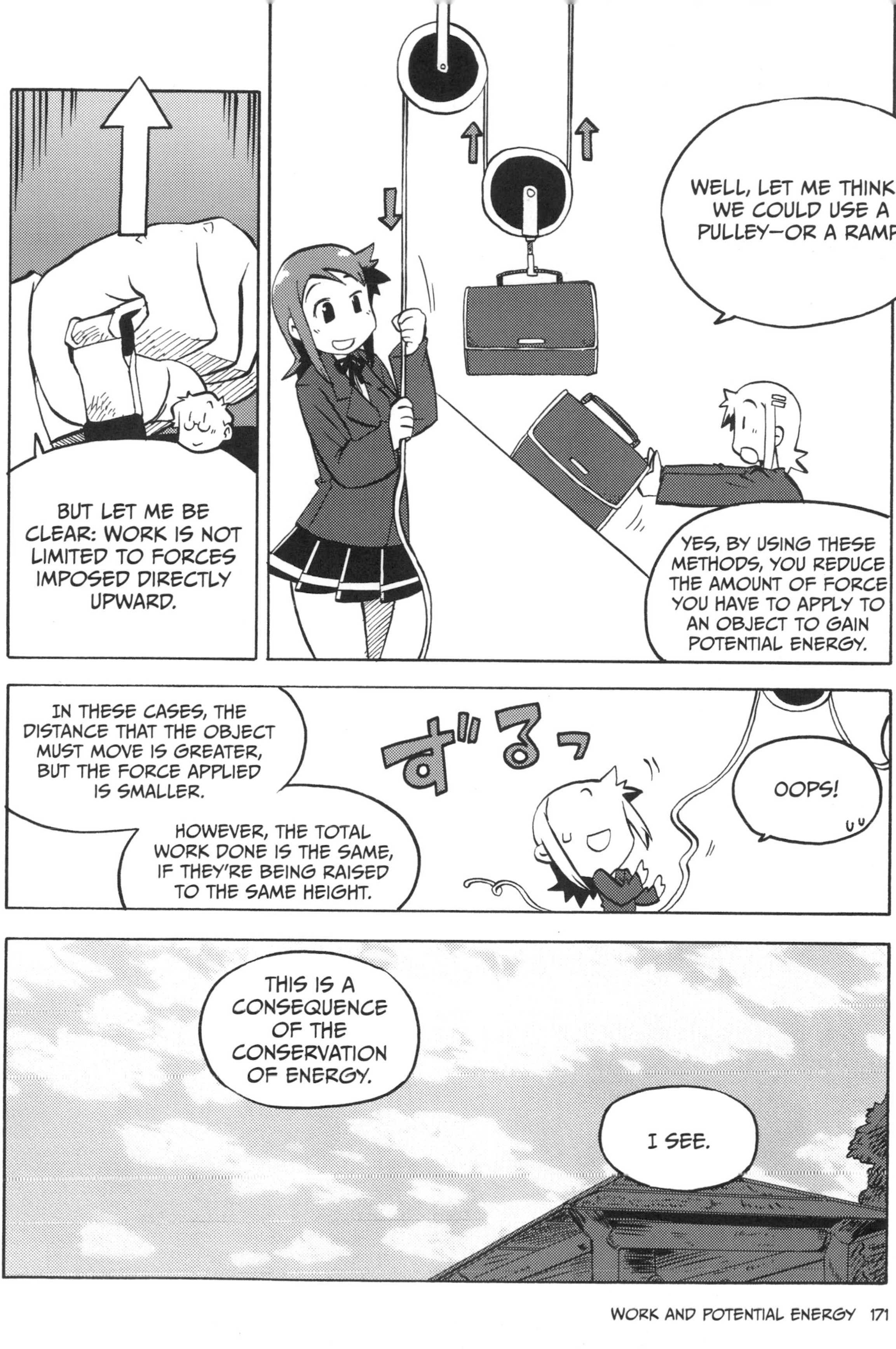
BUT LET ME BE CLEAR: WORK IS NOT LIMITED TO FORCES IMPOSED DIRECTLY UPWARD.
WELL, LET ME THINK... WE COULD USE A PULLEY—OR A RAMP.
YES, BY USING THESE METHODS, YOU REDUCE THE AMOUNT OF FORCE YOU HAVE TO APPLY TO AN OBJECT TO GAIN POTENTIAL ENERGY.
IN THESE CASES, THE DISTANCE THAT THE OBJECT MUST MOVE IS GREATER, BUT THE FORCE APPLIED IS SMALLER.
HOWEVER, THE TOTAL WORK DONE IS THE SAME, IF THEY'RE BEING RAISED TO THE SAME HEIGHT.
ずるっ
OOPS!
THIS IS A CONSEQUENCE OF THE CONSERVATION OF ENERGY.
I SEE.

LABORATORY

WORK AND THE CONSERVATION OF ENERGY

 Let's consider a scenario in which we are lifting a heavy load to a certain height. The simplest way to do this is to lift straight up. The following diagram shows how it looks.

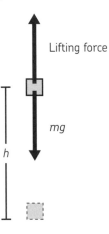

Lifting force

mg

h

 We are lifting a load with mass m to height h.

 Let's consider how much work we must do to lift the load to a height of h by imposing a force equal to the force of gravity of the mass—that is, we'll impose a force upward equivalent to the force downward from gravity. Assuming g for gravitational acceleration, we know that the force downward is mg:

work upward = force of lifting × height h = mgh

Note that for simplicity's sake, we won't take into account friction or air resistance in these examples. But this is a hard way to lift something so heavy!

 Hmm . . . maybe it'd be easier if we pushed the load up a ramp.

 Yes, let's consider the case of pushing the load up an incline.

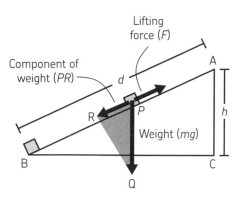

Lifting force (F)

Component of weight (PR)

d

Weight (mg)

R P

A

h

B C

Q

 Look at this diagram. The magnitude of the force needed to push the load up this ramp (F) is equal to the component of the force of gravity parallel to the ramp (PR). So, if the ramp has a length of d, the work required to move the load to height h can be represented as:

work = Fd

Now, you know intuitively that F is smaller than mg, and d is larger than h.

 That makes sense. Is that why it takes the same amount of work to push the load up a ramp as it does to lift the load straight up?

 Yes, indeed. Now let's show why this works, mathematically. $\triangle ABC$ represents the ramp in the figure, and $\triangle PQR$ represents the composition of the force mg. These two triangles are similar—this means that $\angle CAB = \angle RPQ$. This also means that the proportion of their corresponding sides must be the same, as well. Thus, the following must be true:

$$\frac{AB}{AC} = \frac{PQ}{PR}$$

Let's make this a little less abstract. The line segment AB equals d (length of ramp) and AC equals h (height). Similarly, the line segment PQ equals mg (the force downward, due to gravity), while PR equals F (the force applied to offset a portion of that force).

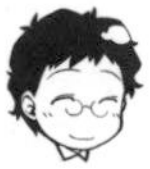

That means:

$$\frac{d}{h} = \frac{mg}{F}$$

Look, with just a little rearranging of this equation we get the following:

$$Fd = mgh$$

Therefore, the work to lift a load using a ramp must be equal to the work to lift that load straight upward.

Also, please note that our results are the same, regardless of the angle of the ramp. Given the conservation of energy, regardless of the lifting route, the work done for lifting an object with mass m to height h is equal to the following:

$$\text{force required to balance gravity} \times \text{height} = mgh$$

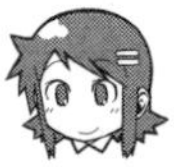

So, whatever method you use to lift something, the amount of work you do is the same.

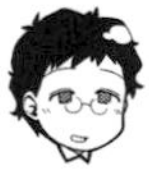

To put it another way, your work increases the potential energy of the load by mgh.

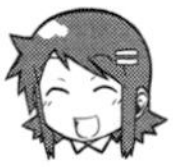

And I bet it works for negative work, too. That is, you'd see a decrease in potential energy of mgh if you lower an object by mgh.

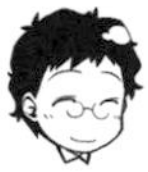

Yep, that's right.

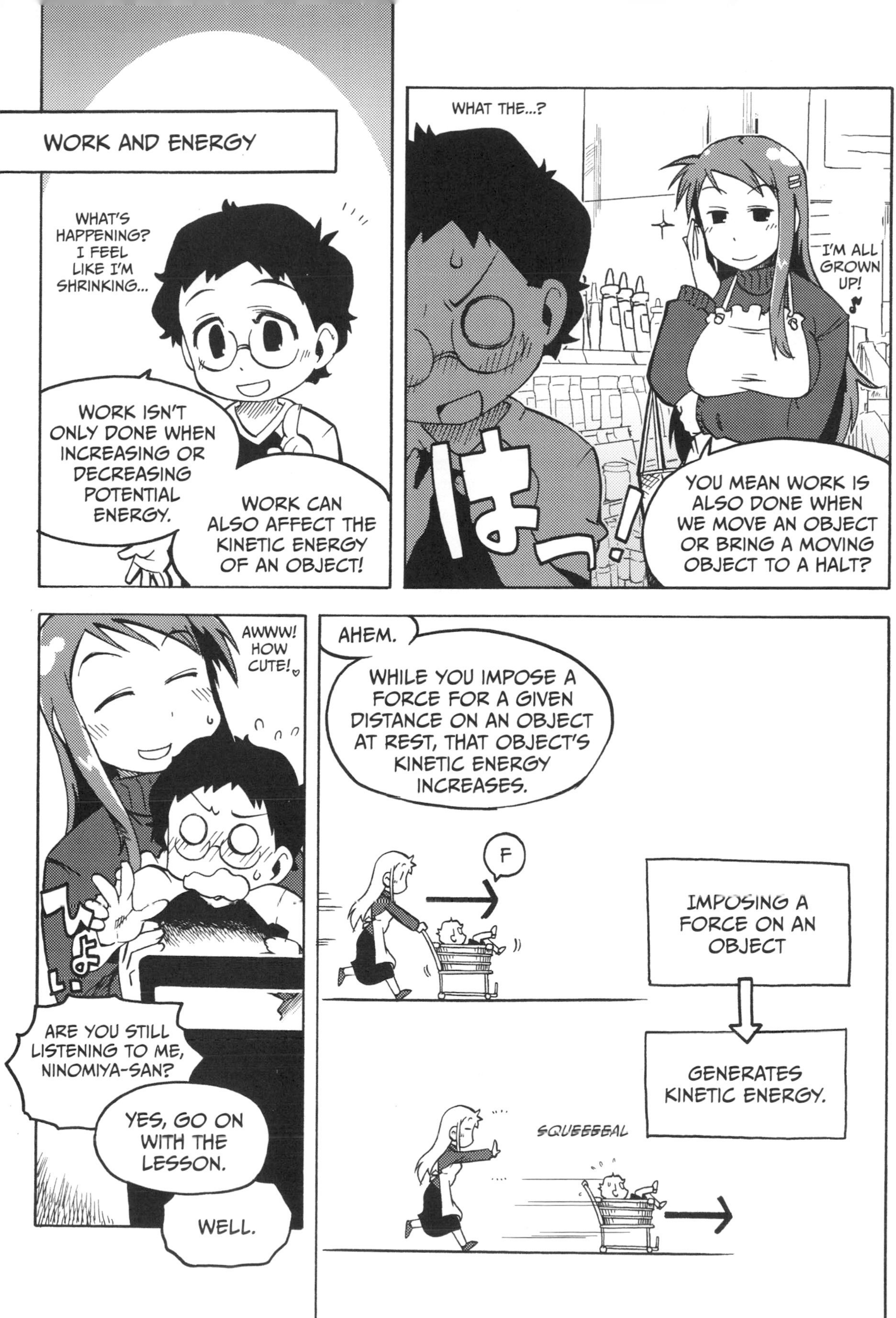
WORK AND ENERGY
WHAT'S HAPPENING? I FEEL LIKE I'M SHRINKING...
WORK ISN'T ONLY DONE WHEN INCREASING OR DECREASING POTENTIAL ENERGY.
WORK CAN ALSO AFFECT THE KINETIC ENERGY OF AN OBJECT!
WHAT THE...?
I'M ALL GROWN UP!
はっ!
YOU MEAN WORK IS ALSO DONE WHEN WE MOVE AN OBJECT OR BRING A MOVING OBJECT TO A HALT?
AWWW! HOW CUTE!
ARE YOU STILL LISTENING TO ME, NINOMIYA-SAN?
YES, GO ON WITH THE LESSON.
WELL.
AHEM.
WHILE YOU IMPOSE A FORCE FOR A GIVEN DISTANCE ON AN OBJECT AT REST, THAT OBJECT'S KINETIC ENERGY INCREASES.
F
IMPOSING A FORCE ON AN OBJECT
GENERATES KINETIC ENERGY.
SQUEEEEAL

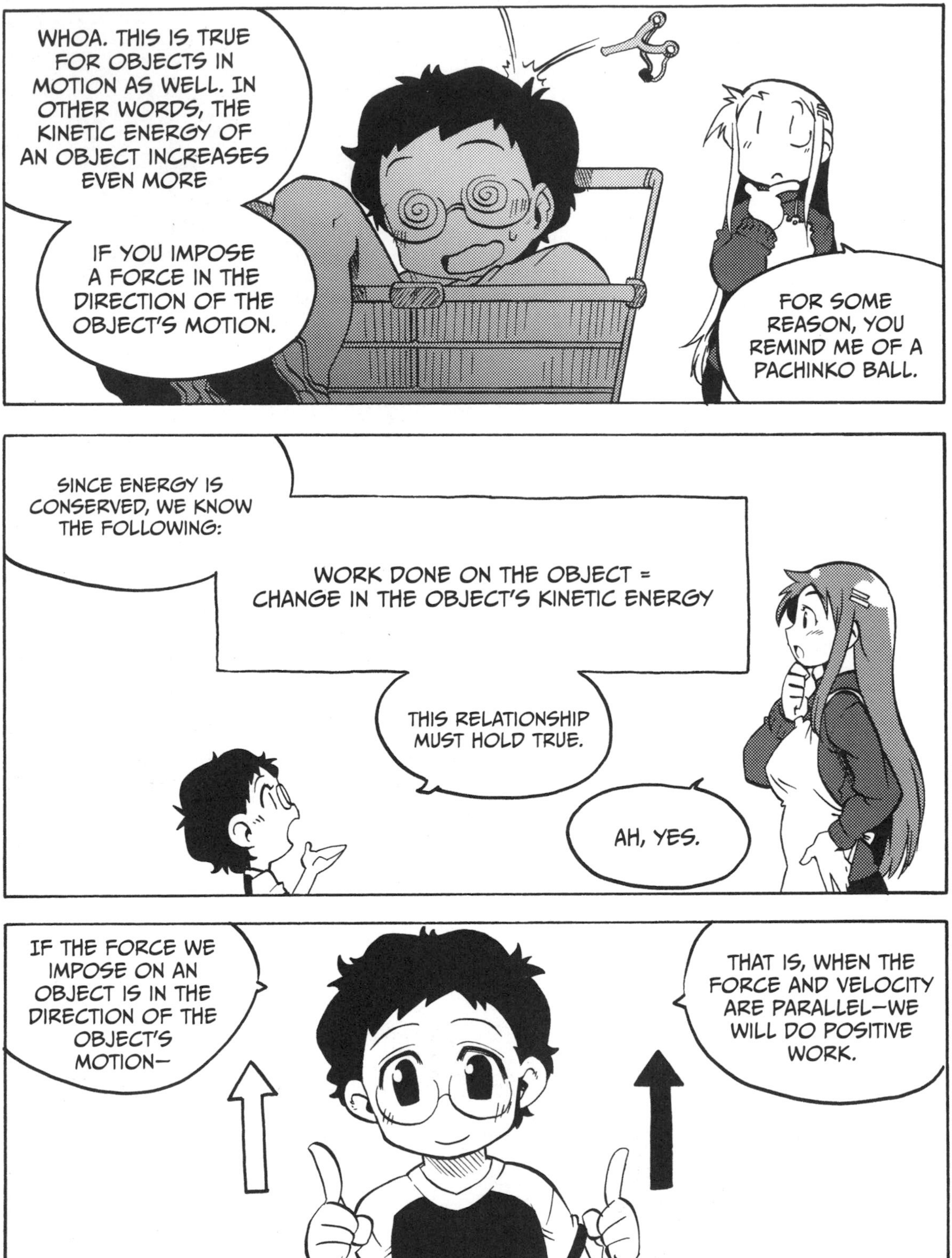
WHOA. THIS IS TRUE FOR OBJECTS IN MOTION AS WELL. IN OTHER WORDS, THE KINETIC ENERGY OF AN OBJECT INCREASES EVEN MORE
IF YOU IMPOSE A FORCE IN THE DIRECTION OF THE OBJECT'S MOTION.
FOR SOME REASON, YOU REMIND ME OF A PACHINKO BALL.
SINCE ENERGY IS CONSERVED, WE KNOW THE FOLLOWING:
WORK DONE ON THE OBJECT = CHANGE IN THE OBJECT'S KINETIC ENERGY
THIS RELATIONSHIP MUST HOLD TRUE.
AH, YES.
IF THE FORCE WE IMPOSE ON AN OBJECT IS IN THE DIRECTION OF THE OBJECT'S MOTION—
THAT IS, WHEN THE FORCE AND VELOCITY ARE PARALLEL—WE WILL DO POSITIVE WORK.

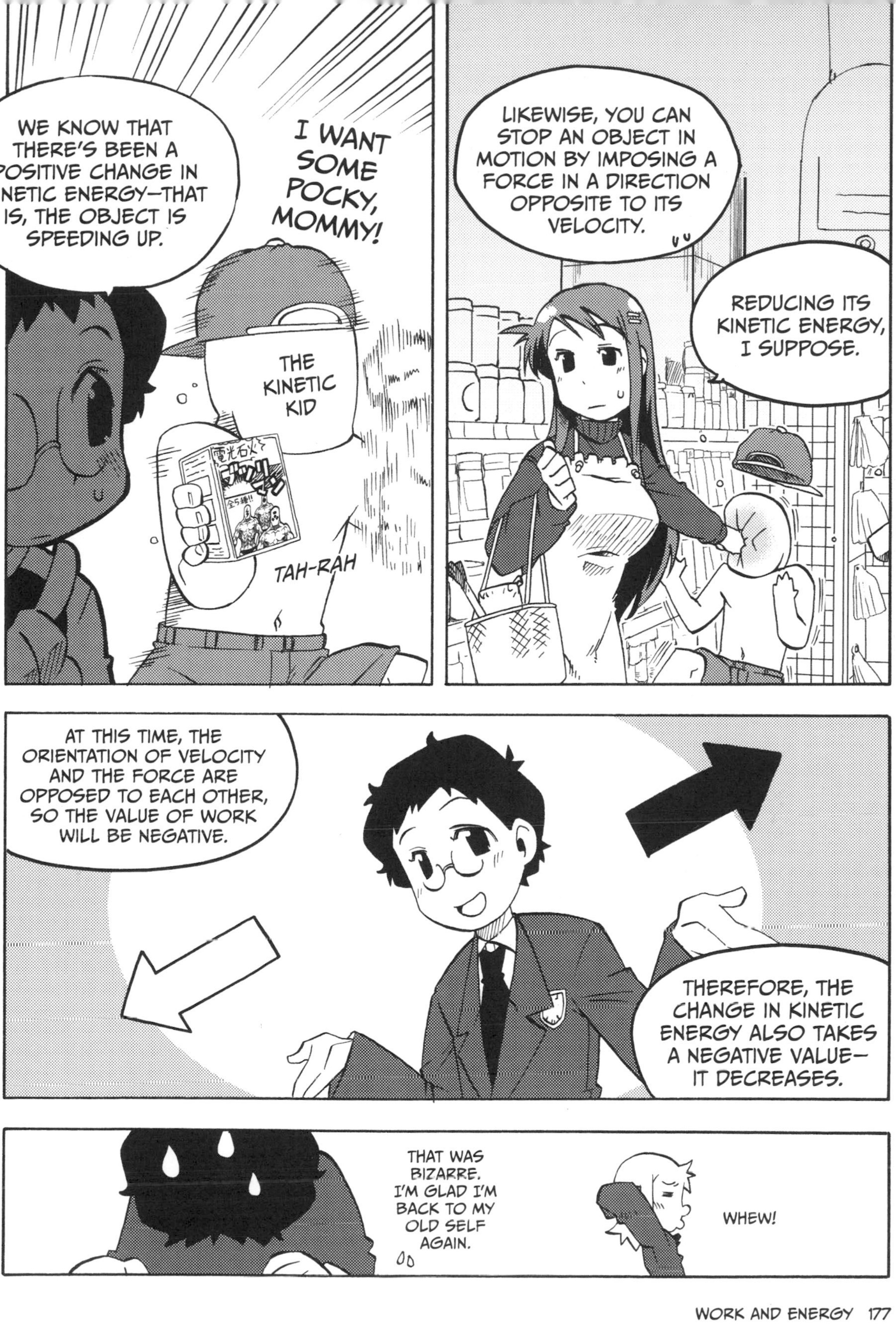
WE KNOW THAT THERE'S BEEN A POSITIVE CHANGE IN KINETIC ENERGY—THAT IS, THE OBJECT IS SPEEDING UP.
I WANT SOME POCKY, MOMMY!
THE KINETIC KID
TAH-RAH
LIKEWISE, YOU CAN STOP AN OBJECT IN MOTION BY IMPOSING A FORCE IN A DIRECTION OPPOSITE TO ITS VELOCITY.
REDUCING ITS KINETIC ENERGY, I SUPPOSE.
AT THIS TIME, THE ORIENTATION OF VELOCITY AND THE FORCE ARE OPPOSED TO EACH OTHER, SO THE VALUE OF WORK WILL BE NEGATIVE.
THEREFORE, THE CHANGE IN KINETIC ENERGY ALSO TAKES A NEGATIVE VALUE—IT DECREASES.
THAT WAS BIZARRE. I'M GLAD I'M BACK TO MY OLD SELF AGAIN.
WHEW!

LABORATORY

THE RELATIONSHIP BETWEEN WORK AND KINETIC ENERGY

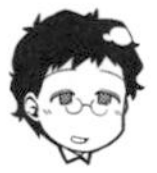

Let's examine how we can derive an equation that expresses the relationship between work and kinetic energy. Suppose we continue to impose force F on a cart in motion, in a direction parallel to that cart's velocity. That cart has mass m and starts with an initial, uniform velocity of v.

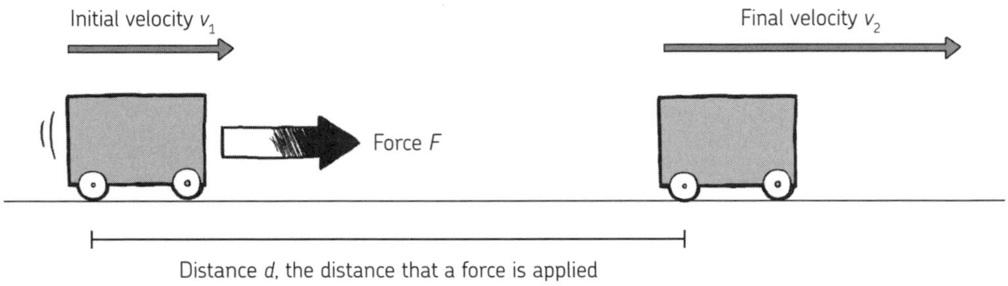
Initial velocity v_1
Final velocity v_2
Force F
Distance d, the distance that a force is applied

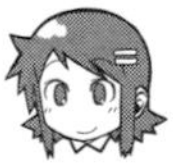

That means an additional force is imposed on the object in motion.

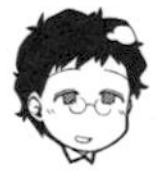

At this time, the following is true:

$$\text{work done on the object} = Fd$$

Also, since we've represented the final velocity as v_2, we can represent the change in the object's kinetic energy as the following:

$$\text{change in kinetic energy} = \tfrac{1}{2}mv_2^{\,2} - \tfrac{1}{2}mv_1^{\,2}$$

And since we already know that the change in kinetic energy is equal to the work done on the object, we can express the following relationship:

$$\tfrac{1}{2}mv_2^{\,2} - \tfrac{1}{2}mv_1^{\,2} = Fd$$

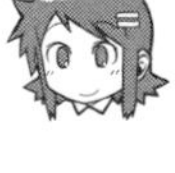

Aha.

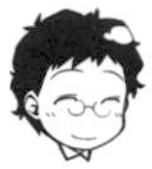

We can also derive this equation another way. Since F is defined as constant, the cart is experiencing uniform acceleration. Therefore, if we represent the cart's acceleration with a, we know that the following must be true:

$$v_2^2 - v_1^2 = 2ad$$

(Why is this so? See expression ❸ on page 85.) To get closer to our original expression, we'll substitute using Newton's second law:

$$F = ma\text{, or rearranged just a little, } a = \frac{F}{m}$$

And we'll get the following:

$$v_2^2 - v_1^2 = \frac{2Fd}{m}$$

Then if you simply multiply both sides by $\frac{1}{2}$, you're there!

$$\tfrac{1}{2}mv_2^2 - \tfrac{1}{2}mv_1^2 = Fd$$

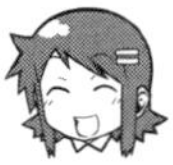

I can get it right if I calculate very carefully.

BRAKING DISTANCE AND SPEED

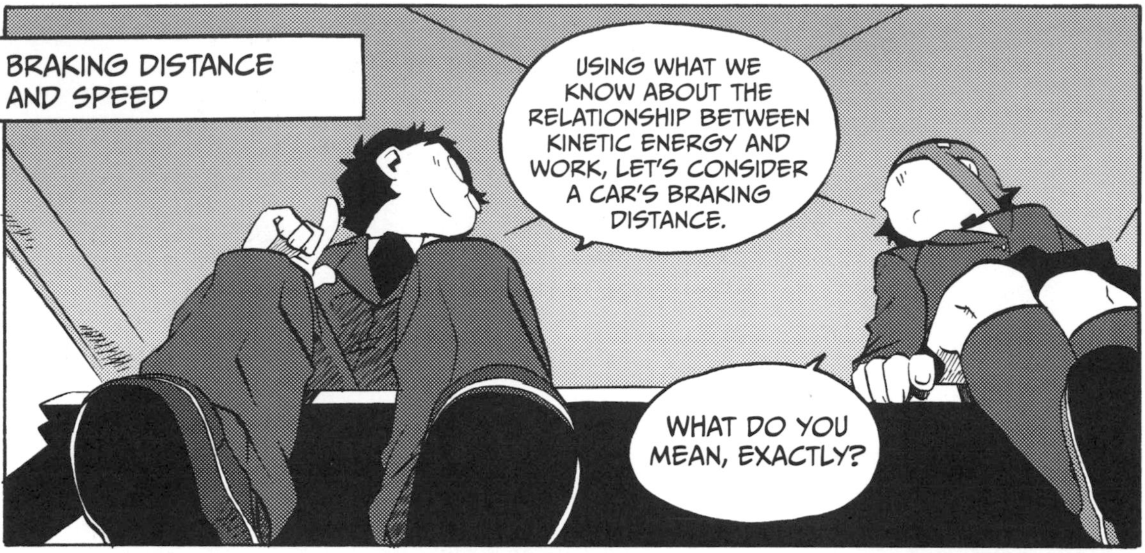
USING WHAT WE KNOW ABOUT THE RELATIONSHIP BETWEEN KINETIC ENERGY AND WORK, LET'S CONSIDER A CAR'S BRAKING DISTANCE.
WHAT DO YOU MEAN, EXACTLY?

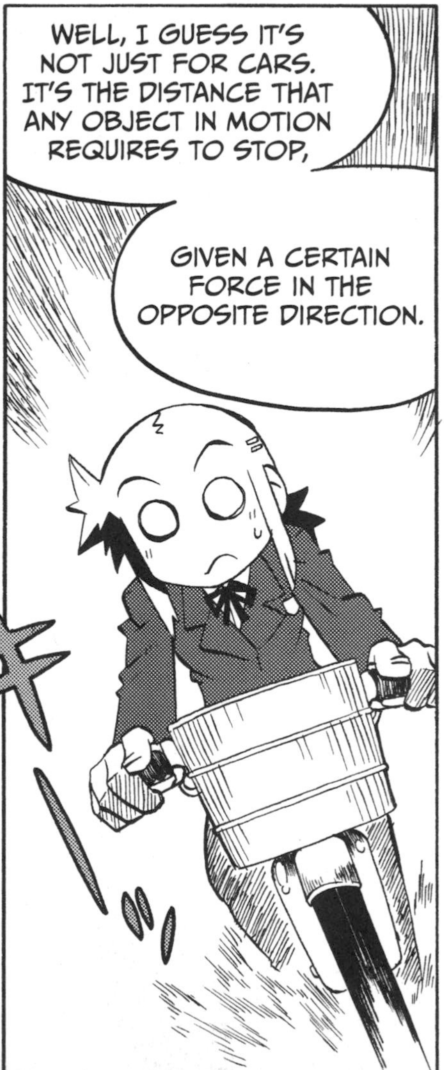
WELL, I GUESS IT'S NOT JUST FOR CARS. IT'S THE DISTANCE THAT ANY OBJECT IN MOTION REQUIRES TO STOP,
GIVEN A CERTAIN FORCE IN THE OPPOSITE DIRECTION.

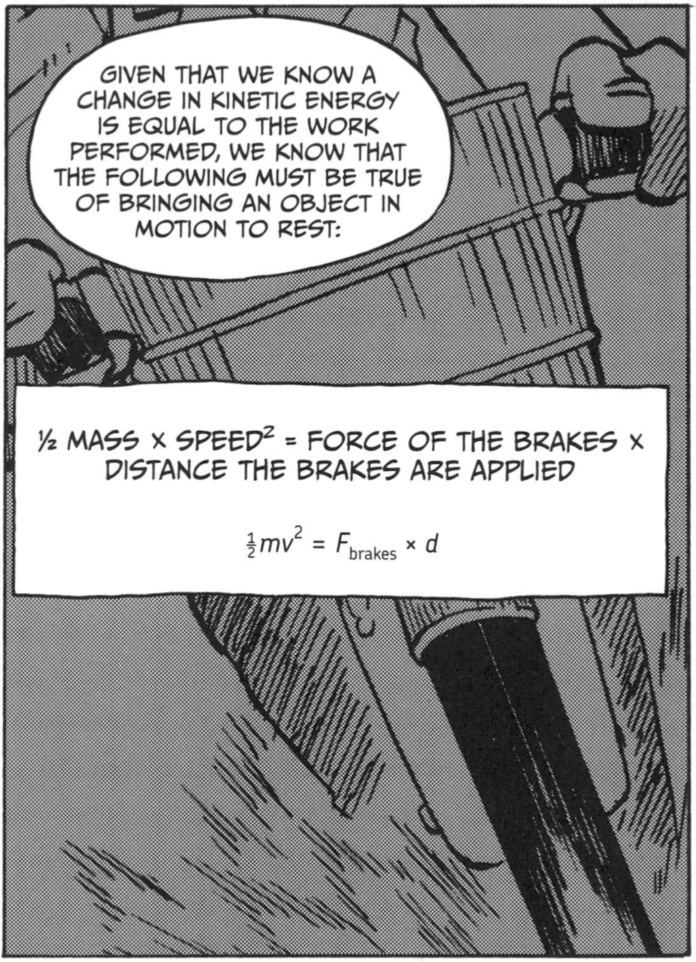
GIVEN THAT WE KNOW A CHANGE IN KINETIC ENERGY IS EQUAL TO THE WORK PERFORMED, WE KNOW THAT THE FOLLOWING MUST BE TRUE OF BRINGING AN OBJECT IN MOTION TO REST:
½ MASS × SPEED² = FORCE OF THE BRAKES × DISTANCE THE BRAKES ARE APPLIED
$\frac{1}{2}mv^2 = F_{brakes} \times d$

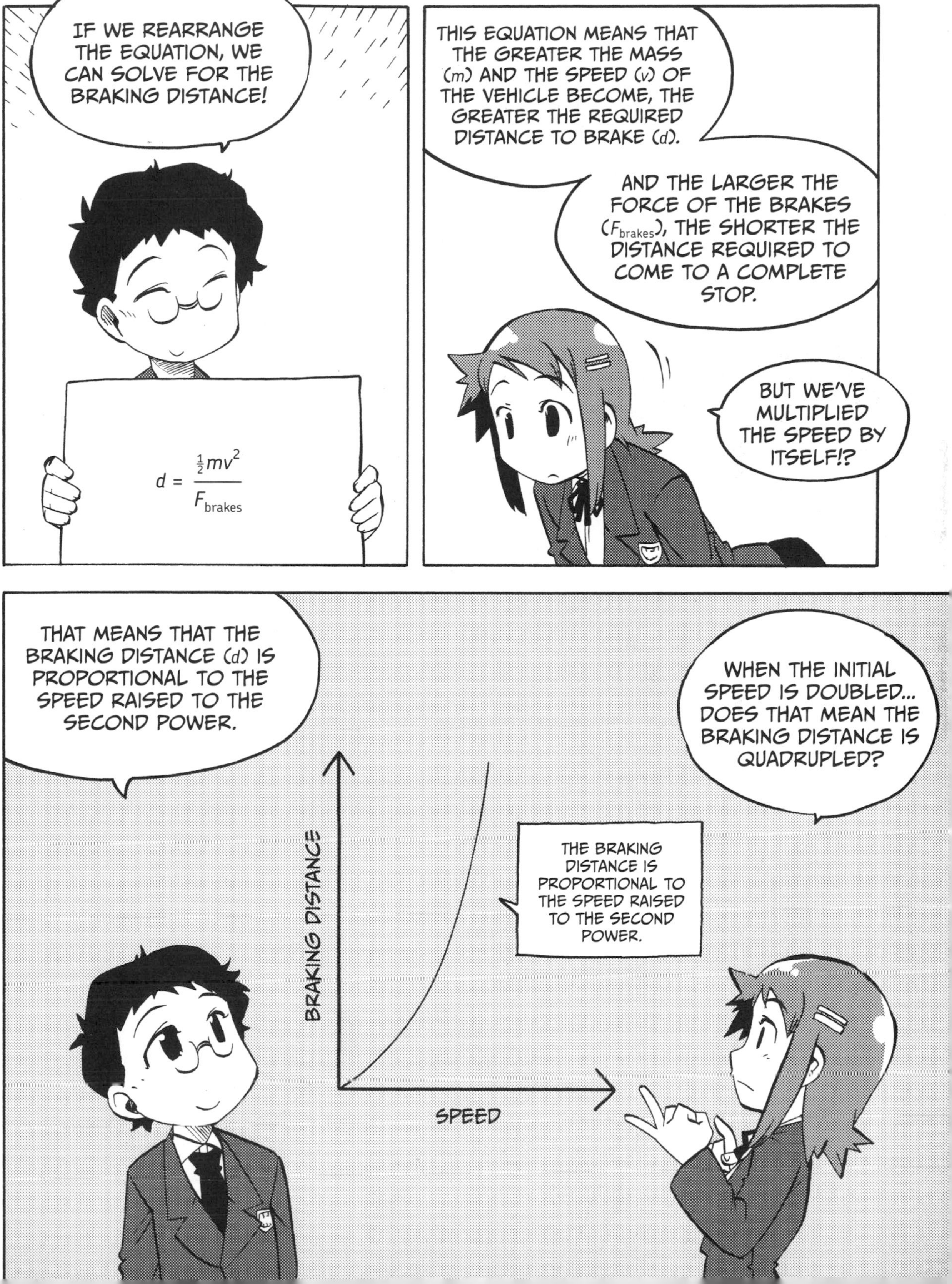
IF WE REARRANGE THE EQUATION, WE CAN SOLVE FOR THE BRAKING DISTANCE!
$d = \frac{\frac{1}{2}mv^2}{F_{brakes}}$
THIS EQUATION MEANS THAT THE GREATER THE MASS (m) AND THE SPEED (v) OF THE VEHICLE BECOME, THE GREATER THE REQUIRED DISTANCE TO BRAKE (d).
AND THE LARGER THE FORCE OF THE BRAKES (F_{brakes}), THE SHORTER THE DISTANCE REQUIRED TO COME TO A COMPLETE STOP.
BUT WE'VE MULTIPLIED THE SPEED BY ITSELF!?
THAT MEANS THAT THE BRAKING DISTANCE (d) IS PROPORTIONAL TO THE SPEED RAISED TO THE SECOND POWER.
WHEN THE INITIAL SPEED IS DOUBLED... DOES THAT MEAN THE BRAKING DISTANCE IS QUADRUPLED?
BRAKING DISTANCE
THE BRAKING DISTANCE IS PROPORTIONAL TO THE SPEED RAISED TO THE SECOND POWER.
SPEED

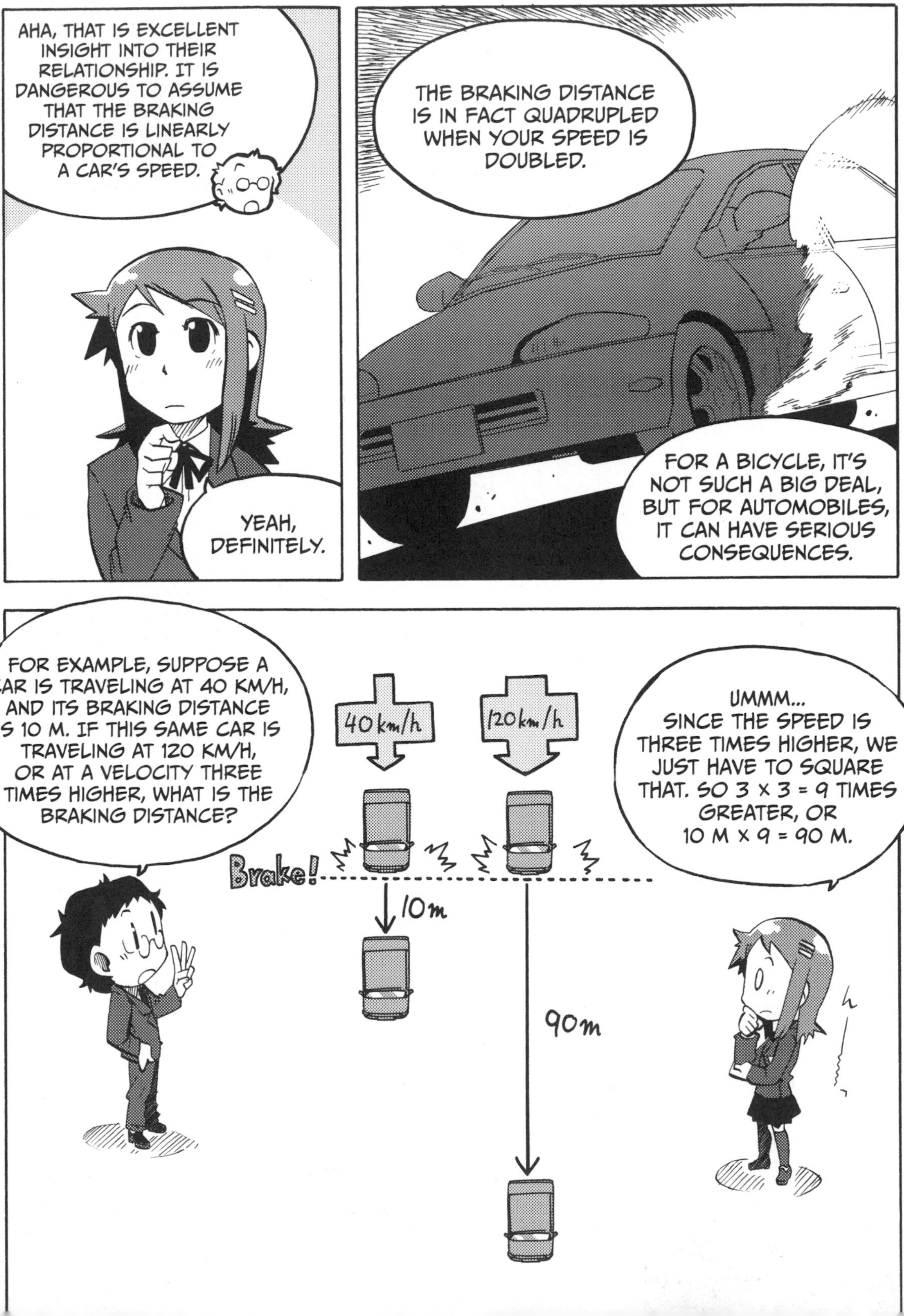
AHA, THAT IS EXCELLENT INSIGHT INTO THEIR RELATIONSHIP. IT IS DANGEROUS TO ASSUME THAT THE BRAKING DISTANCE IS LINEARLY PROPORTIONAL TO A CAR'S SPEED.
YEAH, DEFINITELY.
THE BRAKING DISTANCE IS IN FACT QUADRUPLED WHEN YOUR SPEED IS DOUBLED.
FOR A BICYCLE, IT'S NOT SUCH A BIG DEAL, BUT FOR AUTOMOBILES, IT CAN HAVE SERIOUS CONSEQUENCES.
FOR EXAMPLE, SUPPOSE A CAR IS TRAVELING AT 40 KM/H, AND ITS BRAKING DISTANCE IS 10 M. IF THIS SAME CAR IS TRAVELING AT 120 KM/H, OR AT A VELOCITY THREE TIMES HIGHER, WHAT IS THE BRAKING DISTANCE?
40km/h
120km/h
Brake!
10m
90m
UMMM... SINCE THE SPEED IS THREE TIMES HIGHER, WE JUST HAVE TO SQUARE THAT. SO 3 × 3 = 9 TIMES GREATER, OR 10 M × 9 = 90 M.

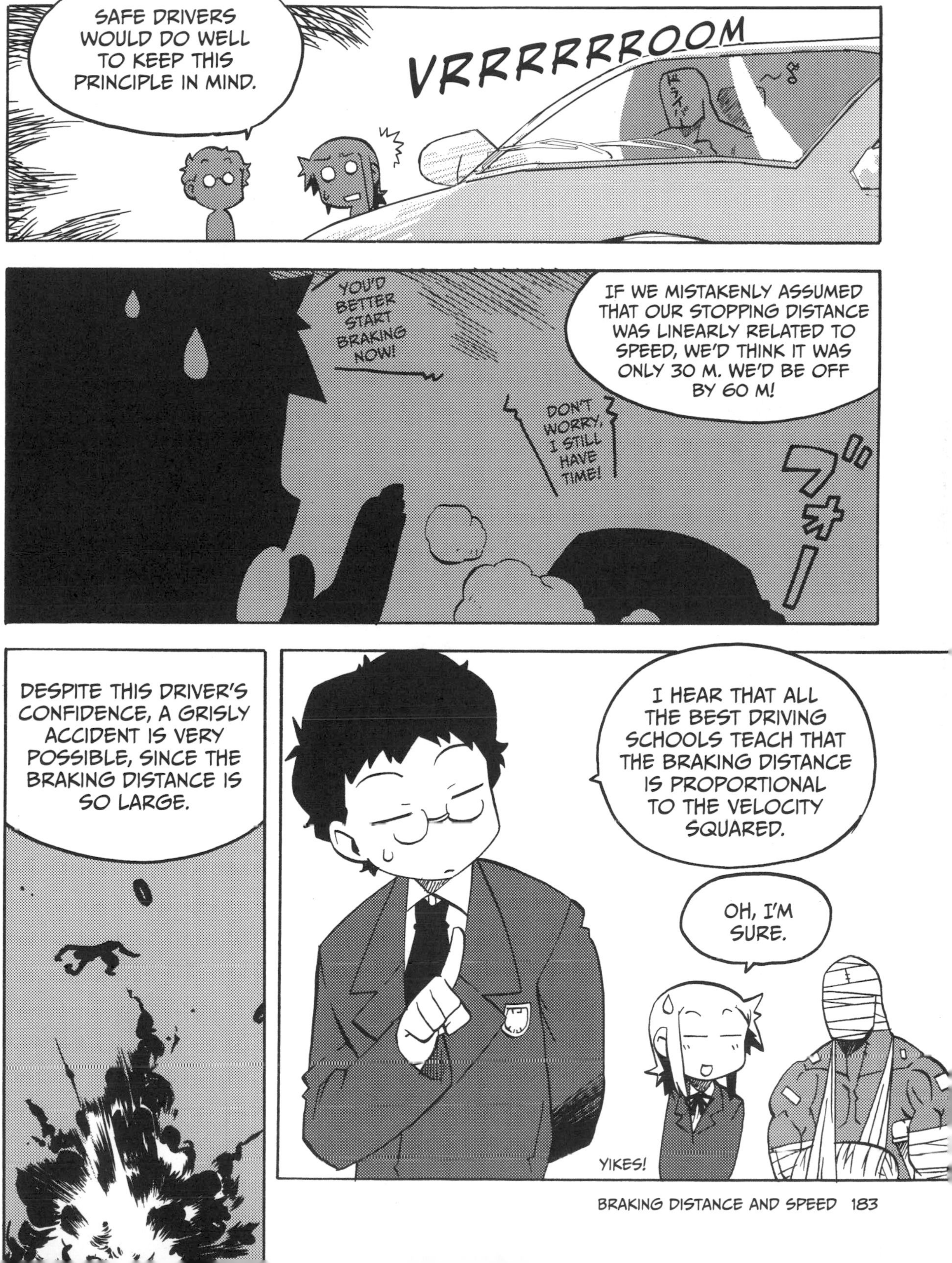
SAFE DRIVERS WOULD DO WELL TO KEEP THIS PRINCIPLE IN MIND.
VRRRRRROOM
YOU'D BETTER START BRAKING NOW!
IF WE MISTAKENLY ASSUMED THAT OUR STOPPING DISTANCE WAS LINEARLY RELATED TO SPEED, WE'D THINK IT WAS ONLY 30 M. WE'D BE OFF BY 60 M!
DON'T WORRY, I STILL HAVE TIME!
DESPITE THIS DRIVER'S CONFIDENCE, A GRISLY ACCIDENT IS VERY POSSIBLE, SINCE THE BRAKING DISTANCE IS SO LARGE.
I HEAR THAT ALL THE BEST DRIVING SCHOOLS TEACH THAT THE BRAKING DISTANCE IS PROPORTIONAL TO THE VELOCITY SQUARED.
OH, I'M SURE.
YIKES!

THE CONSERVATION OF MECHANICAL ENERGY

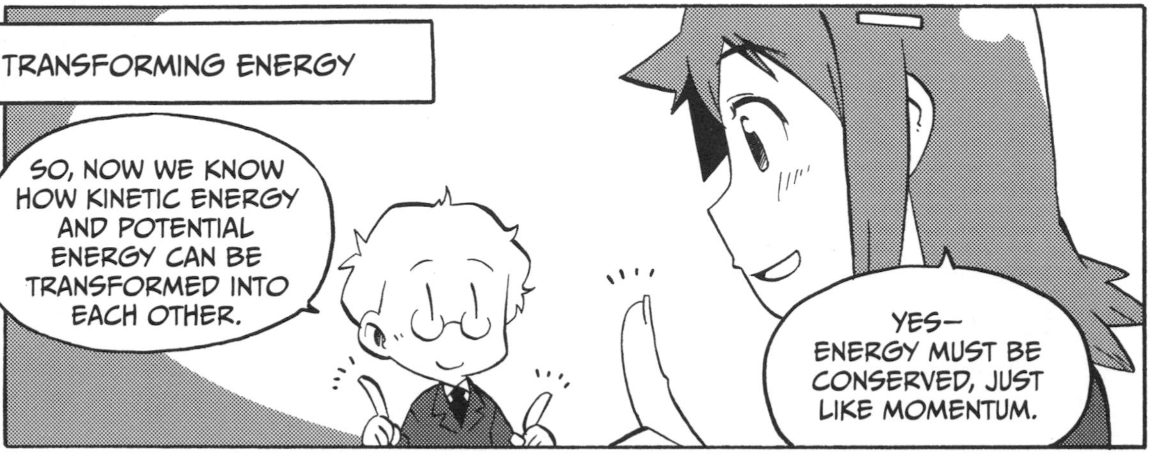
TRANSFORMING ENERGY
SO, NOW WE KNOW HOW KINETIC ENERGY AND POTENTIAL ENERGY CAN BE TRANSFORMED INTO EACH OTHER.
YES— ENERGY MUST BE CONSERVED, JUST LIKE MOMENTUM.

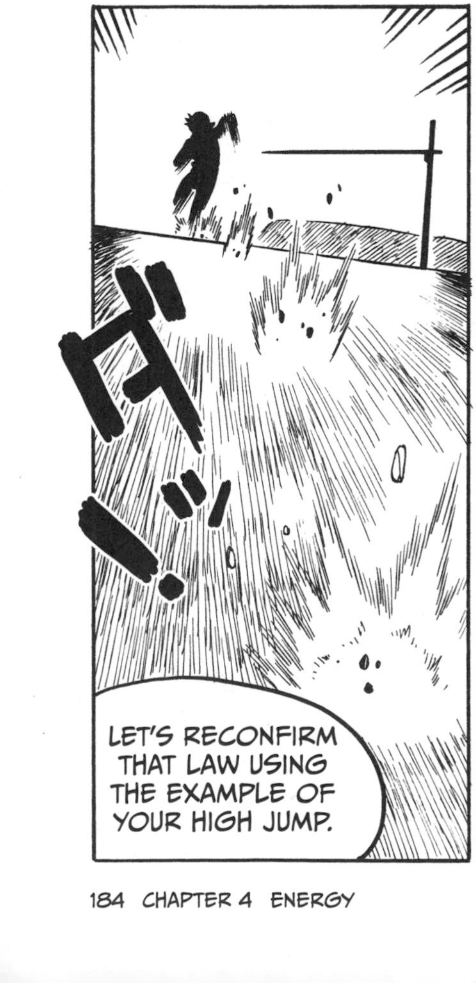
LET'S RECONFIRM THAT LAW USING THE EXAMPLE OF YOUR HIGH JUMP.

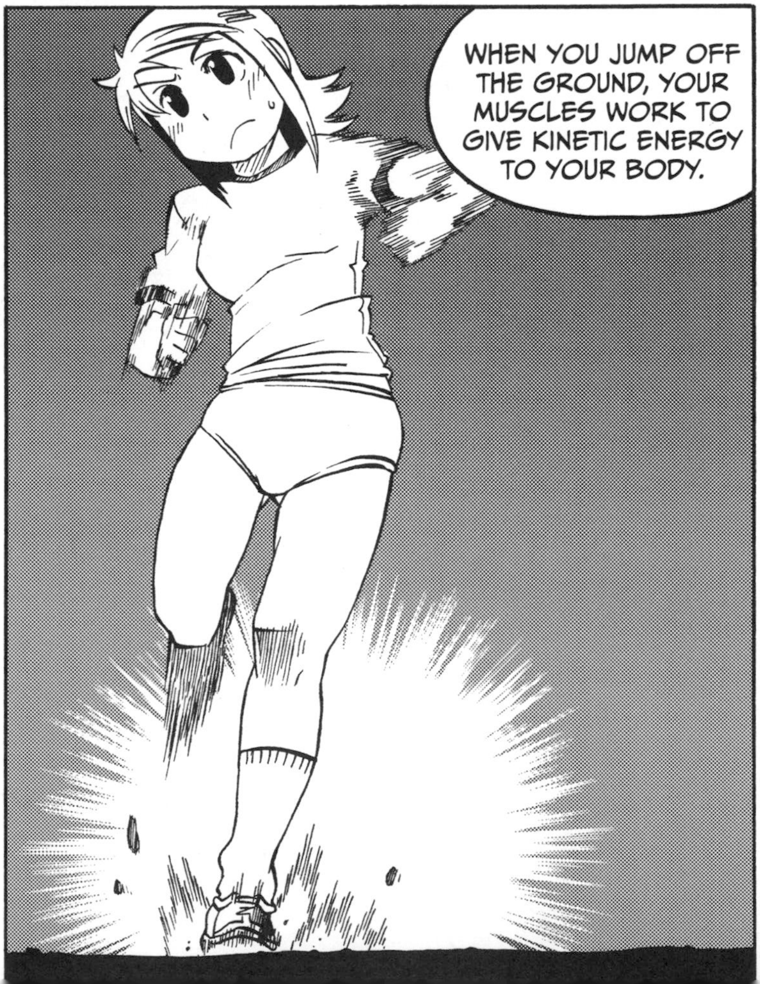
WHEN YOU JUMP OFF THE GROUND, YOUR MUSCLES WORK TO GIVE KINETIC ENERGY TO YOUR BODY.

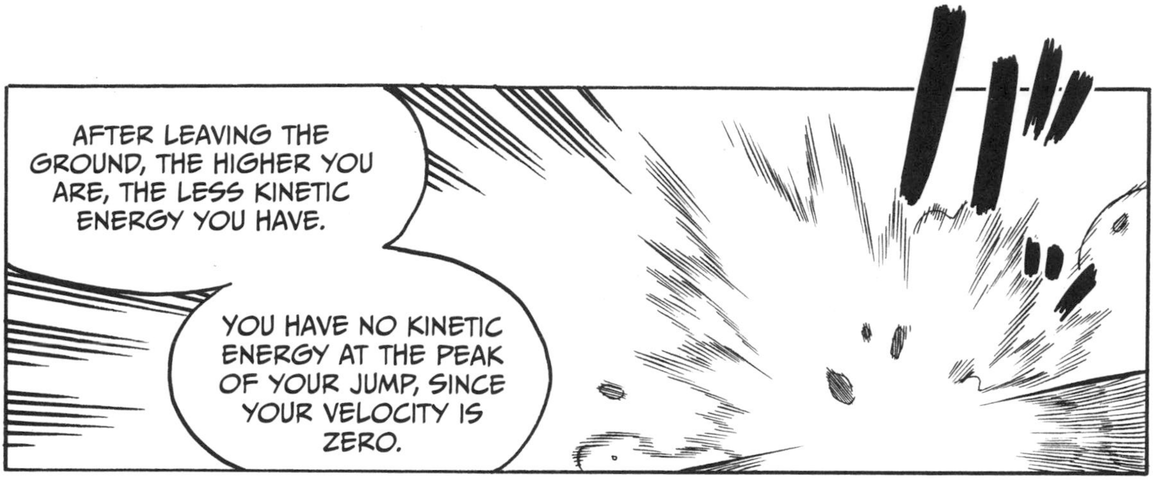
AFTER LEAVING THE GROUND, THE HIGHER YOU ARE, THE LESS KINETIC ENERGY YOU HAVE.
YOU HAVE NO KINETIC ENERGY AT THE PEAK OF YOUR JUMP, SINCE YOUR VELOCITY IS ZERO.

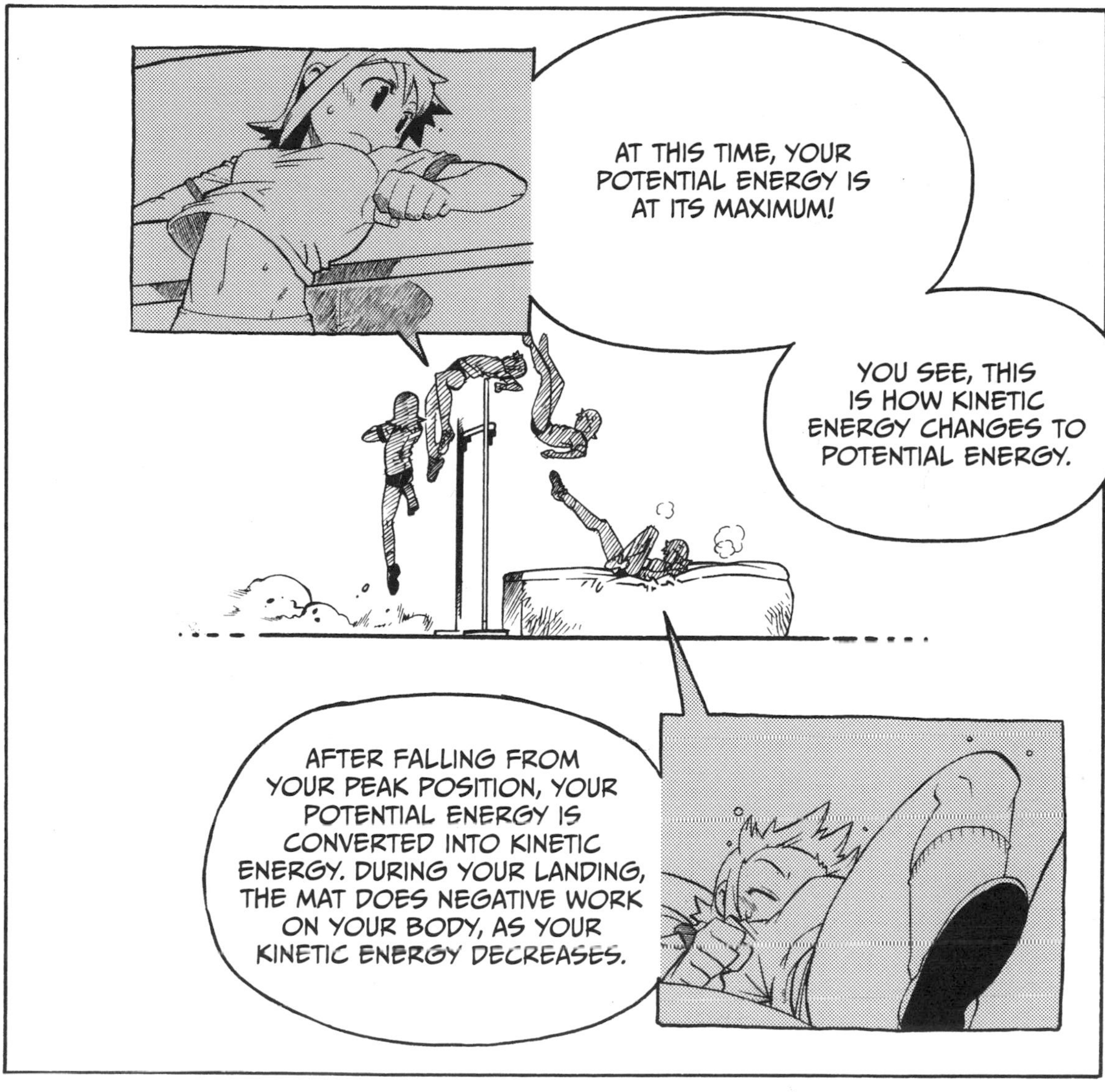
AT THIS TIME, YOUR POTENTIAL ENERGY IS AT ITS MAXIMUM!
YOU SEE, THIS IS HOW KINETIC ENERGY CHANGES TO POTENTIAL ENERGY.
AFTER FALLING FROM YOUR PEAK POSITION, YOUR POTENTIAL ENERGY IS CONVERTED INTO KINETIC ENERGY. DURING YOUR LANDING, THE MAT DOES NEGATIVE WORK ON YOUR BODY, AS YOUR KINETIC ENERGY DECREASES.

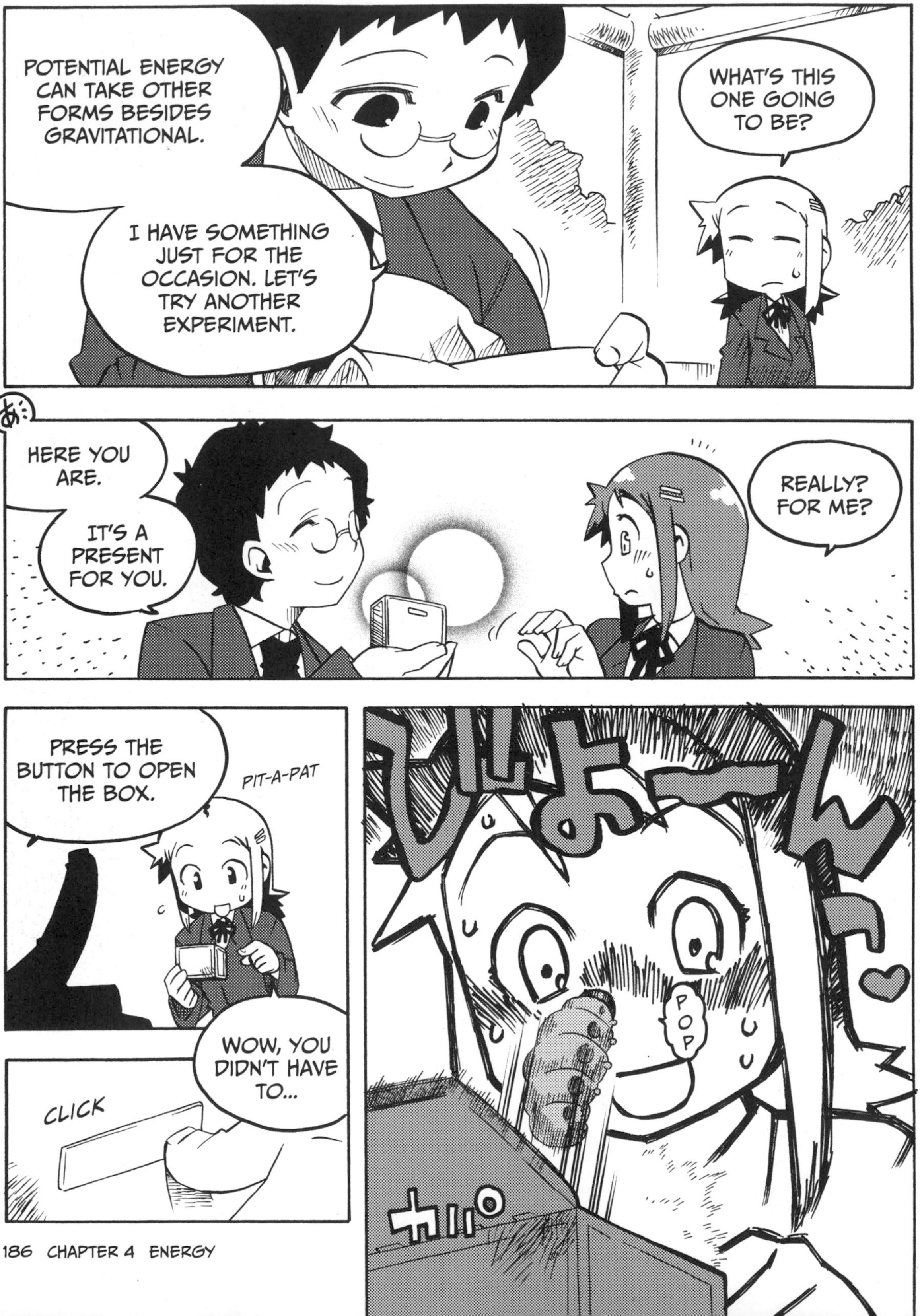
POTENTIAL ENERGY CAN TAKE OTHER FORMS BESIDES GRAVITATIONAL.
I HAVE SOMETHING JUST FOR THE OCCASION. LET'S TRY ANOTHER EXPERIMENT.
WHAT'S THIS ONE GOING TO BE?
HERE YOU ARE.
IT'S A PRESENT FOR YOU.
REALLY? FOR ME?
PRESS THE BUTTON TO OPEN THE BOX.
PIT-A-PAT
WOW, YOU DIDN'T HAVE TO...
CLICK
びよーん
POP
カパ

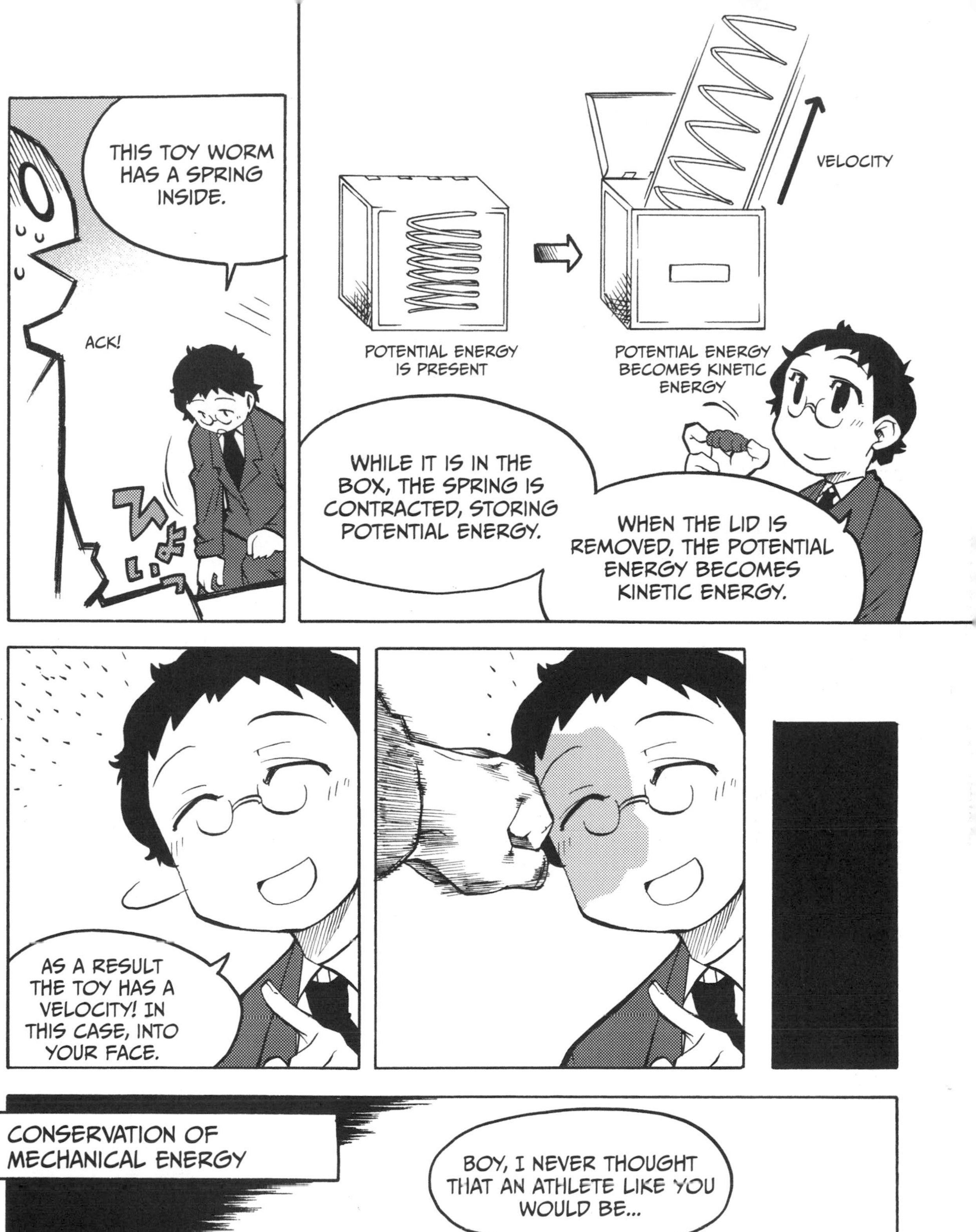
THIS TOY WORM HAS A SPRING INSIDE.
ACK!
VELOCITY
POTENTIAL ENERGY IS PRESENT
POTENTIAL ENERGY BECOMES KINETIC ENERGY
WHILE IT IS IN THE BOX, THE SPRING IS CONTRACTED, STORING POTENTIAL ENERGY.
WHEN THE LID IS REMOVED, THE POTENTIAL ENERGY BECOMES KINETIC ENERGY.
AS A RESULT THE TOY HAS A VELOCITY! IN THIS CASE, INTO YOUR FACE.
CONSERVATION OF MECHANICAL ENERGY
BOY, I NEVER THOUGHT THAT AN ATHLETE LIKE YOU WOULD BE...

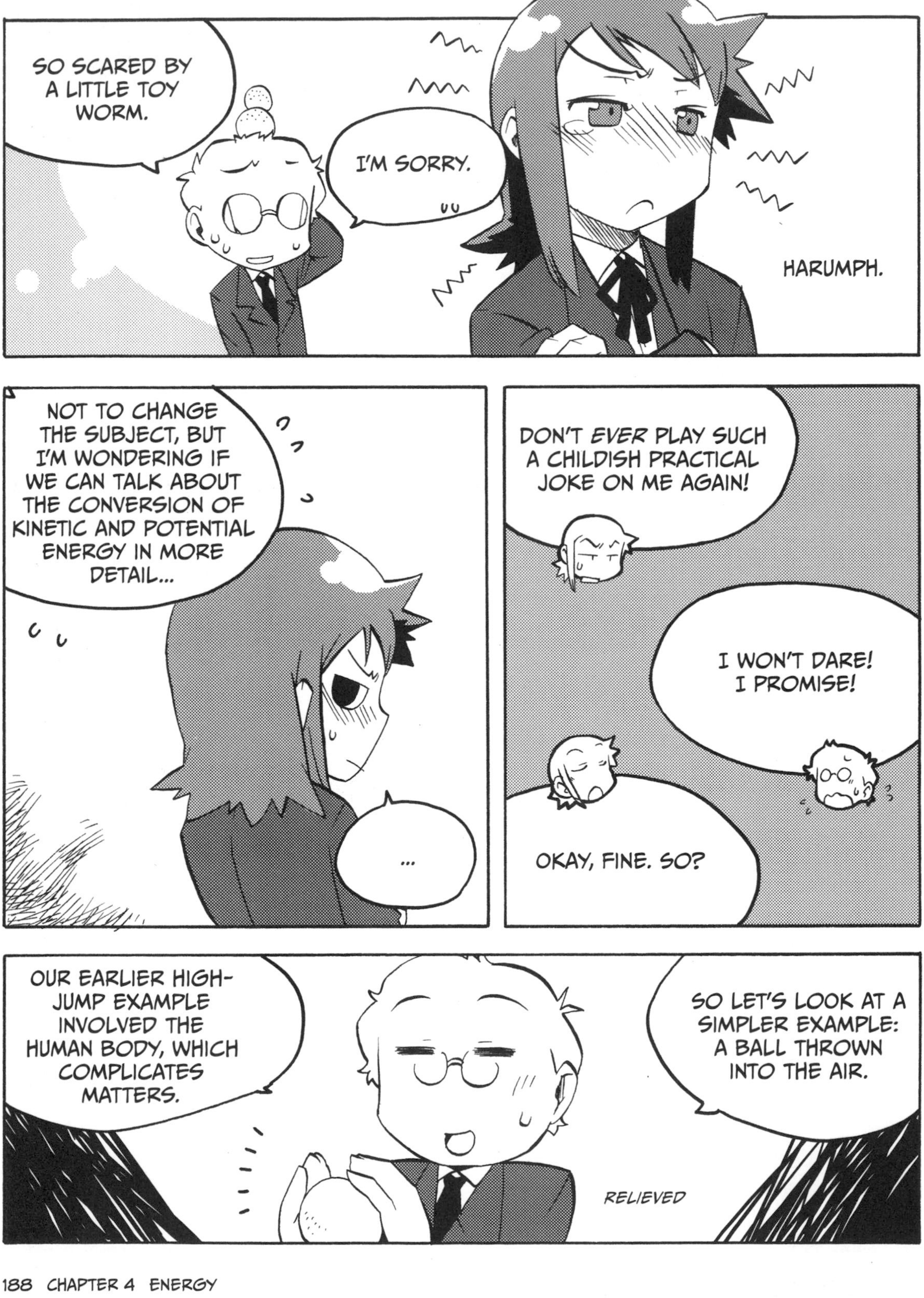
SO SCARED BY A LITTLE TOY WORM.
I'M SORRY.
HARUMPH.
NOT TO CHANGE THE SUBJECT, BUT I'M WONDERING IF WE CAN TALK ABOUT THE CONVERSION OF KINETIC AND POTENTIAL ENERGY IN MORE DETAIL...
...
DON'T *EVER* PLAY SUCH A CHILDISH PRACTICAL JOKE ON ME AGAIN!
I WON'T DARE! I PROMISE!
OKAY, FINE. SO?
OUR EARLIER HIGH-JUMP EXAMPLE INVOLVED THE HUMAN BODY, WHICH COMPLICATES MATTERS.
SO LET'S LOOK AT A SIMPLER EXAMPLE: A BALL THROWN INTO THE AIR.
RELIEVED

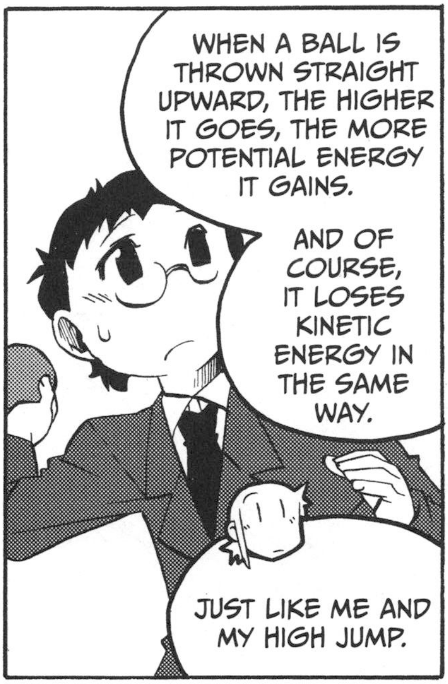
WHEN A BALL IS THROWN STRAIGHT UPWARD, THE HIGHER IT GOES, THE MORE POTENTIAL ENERGY IT GAINS.
AND OF COURSE, IT LOSES KINETIC ENERGY IN THE SAME WAY.
JUST LIKE ME AND MY HIGH JUMP.

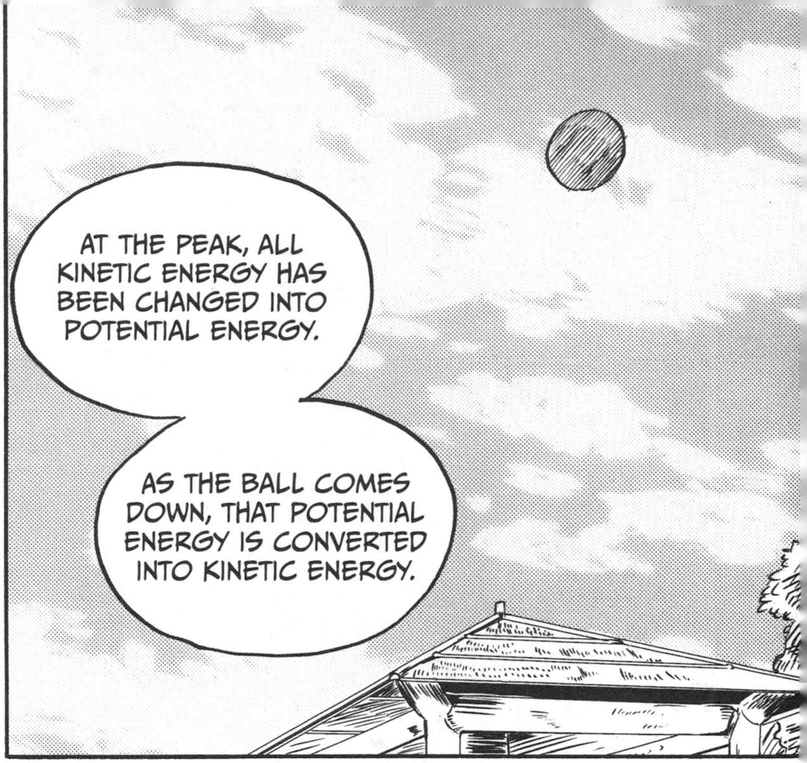
AT THE PEAK, ALL KINETIC ENERGY HAS BEEN CHANGED INTO POTENTIAL ENERGY.
AS THE BALL COMES DOWN, THAT POTENTIAL ENERGY IS CONVERTED INTO KINETIC ENERGY.

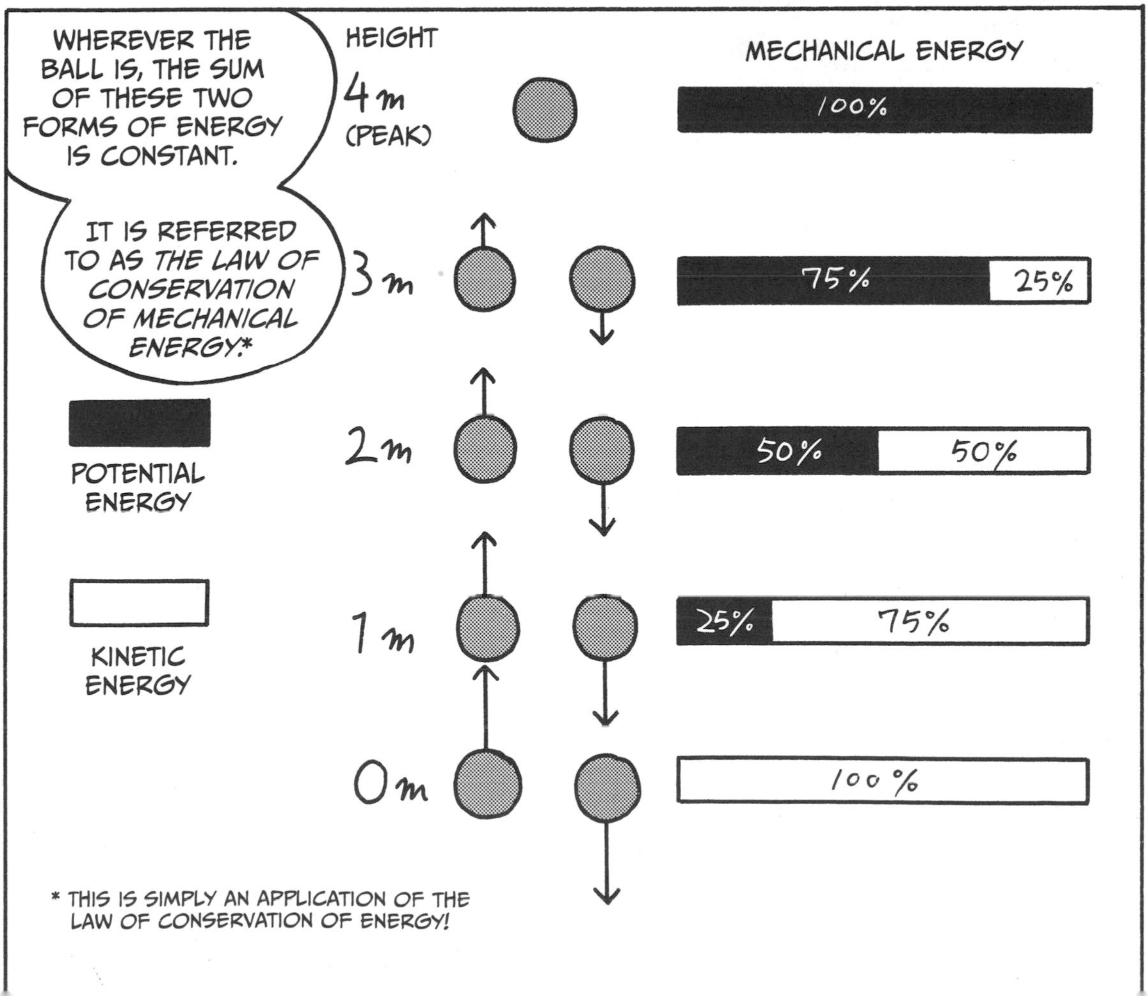
WHEREVER THE BALL IS, THE SUM OF THESE TWO FORMS OF ENERGY IS CONSTANT.
IT IS REFERRED TO AS *THE LAW OF CONSERVATION OF MECHANICAL ENERGY*.*
HEIGHT
MECHANICAL ENERGY
4 m (PEAK)
100%
3 m
75%
25%
2 m
50%
50%
1 m
25%
75%
0 m
100%
POTENTIAL ENERGY
KINETIC ENERGY
* THIS IS SIMPLY AN APPLICATION OF THE LAW OF CONSERVATION OF ENERGY!

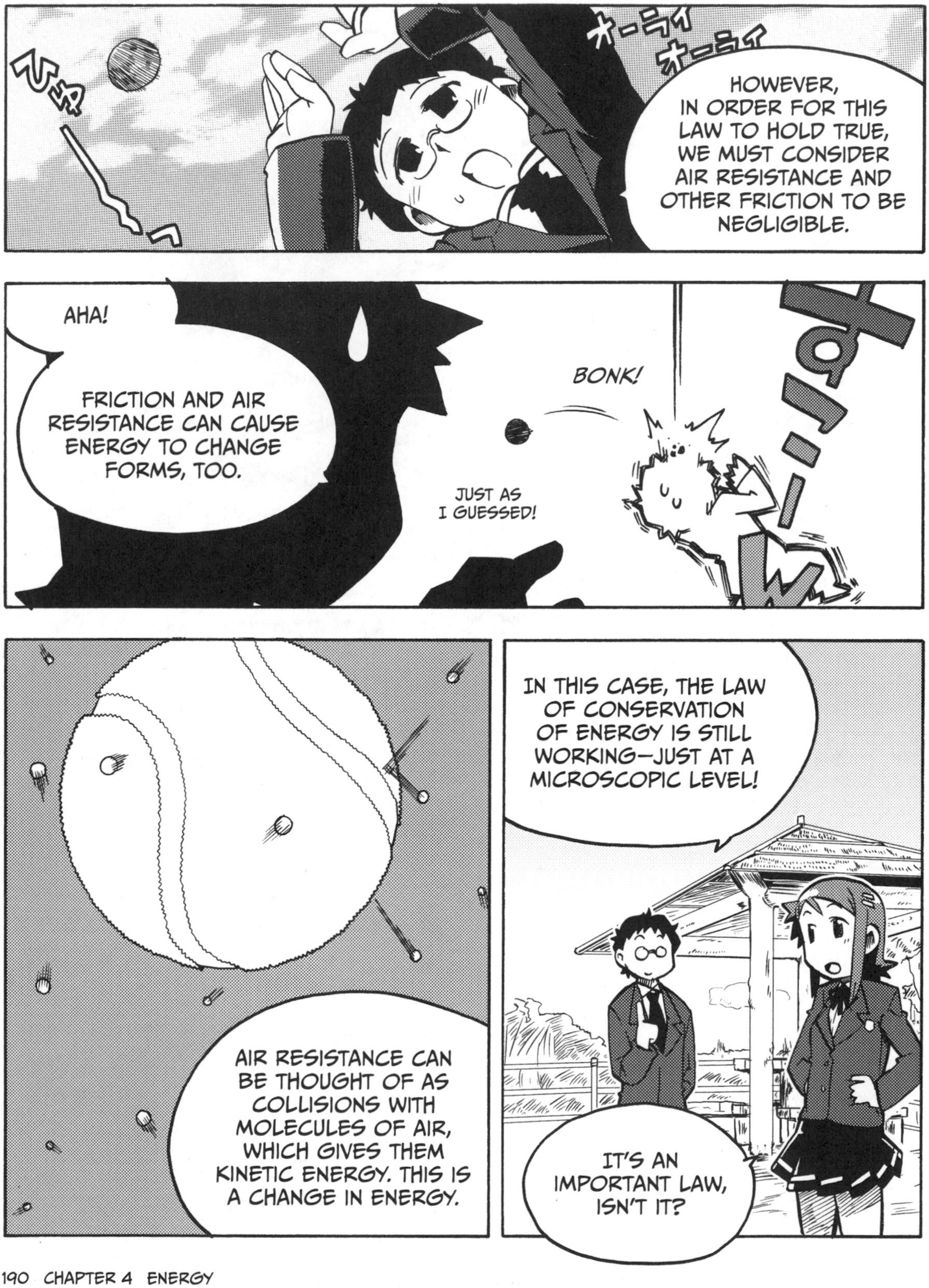
オーライ
オーライ
HOWEVER, IN ORDER FOR THIS LAW TO HOLD TRUE, WE MUST CONSIDER AIR RESISTANCE AND OTHER FRICTION TO BE NEGLIGIBLE.
AHA!
FRICTION AND AIR RESISTANCE CAN CAUSE ENERGY TO CHANGE FORMS, TOO.
BONK!
JUST AS I GUESSED!
AIR RESISTANCE CAN BE THOUGHT OF AS COLLISIONS WITH MOLECULES OF AIR, WHICH GIVES THEM KINETIC ENERGY. THIS IS A CHANGE IN ENERGY.
IN THIS CASE, THE LAW OF CONSERVATION OF ENERGY IS STILL WORKING—JUST AT A MICROSCOPIC LEVEL!
IT'S AN IMPORTANT LAW, ISN'T IT?

LABORATORY

THE LAW OF CONSERVATION OF MECHANICAL ENERGY IN ACTION

Let's prove that the law of conservation of mechanical energy applies when throwing a ball straight upward.

First, we know that the equation for a change in kinetic energy and work is as follows:

❶ $\frac{1}{2}mv_2^2 - \frac{1}{2}mv_1^2 = Fd$

That is:

the change in KE = work

Yes, we confirmed that earlier.

In this case, the work Fd represents the work done by gravity. Assume that the ball starts at height h_1 with velocity v_1. After traveling distance d, it is at height h_2, and its velocity has diminished to v_2. The distance d can be thought of as the change in height—or $h_2 - h_1$.

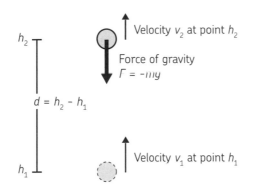

h_2

Velocity v_2 at point h_2

Force of gravity
$F = -mg$

$d = h_2 - h_1$

Velocity v_1 at point h_1

h_1

Yeah, so what's the big deal? Are you trying to show that the force of gravity is doing negative work on the ball?

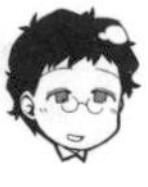

Exactly. The force of gravity is acting against the direction of the velocity. So it's expressed as:

$$F = -mg$$

That means that the work done by the ball (force × distance) is equal to:

$$Fd = -mg(h_2 - h_1)$$

Substituting values from the first equation ❶, we get the following:

$$\tfrac{1}{2}mv_2^{\,2} - \tfrac{1}{2}mv_1^{\,2} = -mg(h_2 - h_1)$$

Now, let's rewrite it a few times, first expanding the terms on the right side:

$$\tfrac{1}{2}mv_2^{\,2} - \tfrac{1}{2}mv_1^{\,2} = mgh_1 - mgh_2$$

Then, make a little switcheroo, and we have something that should be familiar:

$$\tfrac{1}{2}mv_2^{\,2} + mgh_2 = \tfrac{1}{2}mv_1^{\,2} + mgh_1$$

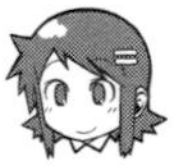

Yes, it is. It's showing that the sum of the kinetic energy and potential energy at both h_1 and h_2 must be the same.

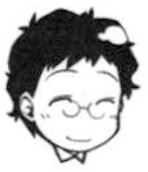

Yes, that's it exactly.

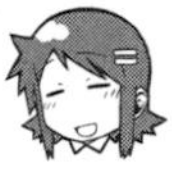

So the left side of this equation is the total mechanical energy at point h_2, and the right side is the total mechanical energy at point h_1.

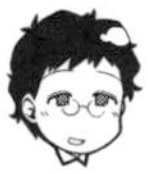

Yes, we've derived an equation that indicates that the sum of the mechanical energy must be equal at any two points of a ball's path, when it is thrown directly into the air.

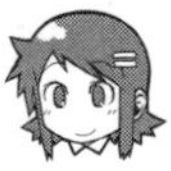

Yes, I see that.

Now, let's use this equation to calculate something a bit different—the velocity (v_1) at which you need to throw a ball to reach a certain *maximum* height (h_2). Since the ball's velocity reaches zero at the peak, we know it has no kinetic energy at that time.

And for simplicity's sake, let's set h_1 equal to 0—that is, we'll measure h_2 from the ball's launching point. That is, h_2 will equal d, the distance the ball travels.

This means that the kinetic energy the ball has at its launching point must equal the potential energy it has at its height.

Therefore, the following is true:

$$PE_2 = KE_1$$

$$mgd = \tfrac{1}{2}mv_1^2$$

Wait, I think I see something interesting here—mass appears on both sides of this equation. That means that the mass does not affect the relationship!

You're right! Let's solve for the initial velocity v_1:

$$mgd = \tfrac{1}{2}mv_1^2$$

$$gd = \tfrac{1}{2}v_1^2$$

$$2gd = v_1^2$$

$$\sqrt{2gd} = v_1$$

If we just use real numbers in this equation, we can find the required initial velocity to reach a particular height!

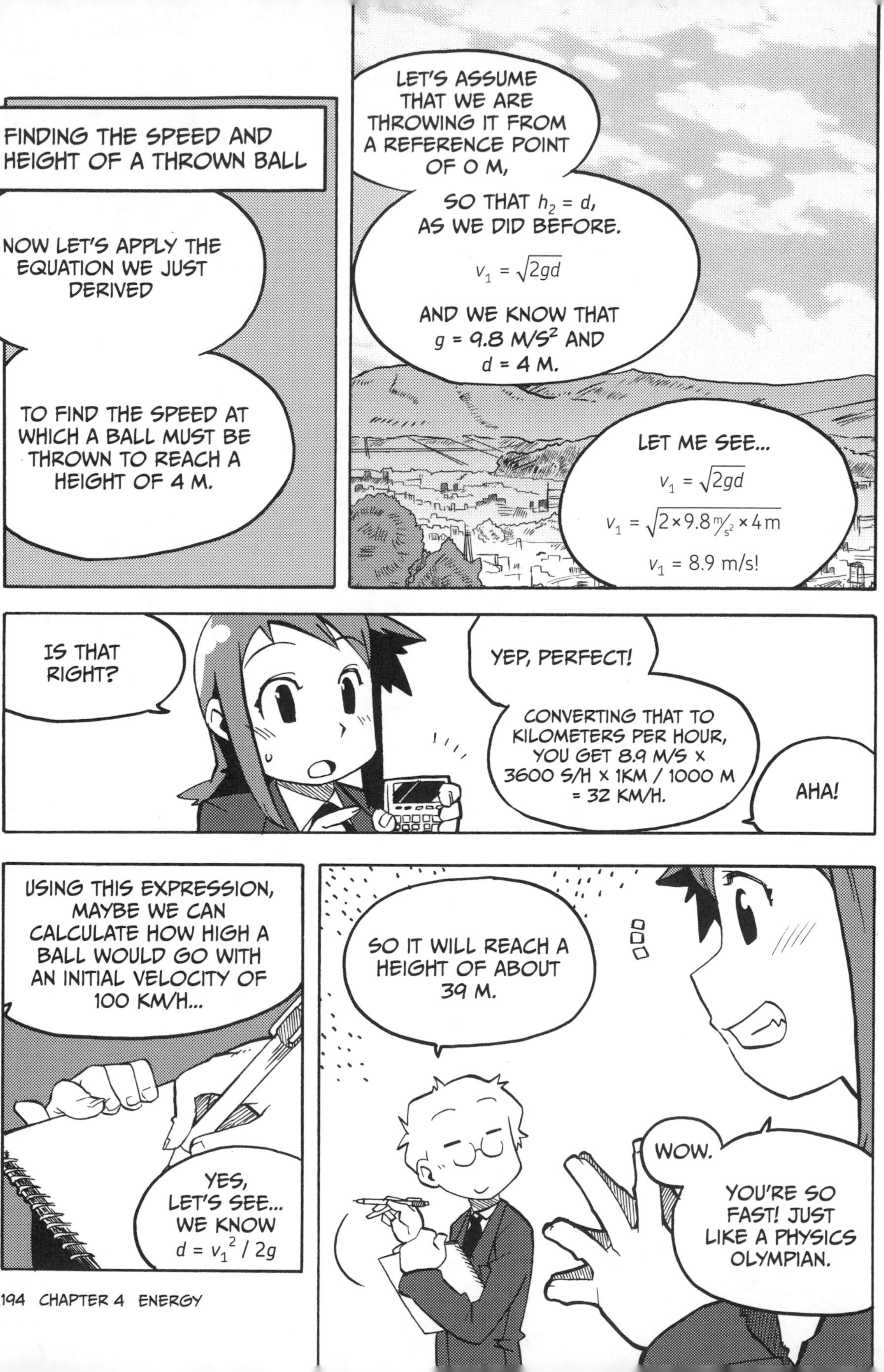

FINDING THE SPEED AND HEIGHT OF A THROWN BALL
NOW LET'S APPLY THE EQUATION WE JUST DERIVED
TO FIND THE SPEED AT WHICH A BALL MUST BE THROWN TO REACH A HEIGHT OF 4 M.
LET'S ASSUME THAT WE ARE THROWING IT FROM A REFERENCE POINT OF 0 M,
SO THAT $h_2 = d$, AS WE DID BEFORE.
$v_1 = \sqrt{2gd}$
AND WE KNOW THAT g = 9.8 M/S² AND d = 4 M.
LET ME SEE...
$v_1 = \sqrt{2gd}$
$v_1 = \sqrt{2 \times 9.8\,\text{m/s}^2 \times 4\,\text{m}}$
v_1 = 8.9 m/s!
IS THAT RIGHT?
YEP, PERFECT!
CONVERTING THAT TO KILOMETERS PER HOUR, YOU GET 8.9 M/S × 3600 S/H × 1KM / 1000 M = 32 KM/H.
AHA!
USING THIS EXPRESSION, MAYBE WE CAN CALCULATE HOW HIGH A BALL WOULD GO WITH AN INITIAL VELOCITY OF 100 KM/H...
YES, LET'S SEE... WE KNOW $d = v_1^2 / 2g$
SO IT WILL REACH A HEIGHT OF ABOUT 39 M.
WOW.
YOU'RE SO FAST! JUST LIKE A PHYSICS OLYMPIAN.

LABORATORY

CONSERVATION OF MECHANICAL ENERGY ON A SLOPE

 The law of conservation of mechanical energy holds true, even when you're not talking about balls in the air, right? Wouldn't it work for lots of other situations, too, like an object on a slope?

 Well, let's examine a case where you slide a box from height h to height 0. On the way down, let's assume that the box attains velocity v_A at height h_A, velocity v_B at height h_B, and so on.

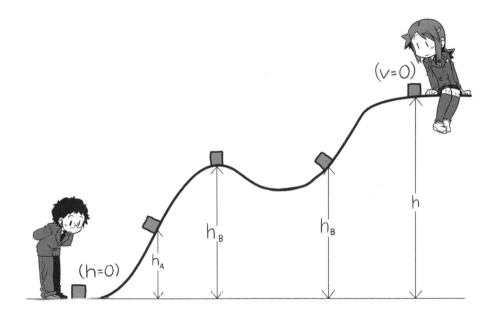

Since $v = 0$ at the highest height, the initial potential energy the box has is equal to all its mechanical energy. But we also know that the potential energy at point h is mgh, so we could express that as:

$$PE_h = mgh$$

Now, how can you express the kinetic energy (KE_0) the box has at point 0?

We already know that kinetic energy is equal to this:

$$KE_0 = \tfrac{1}{2}mv^2$$

Exactly! And we know that kinetic energy at $h = 0$ must equal the potential energy at point h:

$$PE_h = KE_0$$

But furthermore, due to the conservation of energy, we know that the sum of the mechanical energy must stay the same at all intermediate points on this slope. That is:

$$KE_A + PE_A = KE_B + PE_A$$

$$\tfrac{1}{2}mv_A^2 + mgh_A = \tfrac{1}{2}mv_B^2 + mgh_B$$

And this also implies that the potential energy is equivalent at two points of the same height, like point B in the figure. At these two points, the box's kinetic energy is equivalent, even if the orientation of its velocity is different.

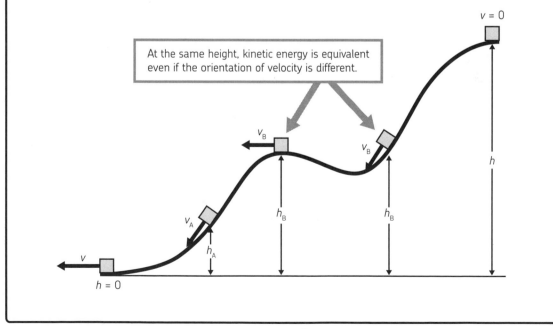

At the same height, kinetic energy is equivalent even if the orientation of velocity is different.

Kinetic energy is not associated with the orientation of velocity!

 Yes, sir! Er, ma'am. Kinetic energy only has a magnitude. Similarly, potential energy only depends on height.

 If we extended this slope, would it be possible for the box to go back up to its original height again?

 Yes, it would be possible, provided that friction and air resistance are negligible. Of course, it'd be impossible to go beyond that original height of h.

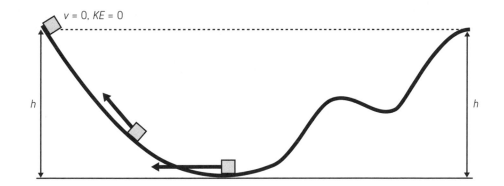

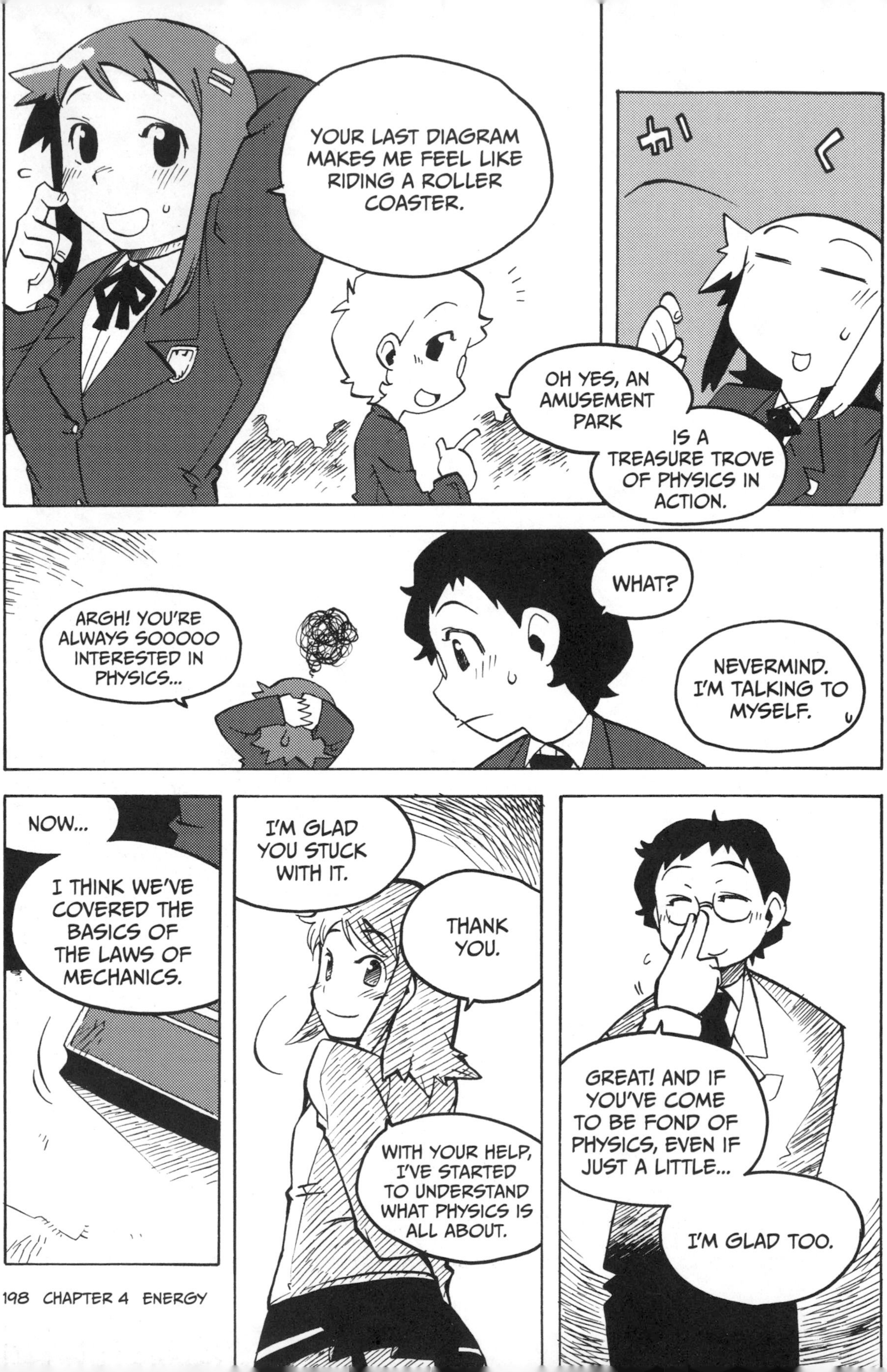
YOUR LAST DIAGRAM MAKES ME FEEL LIKE RIDING A ROLLER COASTER.
OH YES, AN AMUSEMENT PARK
IS A TREASURE TROVE OF PHYSICS IN ACTION.
ARGH! YOU'RE ALWAYS SOOOOO INTERESTED IN PHYSICS...
WHAT?
NEVERMIND. I'M TALKING TO MYSELF.
NOW...
I THINK WE'VE COVERED THE BASICS OF THE LAWS OF MECHANICS.
I'M GLAD YOU STUCK WITH IT.
THANK YOU.
WITH YOUR HELP, I'VE STARTED TO UNDERSTAND WHAT PHYSICS IS ALL ABOUT.
GREAT! AND IF YOU'VE COME TO BE FOND OF PHYSICS, EVEN IF JUST A LITTLE...
I'M GLAD TOO.

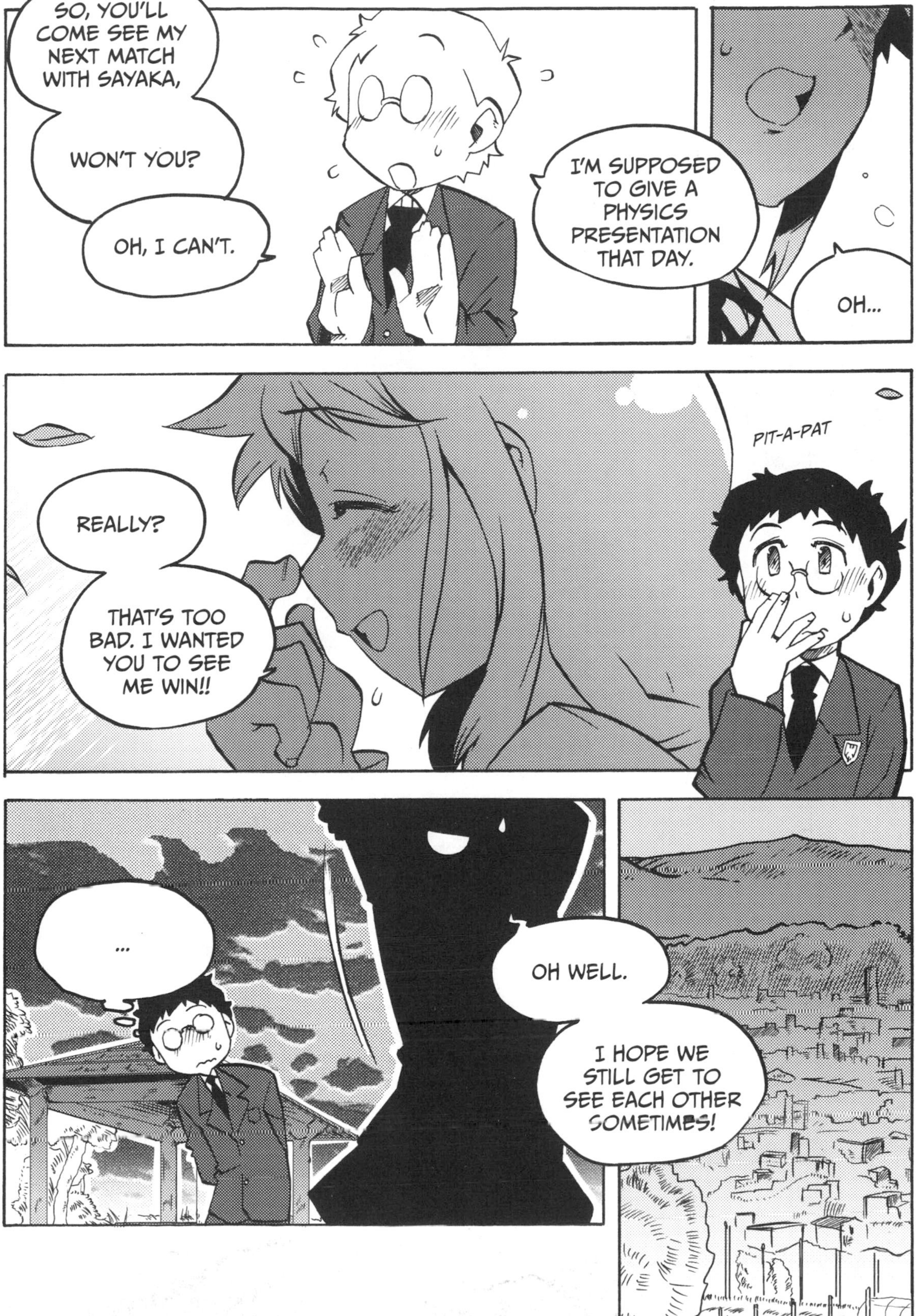
SO, YOU'LL COME SEE MY NEXT MATCH WITH SAYAKA,
WON'T YOU?
OH, I CAN'T.
I'M SUPPOSED TO GIVE A PHYSICS PRESENTATION THAT DAY.
OH...
REALLY?
THAT'S TOO BAD. I WANTED YOU TO SEE ME WIN!!
PIT-A-PAT
...
OH WELL.
I HOPE WE STILL GET TO SEE EACH OTHER SOMETIMES!

UNITS FOR MEASURING ENERGY

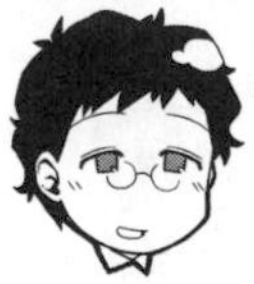

The units for energy can be found by applying the definition of mechanical energy, which is as follows:

$$\text{kinetic energy} = \tfrac{1}{2} \times \text{mass} \times (\text{speed})^2$$

From the expression above, we find the following:

$$\text{units for energy} = \text{units for mass} \times \text{units for speed} \times \text{units for speed}$$

$$1 \text{ joule} = \text{kg} \times \text{m}^2 / \text{s}^2$$

NOTE *As $\frac{1}{2}$ does not affect the units, you can omit it when determining the units.*

Since energy is a very common physical quantity, a special unit, the *joule (J)*, is assigned to it. On the other hand, given the fact that the variation in kinetic energy equals the work done (which we learned on page 176), the following is true:

$$\text{units for energy} = \text{units for work}$$

Therefore, the following expression is also true:

$$\text{units for work} = \text{units for force} \times \text{units for distance} = (\text{N}) \times (\text{m}) = (\text{N} \times \text{m})$$

At first glance, this unit, (N × m), looks different from a joule (kg × m^2/s^2). However, recall that a newton (N) is simply equal to 1 kg × m/s^2. So by multiplying force and distance, we do indeed have the same unit.

To get an idea about how much energy is represented by 1J, it is useful to keep in mind that 1J equals 1 (N × m). In other words, you can say, "1J represents energy generated from work that moves an object by 1 m and continues to impose a force of 1N on it."

Additionally, given that the force of gravity on an object with mass of 1 kg is 9.8N, the mass of an object under exactly 1N of gravity is 1 / 9.8 kg = 0.102 kg = 102 g. This is what I meant when I said, "One joule is equivalent to the energy required to lift a 102 g object directly upward 1 meter" (on page 161).

Besides the joule, another common unit for measuring energy is the *calorie (cal)*, which is used for thermal appliances such as heaters and food. One calorie (1 cal) represents the thermal energy required for increasing the temperature of one gram of water by 1°C under one atmosphere of pressure (1 atm). Relative to a joule, this unit is defined as follows: 1 cal = 4.2J.

When talking about food, the *kilocalorie (kcal)* is used. One kilocalorie is defined as 1,000 calories. Although the term *calorie* is used informally when talking about food and diet, the scientific unit being referred to is in fact the kilocalorie.

For example, the energy in 50 g of ice cream is about 100 kcal. If you convert it into joules, you get the following:

$$100 \text{ kcal} = 100000 \text{ cal} = 4.2 \times 100000\text{J} = 420000\text{J}$$

It seems like quite a high value, but it isn't really, if you compare it to the amount of energy we need to survive. According to data from the Japanese Ministry of Health, Labor and Welfare, the daily energy requirement is about 2,200 kcal for a 17-year-old female and about 2,700 kcal for a 17-year-old male. Kilocalories are converted into joules like this:

$$2200 \text{ kcal} \times 1000 \text{ cal/kcal} \times 4.2 \text{J/cal} = 9240000 \text{J}$$

Let's see how much that is. Since the energy required to lift a load with mass of 1 kg one meter is 9.8J, that value is nearly the amount of energy required to lift a mass of one million kilograms just one meter! That indicates we need a tremendous quantity of energy every day in order to maintain life.

POTENTIAL ENERGY

Kinetic energy resides in an object in motion. In contrast, potential energy is not stored inside an object—it's usually energy that comes from an object's position. Typical forms of potential energy include gravitational potential energy and the potential energy of an electrostatic field, which provides the attractive and repulsive force of electricity.

You can also regard the elastic energy of springs and rubber as a form of potential energy. However, different factors are involved in storing this potential energy in different materials. The resilience of springs comes from the spring's contraction to its original state—a spring wants to recover its stable initial position after gaps between atoms (dependent on the potential energy of the electrostatic field working among atoms) are slightly displaced. A coil-type spring used in the real world is designed to transform minor distortion occurring on a straight metal rod into greater displacement by adopting a coiled shape.

Natural state
Expanded
Distorted
Contracted
Force

State with wider gaps between atoms (unstable, with a high potential energy level)
Natural state of gaps between atoms (stable, with a low potential energy level)
State with smaller gaps between atoms (unstable, with a high potential energy level)

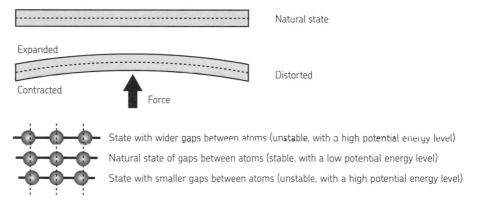

On the other hand, the elasticity of rubber originates in the activity of polymer molecules to recover the initial state with a greater "disorder," where they are very closely curled up, after a state with a lower "disorder," in which molecules are expanded and aligned.

Polymer molecules of rubber

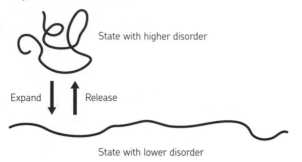

State with higher disorder

Expand | ↑ Release

State with lower disorder

SPRINGS AND THE CONSERVATION OF ENERGY

Let's think about resilience of a spring as an example of the conservation of energy.

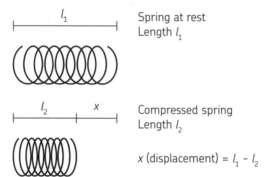

Spring at rest
Length l_1

Compressed spring
Length l_2

x (displacement) = $l_1 - l_2$

When you compress[*] a spring with spring constant k (you can think of k as a measure of how springy your spring is in N/m) by x (the distance beyond its natural length), the potential energy stored in the spring can be expressed as follows:

$$PE = \tfrac{1}{2}kx^2$$

This stored energy is called *elastic potential energy*. If we place a mass m next to this spring and the opposite side is fixed, what is the force it will receive? And what will its velocity be?

---

* Note that springs work the same way if you stretch them, as well. These equations will hold true in cases of stretching and compression.

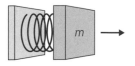

The spring wants to return to its natural state and will exert a force on mass m.

Well, we know that due to the conservation of energy, that mass's kinetic energy must be equal to the spring's potential energy. That means that the following must hold true:

$$PE_{spring} = KE_{mass}$$

$$\tfrac{1}{2}kx^2 = \tfrac{1}{2}mv^2$$

Solving for v, we get:

$$v = \sqrt{\frac{kx^2}{m}}$$

WARNING: CALCULUS AHEAD!

Also, as the spring expands, we know that the object is subject to force:

$$F = \frac{-d}{dx}\left(\tfrac{1}{2}kx^2\right) = -kx$$

Without saying, calculation of work done when the spring with resilient force $F = -kx$ expands by amount x relative to its natural length gives us the following:

$$W = \int_{-x}^{0}(-kx)dx = \tfrac{1}{2}kx^2$$

This matches potential energy. This is only reasonable, given the conservation of energy.

VELOCITY FOR THROWING UPWARD AND HEIGHT ATTAINED

On page 194, in response to Megumi's question about how high a ball would go if it was thrown with an initial velocity of 100 km/h, I answer that it is 39 m.

Let's find out why. Given that we know that the following equation holds, you can solve for h, the height attained by the object thrown·

$$v_1^2 = 2gh$$

$$h = \frac{v_1^2}{2g}$$

Now, plugging in some real numbers for this, we know that 100 km/h equals the following:

$$100\,\frac{\text{km}}{\text{h}} \times 1000\,\frac{\text{m}}{\text{km}} \times \frac{1\text{ h}}{3600\text{ s}} = 27.78\,\frac{\text{m}}{\text{s}}$$

Now let's put that value into our equation and see what we find:

$$h = \frac{v_1^{2}}{2g}$$

$$h = \frac{27.782}{2 \times 9.8\text{ m/s}^2}$$

$$h = 39.36\text{ m}$$

THE ORIENTATION OF FORCE AND WORK

As you know, we represent work in terms of force and the distance (or the displacement) a force is applied to an object. Let's consider an object being moved over displacement *d*, undergoing a force *F* as shown below.

Force $\vec{F}$
d_x
Displacement

$\vec{F}$
θ
$\vec{F}\sin\theta = \vec{F}_y$
$\vec{F}\cos\theta = \vec{F}_x$

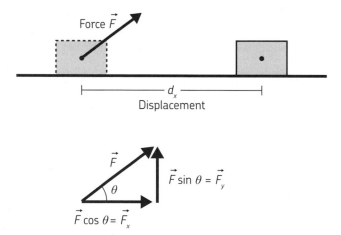

When the orientations of a force and displacement do not match, we must take this into account. In the example above, work (W) is represented as follows:

$$W = \vec{F}_x d_x + \vec{F}_y d_y$$

We've split the forces and displacements into their horizontal (x) and vertical (y) components. However, in this case, we know that the vertical displacement of the box is 0, as the box is moving on level ground. Therefore, we can neglect that term in our calculation of the total work done on the box:

$$W = \vec{F}_x d_x$$

$$W = \vec{F} \cos \theta \times d_x$$

It's also worth noting that we just performed a dot product. So . . . what's that? Well, work and energy are scalars—they have no orientation. But force and displacement are both vectors, as they have an orientation. The multiplication of two vectors in this way is called a *dot product*.

In a case where the force is in a direction opposite to the displacement, the work is said to be negative. This kind of work results in a deceleration.

Additionally, when the orientation of force is perpendicular to the displacement, given that cos 90 = 0, no work is done. A typical case in which the orientation of force is perpendicular to that of displacement is uniform circular motion. While force is working toward the center of the circle (centripetal force), kinetic energy does not change because the value of work is zero. Because of this, an object can move in a circular direction at a uniform speed.

The orientation of the velocity matches that of the displacement.

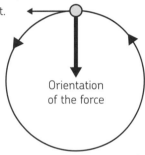

Orientation of the force

FINDING AN AMOUNT OF WORK WITH NONUNIFORM FORCE (ONE-DIMENSIONAL)

WARNING: CALCULUS AHEAD!

In the case of a uniform force, we can express work as the product of the displacement and the force in the direction of the displacement. But many times, forces are not constant.

To deal with nonuniform forces, we can break up the force into short segments. If we break it up into tiny enough pieces, we can say the force is constant during each segment. We can look at any one of these segments, which we will label with a subscript i, and the work can still be expressed as the product we have seen before:

$$\tfrac{1}{2}mv_{i+1}^{2} - \tfrac{1}{2}mv_i^{2} = F\Delta x$$

Of course, this is true for every segment i, so we can add up all of these to find the work done across the entire displacement, from x_1 to x:

$$\left(\tfrac{1}{2}mv_2^{\,2} - \tfrac{1}{2}mv_1^{\,2}\right) + \left(\tfrac{1}{2}mv_3^{\,2} - \tfrac{1}{2}mv_2^{\,2}\right) + \left(\tfrac{1}{2}mv_4^{\,2} - \tfrac{1}{2}mv_3^{\,2}\right) + \ldots =$$
$$F_1\Delta x + F_2\Delta x + F_3\Delta x + \ldots$$

By looking closely at the left-hand side, you can see that most of the terms will cancel! We are left with just two terms:

$$\tfrac{1}{2}mv_n^{\,2} - \tfrac{1}{2}mv_1^{\,2}$$

So we can rewrite this equation as:

$$\tfrac{1}{2}mv_n^{\,2} - \tfrac{1}{2}mv_1^{\,2} = \sum_{i=1}^{n} F_i\Delta x$$

We have added up the little pieces of work done at each instant to get the total change in energy—the work W. You'll see that this looks awfully similar to the definition of an integral. It turns out that if we make the sections infinitely small by making n go to infinity, then we can change the summation to an integral by the rules of calculus:

$$W = \lim_{\substack{\Delta x \to 0 \\ n \to \infty}} \sum_{i=1}^{n} F_i\Delta x$$

$$W = \int_{x_1}^{x} F(x)dx$$

NOTE F *here does not denote a function. Remember, that* F *stands for* force!

This is much easier to see graphically, as all we are doing is adding up the area under the curve on a plot of F vs. x. An integral is exactly this, in the limit of making the width of the segment go to zero.

F
W = Σ_{i=1}^{n} F_i Δx
x_1
Δx
x
F
W = ∫_{x_1}^{x} F(x)dx
x_1
x

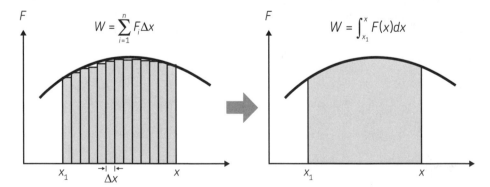

In conclusion, the statement, the change in kinetic energy between two points is equivalent to the work done on the object within that segment, means the following:

$$W = \int_{x_1}^{x} F(x)dx$$

When the above is assumed, the statement can be expressed as:

$$\frac{1}{2}mv^2 - \frac{1}{2}mv_0^2 = W$$

Note that v in the equation above equals the object's final velocity, v_n.

NONCONSERVATIVE FORCE AND THE LAW OF CONSERVATION OF ENERGY

Not all forces can be expressed as having a potential. Forces such as these are said to be *nonconservative*. Friction is a typical nonconservative force. When a nonconservative force is doing work, the energy of a system goes down. For example, if you push a book across a table, it will slide to a stop. This doesn't mean that energy is not conserved—just that it went somewhere that you can't easily get it back. For example, the book gave kinetic energy to the molecules of the table in the form of heat.

FRICTION: A NONCONSERVATIVE FORCE

Now let's examine the force of friction, an example of a nonconservative force. First, let's assume a mass of m is in motion with velocity v_1.

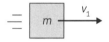

If the object had no forces working on it, it would continue to travel with velocity v_1 forever—that's just Newton's first law in action. But life isn't so simple. Let's assume this object's motion is impeded by the force of friction between the bottom of the object and the surface that it's traveling on.

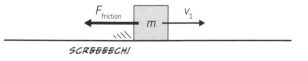

SCREEEECH!

The magnitude of this force depends on two factors: the normal force and the coefficient of friction. But what are those, you ask? Well, the *normal force* is simply the force *perpendicular* to the surface a body travels on. The larger an object's mass, the larger the normal force, and the larger the force of friction itself. In the example above, the normal force is simply the weight of the mass, ($F = ma$, so in this case, $F_{normal} = m \times g$). We'll look at a more complicated example of normal forces shortly to see how the normal force differs from the weight of an object.

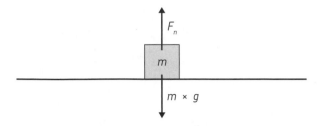

The *coefficient of friction* is simply a measure of how "sticky" two surfaces are. Rubber on concrete, for example, has a very high coefficient of friction. But the coefficient of friction between ice and an ice skate is very low. We use the following formula to determine the frictional force working on an object:

$$F = \mu \times F_n$$

force of friction = coefficient of friction × normal force

Since $F = ma$, we know that the normal force is simply the mass times the acceleration due to gravity. That is, $F_n = m \times g$:

$$F = \mu \times m \times g$$

The variable μ we use to represent the coefficient of friction is the Greek letter mu (pronounced "mew"). Scientists can determine the coefficient of friction of two objects through direct observation and experimentation. The coefficient of friction ranges from very close to zero to greater than one.

But wait, how do we determine the direction of that frictional force? And what happens when the object finally comes to rest? Well, let's use common sense: Friction works to oppose motion. It's always in the opposite direction of velocity or an imposed force (including cases when the object is at rest). And the equation above isn't true in every case. This is simply the *maximum possible force* exerted by friction on the object. When it's at rest with no outside forces imposed, there will be no frictional force. Friction won't move the object backwards, of course!

FRICTION ON A SLOPE

Now let's consider a more complicated scenario. A small mass of m is on a ramp with angle θ. The mass m is attached to a larger mass M by a rope, which exerts a force on the smaller mass, in a direction parallel to the ramp.

m
M
θ

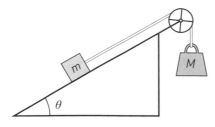

If there are no other forces to consider, the only forces on mass *m* are the force of gravity, $m \times g$, and the force of the tension of the string, $M \times g$. To determine the acceleration of the mass *m*, we'll decompose the force of gravity into a force that opposes the direction of motion (that is, one parallel to the direction of the ramp and the tension of the rope attached to mass *M*), and a force perpendicular to the ramp itself.

mg sin θ
m
F gravity = mg
θ
θ
mg cos θ

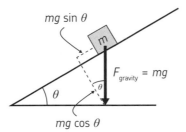

We know that the right triangle formed by the decomposition of this force is similar to the triangle formed by the ramp (that is, it has the same angle, θ). This means that the force opposing the tension of the rope is equal to $mg \sin \theta$. The force that's perpendicular to the ramp and the motion of mass *m* is equal to $mg \cos \theta$. If there's no friction at work, we can ignore this force, as it's offset by a force perpendicular to it, imposed by the ramp itself. This is simply Newton's third law in action.

F normal
Mg
mg sin θ
mg cos θ

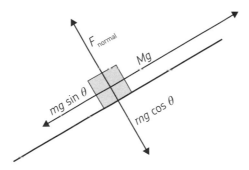

Now that we know all this, can you determine how this system works if we take into account the friction between mass *m* and the ramp? First, let's think about the normal force. Earlier, I said that it's the force perpendicular to the surface. That means that the force of the object perpendicular to the ramp, $mg \cos \theta$, is equal to our normal force. The force of friction for this object is as follows:

$$F_{\text{friction}} = \mu\, mg \cos \theta$$

Taking into account all forces on the object parallel to the ramp ($mg \cos \theta$ is offset by the normal force), we have the following relationship:

$$F_{\text{net}} = Mg - mg \sin \theta - \mu\, mg \cos \theta$$

net force = weight of *M* – component force of gravity – force of friction

Mg
mg sin θ
μ mg cos θ

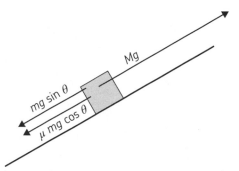

Knowing all this, we can determine how quickly object *m* will accelerate up the ramp!

COLLIDING COINS AND THE CONSERVATION OF ENERGY

WARNING: SERIOUS ALGEBRA AHEAD!

In Chapter 3, we examined collision of coins, the conservation of momentum, and how it must hold true in two dimensions (page 144). In this example, we know that the 100 yen coin's initial momentum in the x direction must match the final momentum of both coins in the x direction. In the following equations the 100 yen coin has mass *m* and the 500 yen coin has mass *M*:

❶ $mv_1 = mv_2 \cos \theta + MV_2 \cos \varphi$

And because the 100 yen coin has no initial momentum in the y direction, we know that the momentum of both coins in the y direction must offset each other:

❷ $0 = mv_2 \sin \theta - MV_2 \sin \varphi$

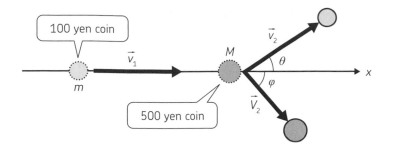

Assuming that this is a completely elastic collision (that is, kinetic energy is conserved), we also know the following:

initial kinetic energy = final kinetic energy

❸ $\quad \frac{1}{2}mv_1^2 = \frac{1}{2}mv_2^2 + \frac{1}{2}MV_2^2$

For these three equations (**❶**, **❷**, and **❸**) we have four unknowns—v_2, V_2, θ, and φ. It's not possible to find exact solutions, since we have too many variables to solve for and not enough equations. However, we can explore the relationships between these variables. So let's examine the 100 yen coin and the relationship between the ratio of its initial and final velocity (v_2 / v_1) to the subsequent scattering angle (θ). For simplicity, we'll assume that $m < M$. (The collision of the 100 yen and 500 yen coins satisfies this condition.)

First, let's solve our equations to get rid of the variable φ. We'll solve for sin φ and cos φ, for simplicity's sake. First, let's consider equation **❶**, where it looks like we can easily solve for cos φ:

$$MV_2 \cos\varphi = mv_1 - mv_2 \cos\theta$$

❹ $\qquad \cos\varphi = \dfrac{mv_1 - mv_2 \cos\theta}{MV_2}$

Now let's solve for sin φ in equation **❷**:

$$MV_2 \sin\varphi = mv_2 \sin\theta$$

❺ $\qquad \sin\varphi = \dfrac{mv_2 \sin\theta}{MV_2}$

With these two relationships in hand, we can substitute these equations (**❹** and **❺**) into a basic trigonometric relationship, which holds true for any angle:

❻ $\quad \sin^2\varphi + \cos^2\varphi = 1$

Be warned that the algebra required in this section is tricky! After solving for V_2^2, you should get the following:

❼ $\quad V_2^2 = \left(\frac{m}{M}\right)^2 \left(v_1^2 - 2v_1 v_2 \cos\theta + v_2^2\right)$

We know energy has been conserved, so we know that equation ❸ must hold true. So let's substitute equation ❼ into equation ❸. Then we'll only have three variables to consider: v_1, v_2, and θ, just as we wanted. Try solving for v_2. (Hint: You may need to use the quadratic formula.)

After all your work, you'll discover the following relationship:

$$\text{❽} \quad v_2 = \frac{\left(\frac{m}{M}\right)\cos\theta + \sqrt{1 - \left(\frac{m}{M}\right)^2 \sin^2\theta}}{1 + \frac{m}{M}} v_1$$

Additionally, assign $\theta = 0$ for this expression, and you'll find $v_1 = v_2$. This is relevant to a case where object 1 passes by object 2 without colliding.

On the other hand, assuming a case where the objects bounce back in opposing directions and $\theta = 180°$, you get the following:

$$\text{❾} \quad v_2 = \frac{1 - \frac{m}{M}}{1 + \frac{m}{M}} v_1$$

This equation indicates that as mass M becomes much larger than mass m, the following relationship holds true: $v_2 = v_1$. (This is because the term (m / M) approaches zero.) This means that an object with smaller mass having a head-on collision with a huge object bounces back at the same velocity it had before it hit the larger object. On the other hand, when $M = m$, $v_2 = 0$. You can reconfirm this relation by causing a head-on collision of two 100 yen coins by replacing the 500 yen coin with another 100 yen coin, taking care not to allow an oblique course. After the collision, the 100 yen coin halts and the other 100 yen coin previously in a stationary state starts traveling at the same speed. In this case, we can easily find that $V_2 = v_1$ from equation ❼. The two coins essentially swap velocities.

Now let's plot on a graph the relationship between the scattering angle (θ) and the velocity ratio v_2 / v_1 for the 100 yen coins before and after the collision. Since the mass of a 100 yen coin is 4.8 g, while that of a 500 yen coin is 7.0 g, we get $m / M = 4.8 / 7.0 = 0.69$. We'll use this result in equation ❽, then solve for v_2 / v_1, and plot the results. Here's the actual equation that we'll graph:

$$\text{❿} \quad \frac{v_2}{v_1} = \frac{0.69\cos\theta + \sqrt{1 - 0.69^2 \sin^2\theta}}{1 + 0.69}$$

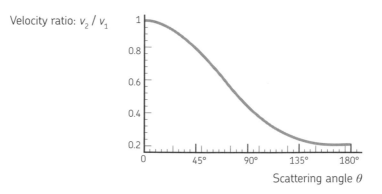
Velocity ratio: v_2 / v_1

Scattering angle θ

This graph should make intuitive sense to you after some consideration. If the scattering angle is greater (that is, the coins' collision is a glancing strike), the secondary velocity of the coin (v_2) will be smaller, thus the relationship of v_2 / v_1 will be smaller as well. Note that if you use objects of different masses, this relationship (and the graph that represents it) will change.

I AM *SO* READY FOR THIS REMATCH!

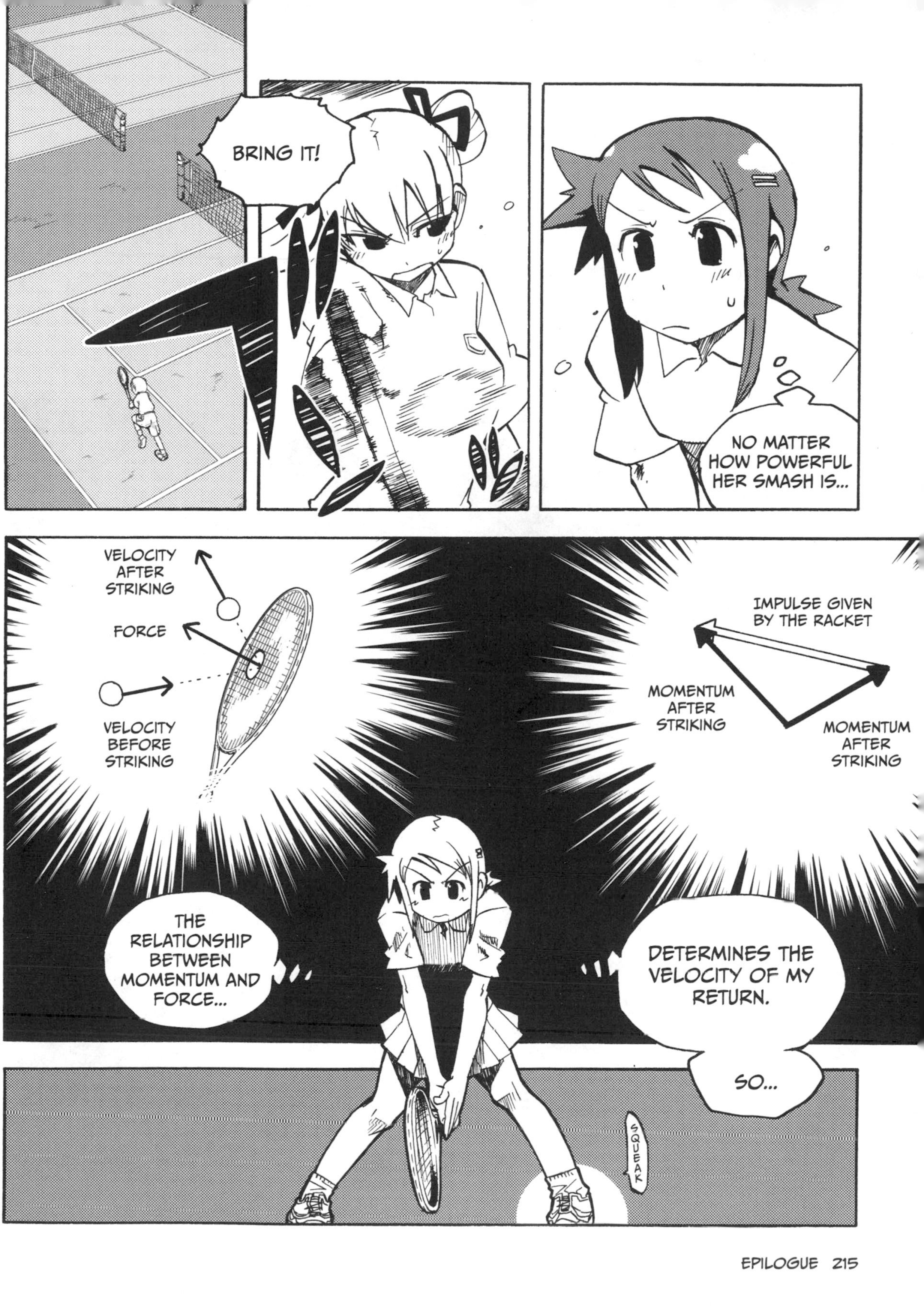
BRING IT!
NO MATTER HOW POWERFUL HER SMASH IS...
VELOCITY AFTER STRIKING
FORCE
VELOCITY BEFORE STRIKING
IMPULSE GIVEN BY THE RACKET
MOMENTUM AFTER STRIKING
MOMENTUM AFTER STRIKING
THE RELATIONSHIP BETWEEN MOMENTUM AND FORCE...
DETERMINES THE VELOCITY OF MY RETURN.
SO...
SQUEAK

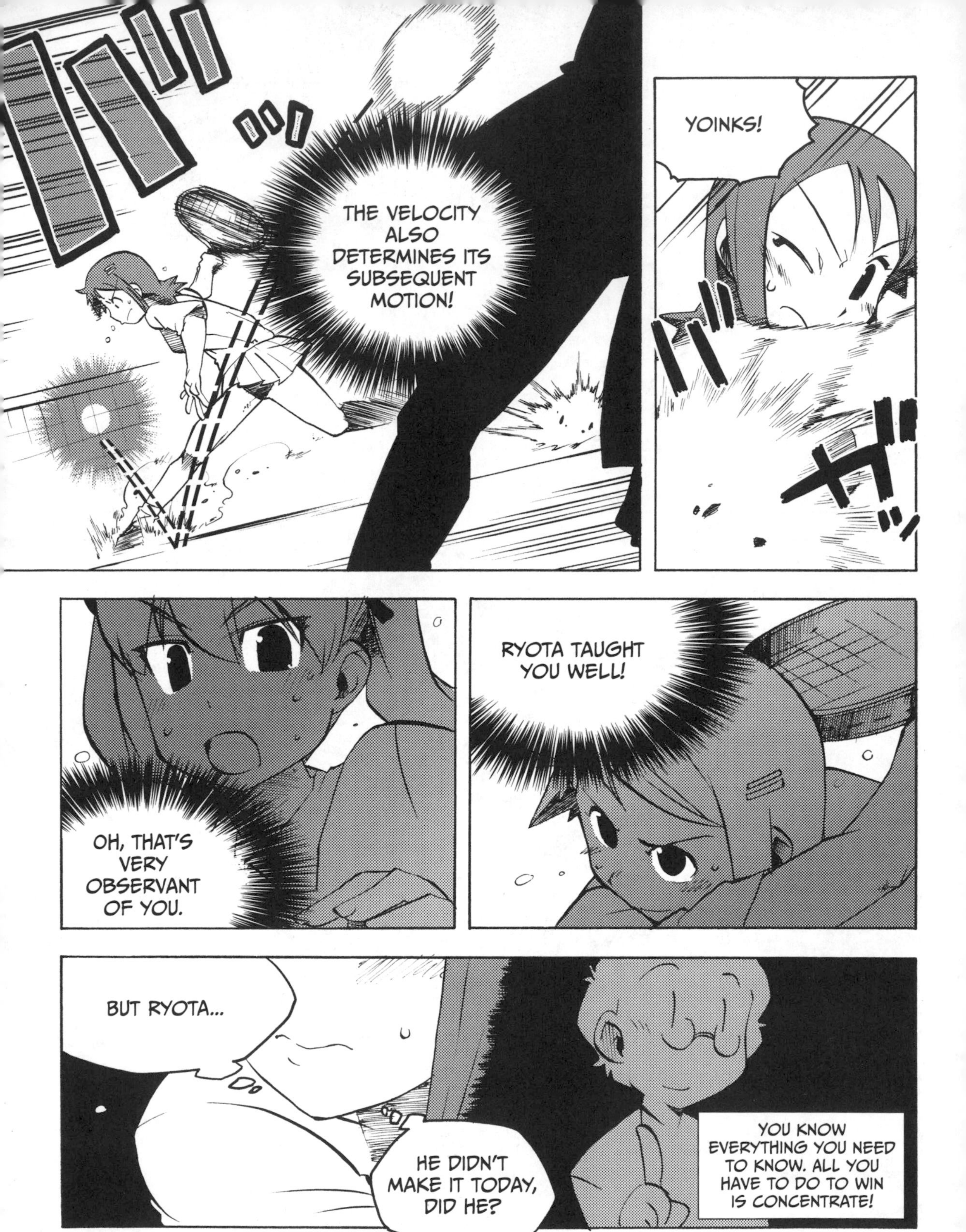
THE VELOCITY ALSO DETERMINES ITS SUBSEQUENT MOTION!
YOINKS!
OH, THAT'S VERY OBSERVANT OF YOU.
RYOTA TAUGHT YOU WELL!
BUT RYOTA...
HE DIDN'T MAKE IT TODAY, DID HE?
YOU KNOW EVERYTHING YOU NEED TO KNOW. ALL YOU HAVE TO DO TO WIN IS CONCENTRATE!

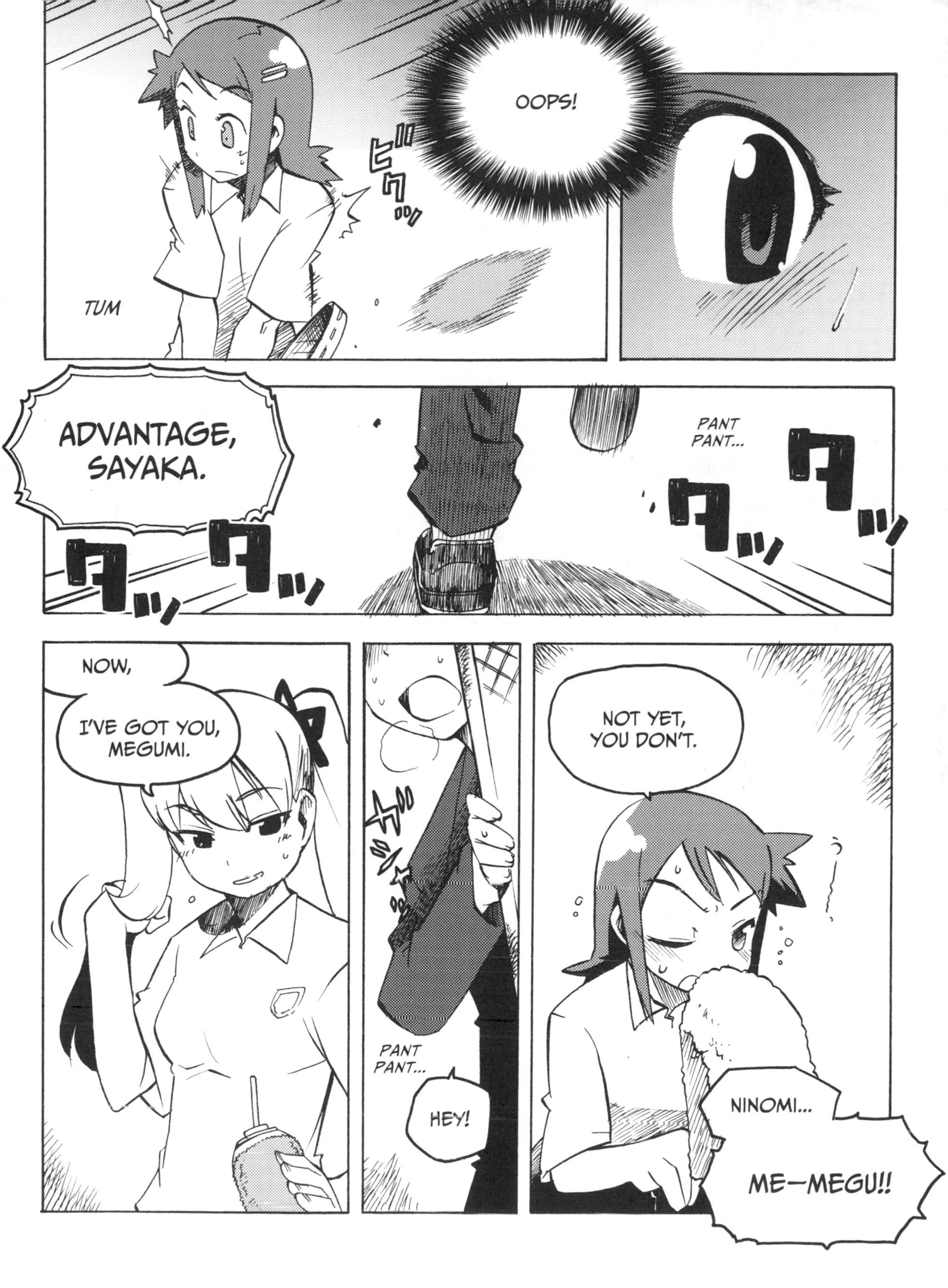
TUM
OOPS!
ADVANTAGE, SAYAKA.
PANT PANT...
NOW,
I'VE GOT YOU, MEGUMI.
PANT PANT...
HEY!
NOT YET, YOU DON'T.
NINOMI...
ME—MEGU!!

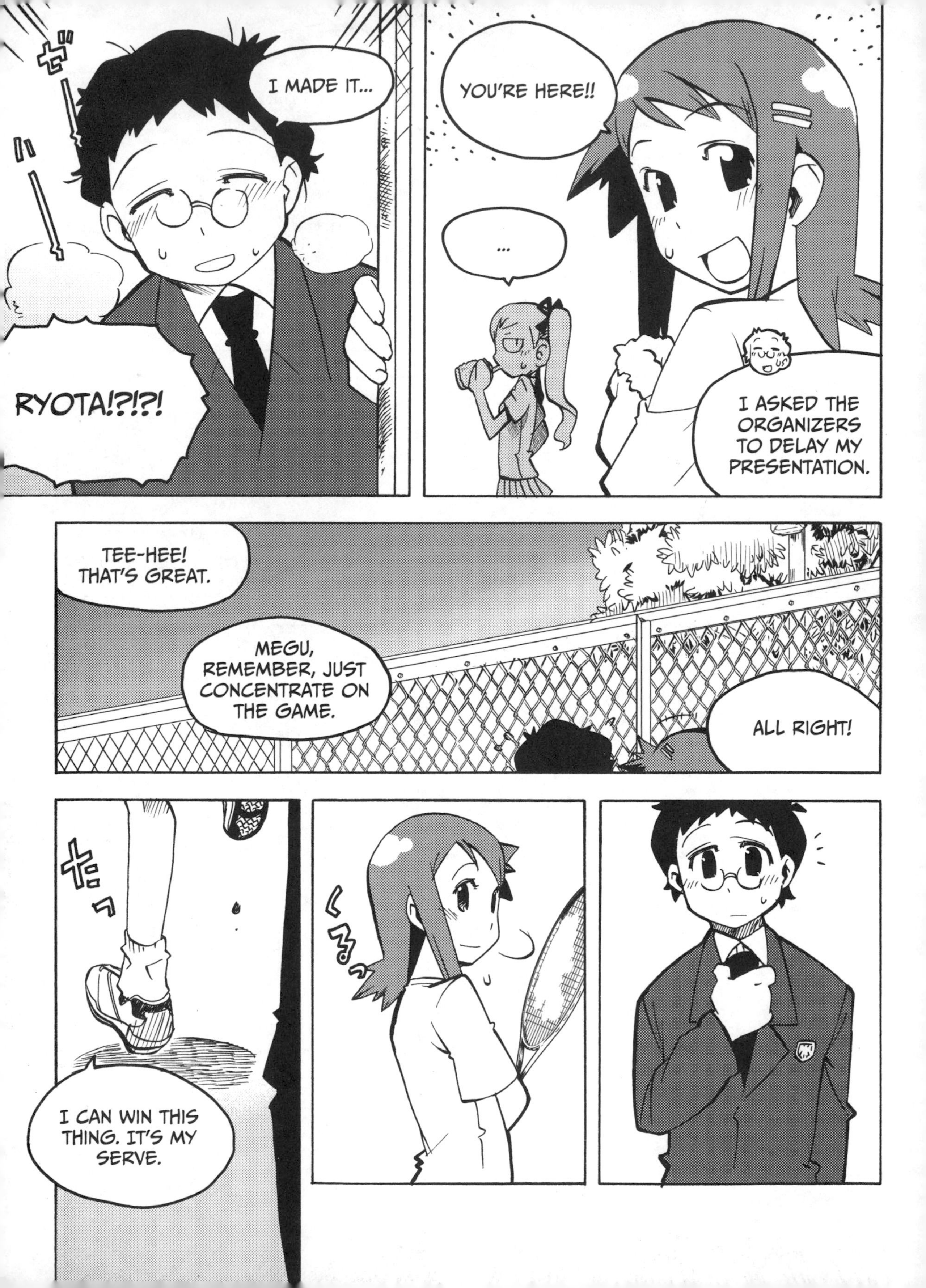
I MADE IT...
RYOTA!?!?!
YOU'RE HERE!!
...
I ASKED THE ORGANIZERS TO DELAY MY PRESENTATION.
TEE-HEE! THAT'S GREAT.
MEGU, REMEMBER, JUST CONCENTRATE ON THE GAME.
ALL RIGHT!
I CAN WIN THIS THING. IT'S MY SERVE.

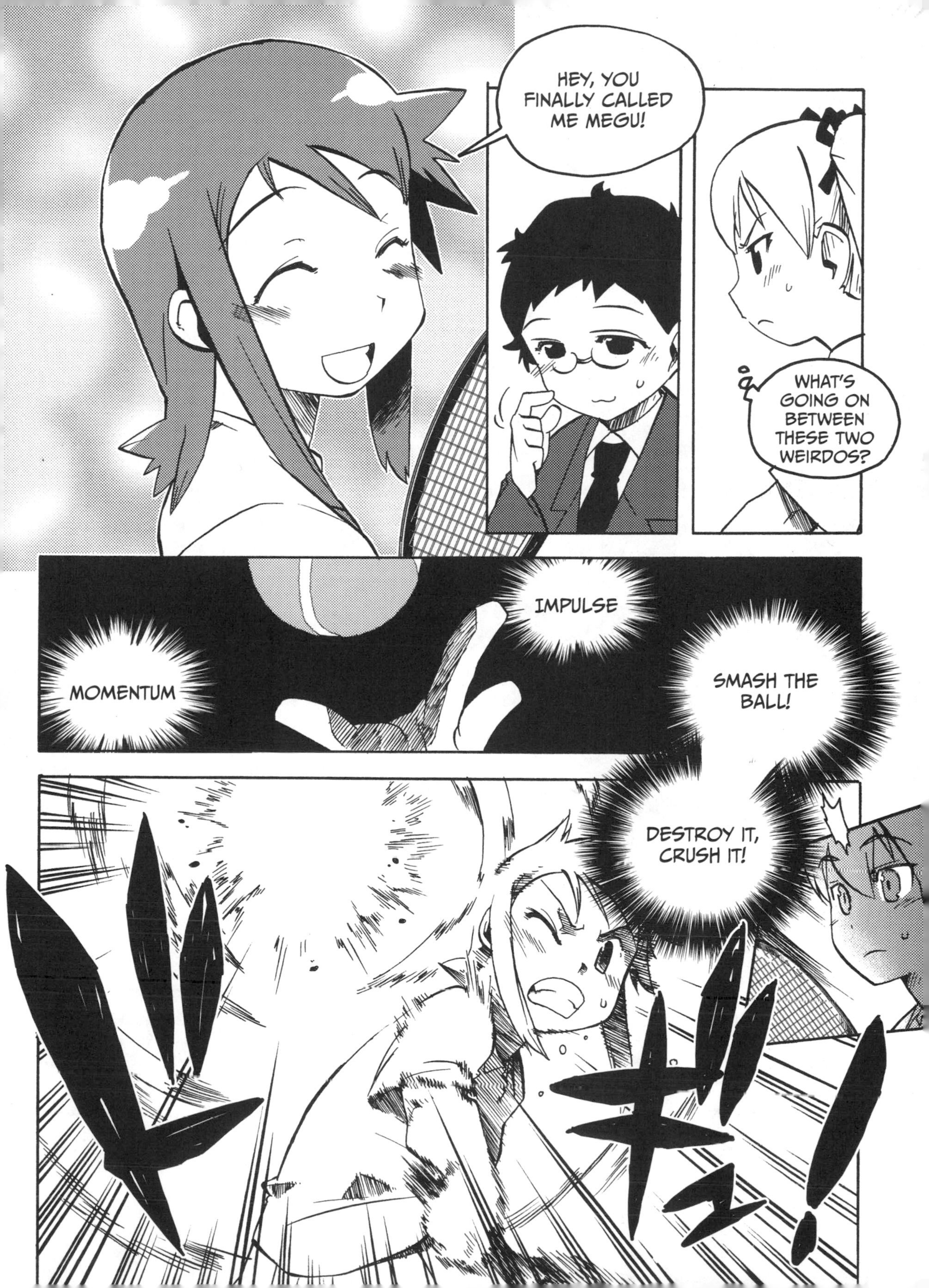
HEY, YOU FINALLY CALLED ME MEGU!
WHAT'S GOING ON BETWEEN THESE TWO WEIRDOS?
IMPULSE
MOMENTUM
SMASH THE BALL!
DESTROY IT, CRUSH IT!

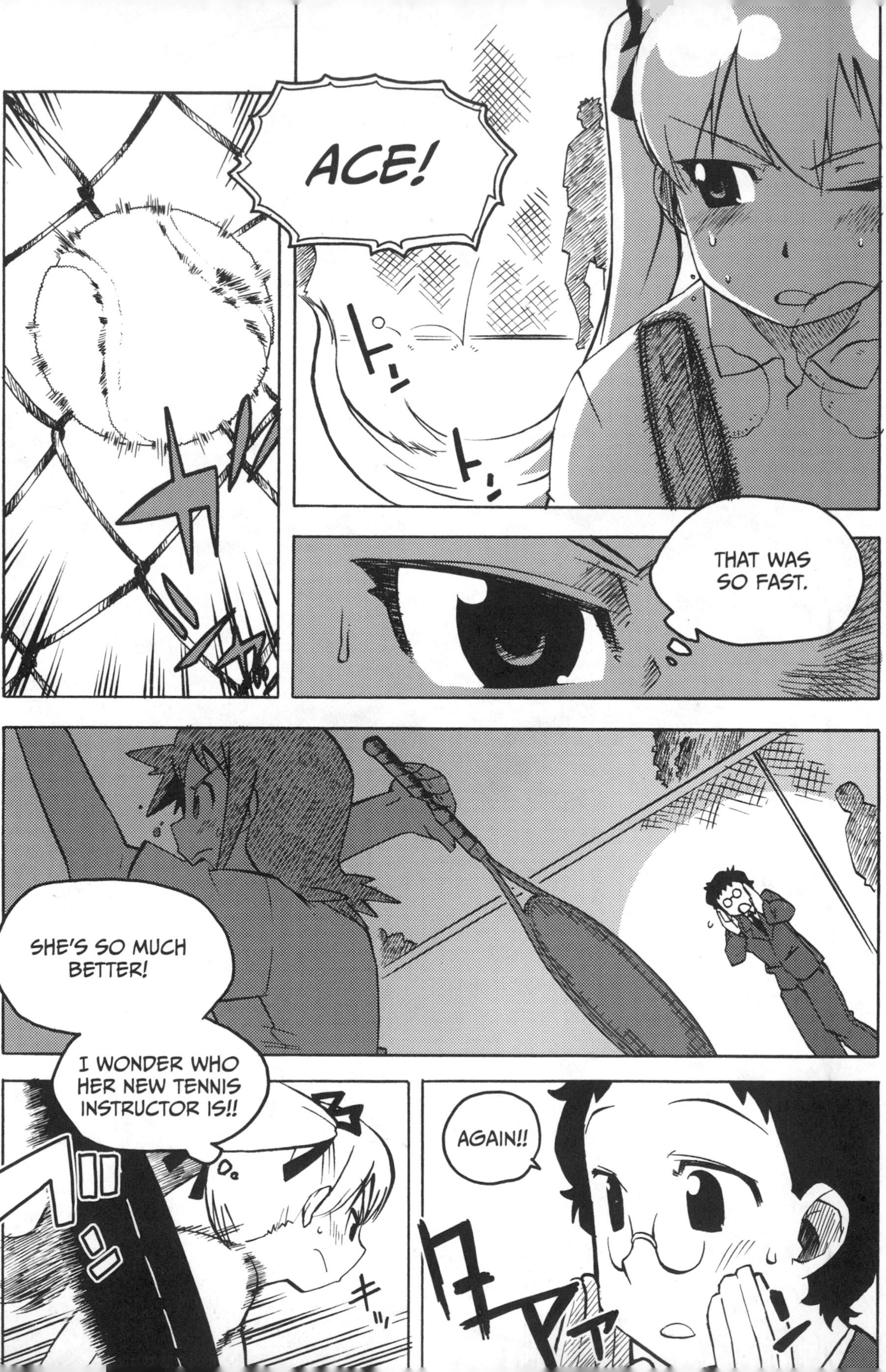
ACE!
THAT WAS SO FAST.
SHE'S SO MUCH BETTER!
I WONDER WHO HER NEW TENNIS INSTRUCTOR IS!!
AGAIN!!

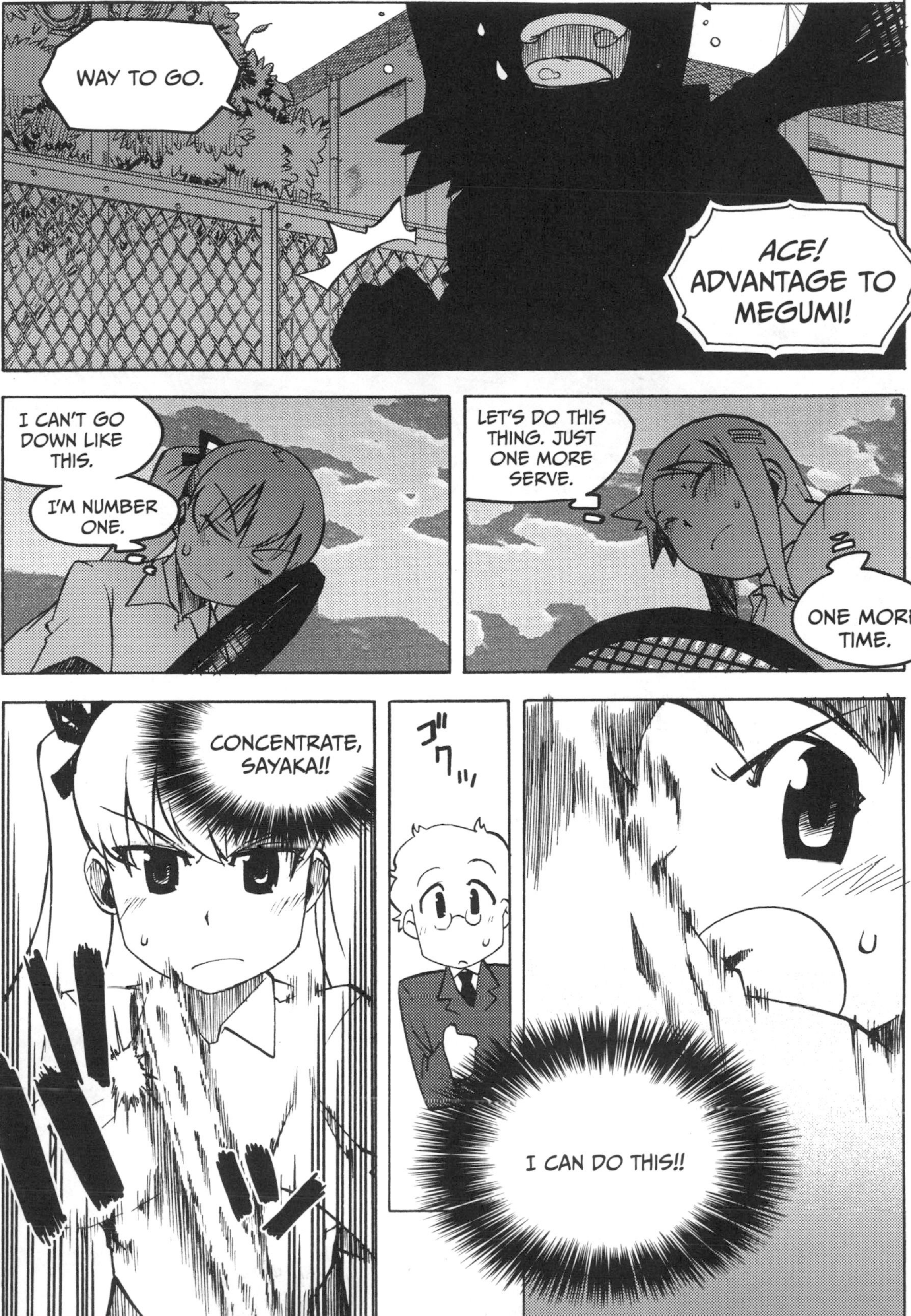
WAY TO GO.
ACE! ADVANTAGE TO MEGUMI!
I CAN'T GO DOWN LIKE THIS.
I'M NUMBER ONE.
LET'S DO THIS THING. JUST ONE MORE SERVE.
ONE MORE TIME.
CONCENTRATE, SAYAKA!!
ゴクッ
I CAN DO THIS!!

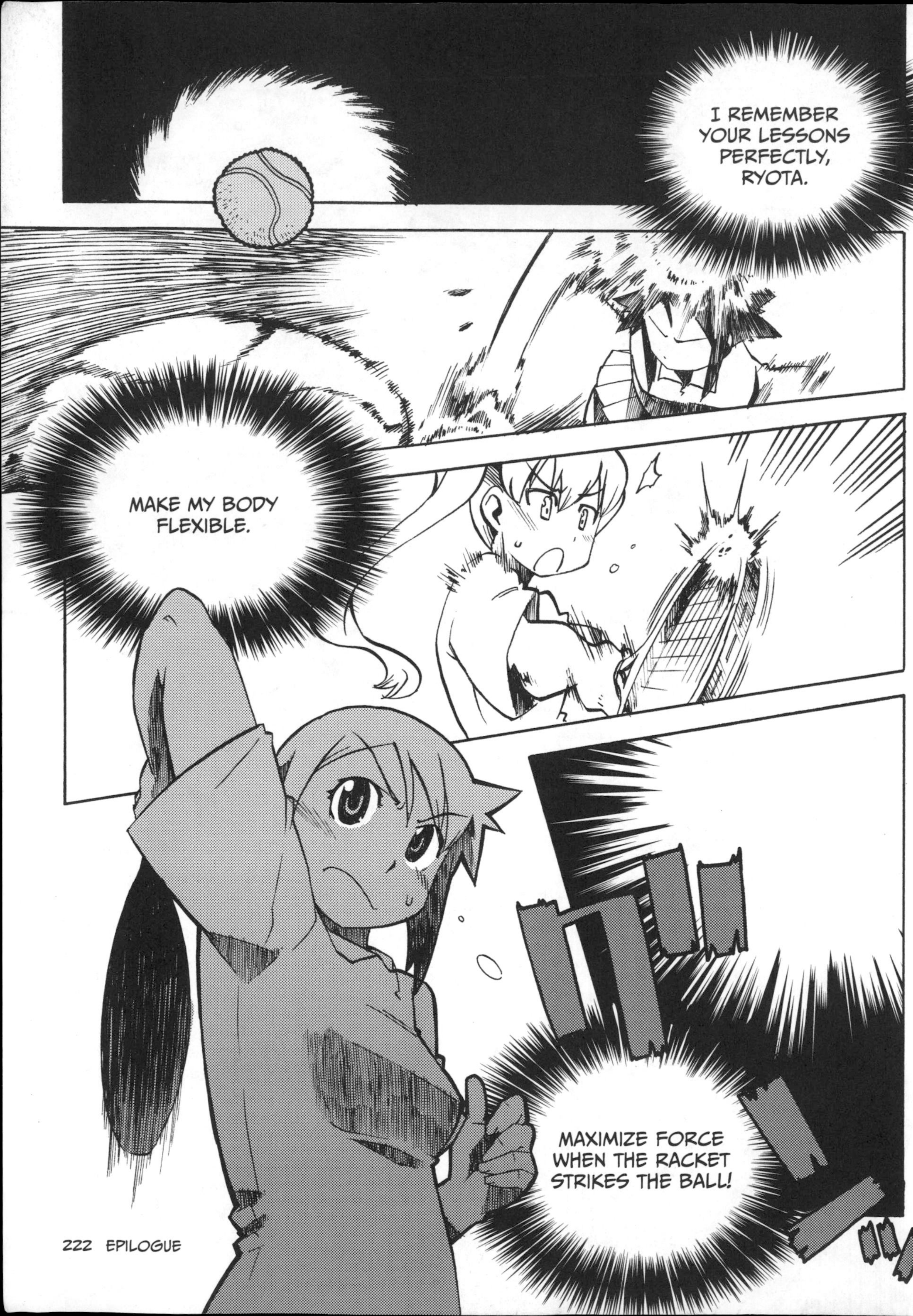
I REMEMBER YOUR LESSONS PERFECTLY, RYOTA.
MAKE MY BODY FLEXIBLE.
MAXIMIZE FORCE WHEN THE RACKET STRIKES THE BALL!

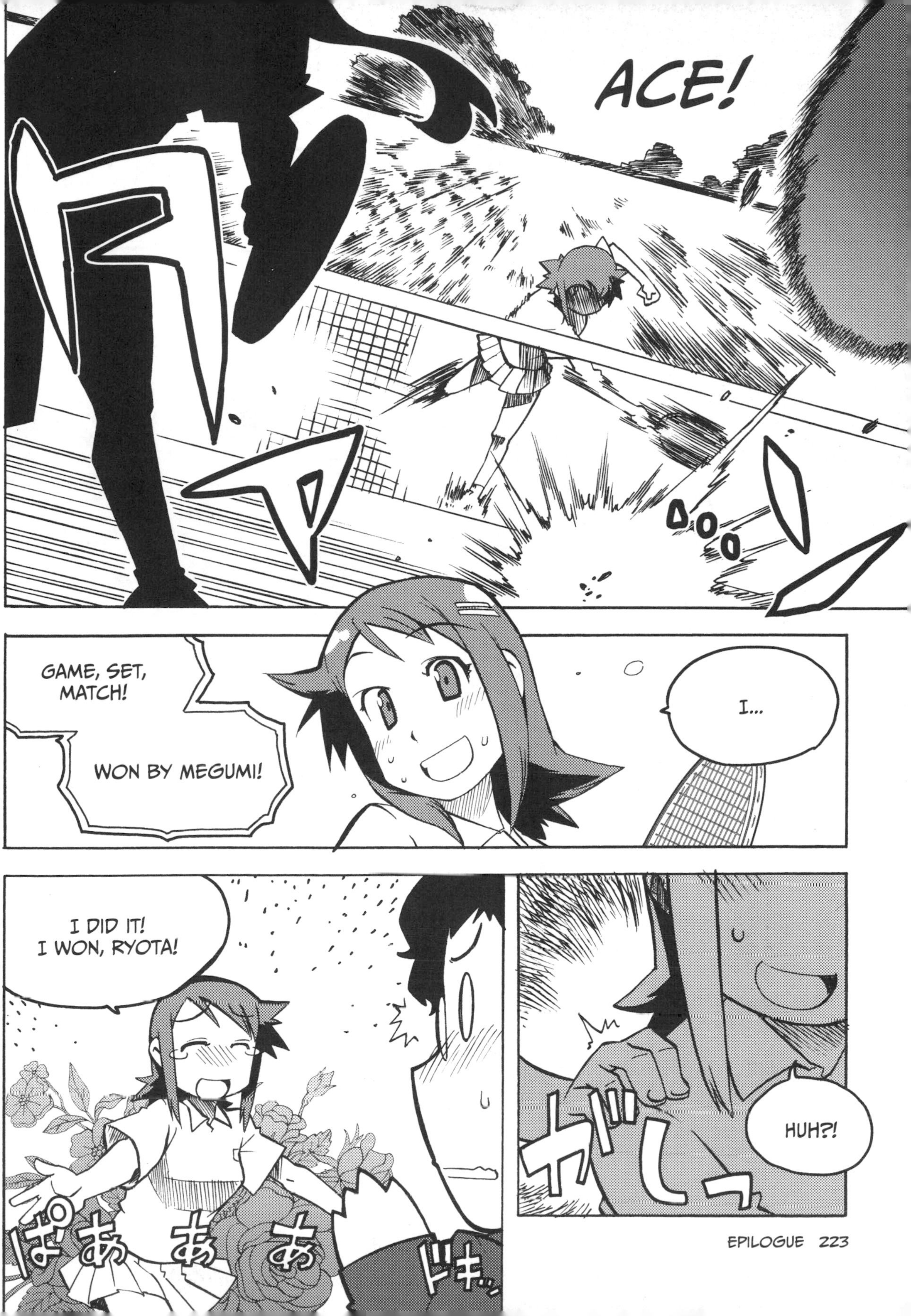
ACE!
GAME, SET, MATCH!
WON BY MEGUMI!
I...
I DID IT! I WON, RYOTA!
ぱあああ
ドキッ
ガしっ
HUH?!

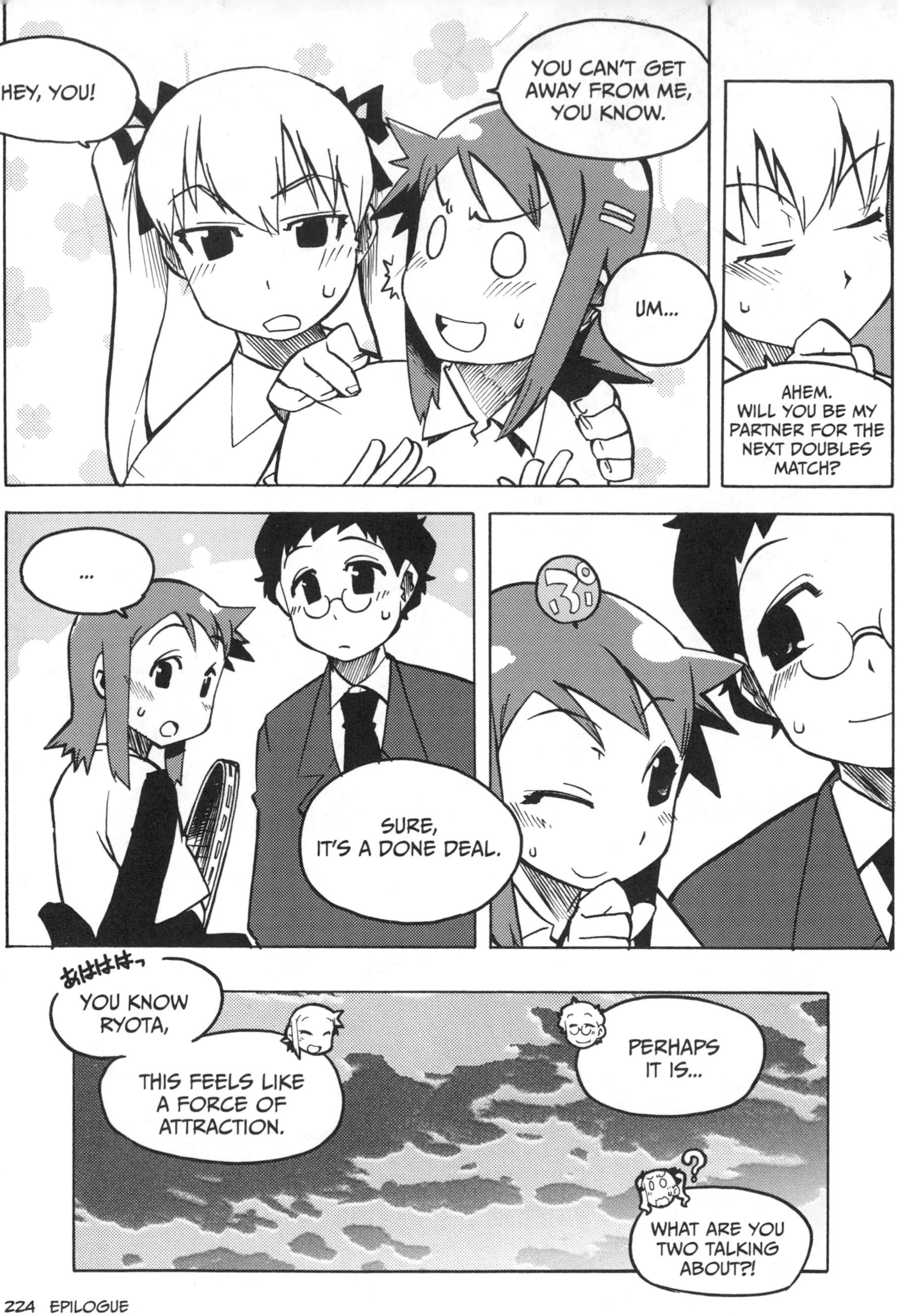

HEY, YOU!
YOU CAN'T GET AWAY FROM ME, YOU KNOW.
UM...
AHEM. WILL YOU BE MY PARTNER FOR THE NEXT DOUBLES MATCH?
...
SURE, IT'S A DONE DEAL.
あははは っ
YOU KNOW RYOTA,
THIS FEELS LIKE A FORCE OF ATTRACTION.
PERHAPS IT IS...
WHAT ARE YOU TWO TALKING ABOUT?!

MAKING SENSE OF UNITS

When it comes to classical mechanics, there are only three *base units*. Using these three simple measurements, you can derive more complicated units of measure like the newton and the joule. The three base units are as follows:

meters, m
(which measure *distance*)

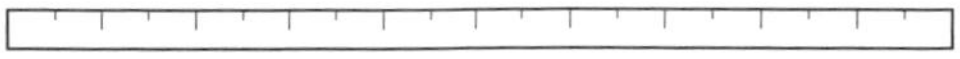

seconds, s
(which measure *time*)

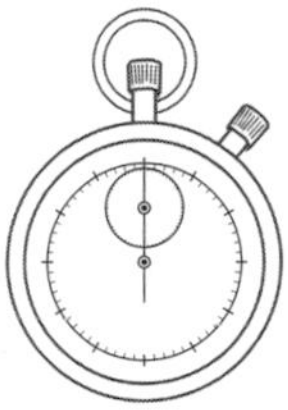

kilograms, kg
(which measure *mass*)

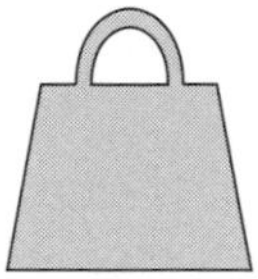

VELOCITY AND ACCELERATION

Let's explore how we can combine these three units to derive new ones. First, let's explore velocity and acceleration:

$$\text{velocity} = \frac{\text{change in distance (m)}}{\text{time (s)}} = \text{m/s}$$

$$\text{acceleration} = \frac{\text{change in velocity (m/s)}}{\text{time (s)}} = \text{m/s}^2$$

Given these relationships, you can see that *velocity* is defined as a change in distance, and *acceleration* is simply the change of that change! Students of calculus know that this means that velocity is the *first* derivative of distance, and acceleration is the *second* derivative of distance (both with respect to time).

FORCE

Given Newton's second law, we know force equals mass times acceleration ($F = ma$):

$$\text{force} = \text{mass (kg)} \times \text{acceleration (m/s}^2) = \text{kg} \times \text{m/s}^2 = \text{N}$$

To save ourselves a headache, we call a kg × m/s^2 a *newton* (N). Remember this relationship, as it will be important in deriving other units!

$$1\ \text{kg} \times \text{m/s}^2 = 1\text{N}$$

MOMENTUM AND IMPULSE

Momentum is an important physical quantity to measure, especially when considering collisions, landings, and impact. It is defined as:

$$\text{momentum} = \text{mass (kg)} \times \text{velocity (m/s)} = \text{kg} \times \text{m/s}$$

Impulse, as you've already read in Chapter 3, is just a change in momentum, and it can be calculated like so:

$$\text{impulse} = \text{force (N)} \times \text{time (s)} = \text{kg} \times \text{m/s}$$

Why does this calculation work? Remember that $1\text{N} = 1 \text{ kg} \times \text{m/s}^2$. Note that the unit for momentum, kg × m/s, has no shorter name.

ENERGY AND WORK

Kinetic energy is defined like so:

$$\text{kinetic energy} = \tfrac{1}{2} \times \text{mass (kg)} \times \text{velocity}^2 \text{ (m/s)} = \text{kg} \times \text{m}^2/\text{s}^2 = \text{J}$$

Just as we did with force, we'll use a simpler name for the unit of energy—the *joule* (J), which is named after English physicist James Prescott Joule. *Gravitational potential energy* can be calculated like this:

$$\text{potential energy} = \text{weight (N)} \times \text{height (m)} = \text{kg} \times \text{m}^2/\text{s}^2 = \text{J}$$

And naturally, our units match those of kinetic energy. *Work* is a measure of the energy transferred by a force over a distance. Notice the similarities in this equation to the previous one:

$$\text{work} = \text{force (N)} \times \text{distance (m)} = \text{kg} \times \text{m}^2/\text{s}^2 = \text{J}$$

The result of all these calculations is the joule, our unit for energy—just as it should be!

SI PREFIXES

You can add a prefix to a unit in order to increase or decrease its magnitude. These prefixes for different powers of 10 are called *SI prefixes*, and they come from internationally determined rules for units called the International System of Units (SI units). For example, 1 kilometer (km) is equal to 1,000 meters, 7 megajoules (MJ) are equal to 7,000,000 joules, and 3 nanograms (ng) are equal to 0.000 000 003 grams.

NOTE *The symbols for prefixes higher than kilo are capitalized.*

Symbol	Name	Power of ten
y	yocto–	10^{-24}
z	zepto–	10^{-21}
a	atto–	10^{-18}
f	femto–	10^{-15}
p	pico–	10^{-12}
n	nano–	10^{-9}
μ	micro–	10^{-6}
m	milli–	1/1000
c	centi–	1/100
d	deci–	1/10
da	deka–	10
h	hecto–	100
k	kilo–	1000
M	mega–	10^{6}
G	giga–	10^{9}
T	tera–	10^{12}
P	peta–	10^{15}
E	exa–	10^{18}
Z	zeta–	10^{21}
Y	yotta–	10^{24}

INDEX

N

O

P

Q

R

S

T

U

V

W

Z

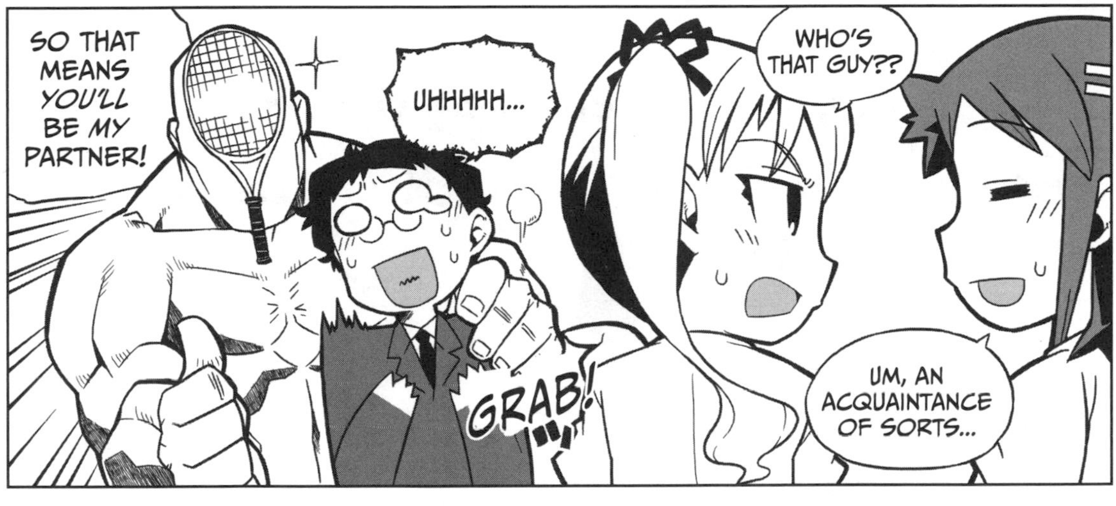
SO THAT MEANS *YOU'LL* BE *MY* PARTNER!
UHHHHH...
WHO'S THAT GUY??
GRAB!
UM, AN ACQUAINTANCE OF SORTS...

ABOUT THE AUTHOR

Hideo Nitta, PhD, is a professor in the Department of Physics at Tokyo Gakugei University. He has had many papers and books published by Japanese and overseas publishers on subjects including quantum dynamics and radiation physics. He also has a strong interest in physics education. He is a member of the International Commission on Physics Education (ICPE), which is a commission of the International Union of Pure and Applied Physics (IUPAP).

PRODUCTION TEAM FOR THE JAPANESE EDITION

Production: TREND-PRO Co., Ltd.

Founded in 1988, TREND-PRO produces newspaper and magazine advertisements incorporating manga for a wide range of clients from government agencies to major corporations and associations. Recently, TREND-PRO is actively participating in advertisement and publishing projects using digital content. Some results of past creations are publicly available at the company's website, *http://www.ad-manga.com/*.

Ikeden Bldg., 3F, 2-12-5 Shinbashi, Minato-ku, Tokyo, Japan

Telephone: 03-3519-6769; Fax: 03-3519-6110

Scenario writer: re_akino

Artist: Keita Takatsu

DTP: Move Co., Ltd.

MORE MANGA GUIDES

The *Manga Guide* series is a co-publication of No Starch Press and Ohmsha, Ltd. of Tokyo, Japan, one of Japan's oldest and most respected scientific and technical book publishers. Each title in the best-selling *Manga Guide* series is the product of the combined work of a manga illustrator, scenario writer, and expert scientist or mathematician. Once each title is translated into English, we rewrite and edit the translation as necessary and have an expert review each volume for technical correctness. The result is the English version you hold in your hands.

Find more *Manga Guides* at your favorite bookstore, and learn more about the series at *http://www.edumanga.me/*.

THE MANGA GUIDE TO
STATISTICS
SHIN TAKAHASHI
TREND-PRO, CO., LTD.

THE MANGA GUIDE TO
DATABASES
MANA TAKAHASHI
SHOKO AZUMA
TREND-PRO CO., LTD.

THE MANGA GUIDE TO
ELECTRICITY
KAZUHIRO FUJITAKI
MATSUDA
TREND-PRO CO., LTD.

THE MANGA GUIDE TO
CALCULUS
HIROYUKI KOJIMA
SHIN TOGAMI
BECOM CO., LTD.

THE MANGA GUIDE TO
MOLECULAR BIOLOGY
MASAHARU TAKEMURA
SAKURA
BECOM CO., LTD.

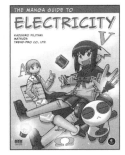

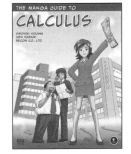

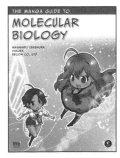

COLOPHON

The Manga Guide to Physics was laid out in Adobe InDesign. The fonts are CCMeanwhile and Chevin.

The book was printed and bound at Malloy Incorporated in Ann Arbor, Michigan. The paper is Glatfelter Spring Forge 60# Eggshell, which is certified by the Sustainable Forestry Initiative (SFI).

UPDATES

Visit *http://www.nostarch.com/mg_physics.htm* for updates, errata, and other information.